Bionic Technology—Robotic Exoskeletons and Prostheses: 2nd Edition

Bionic Technology—Robotic Exoskeletons and Prostheses: 2nd Edition

Guest Editor

Rafhael Milanezi de Andrade

MDPI

Basel • Beijing • Wuhan • Barcelona • Belgrade • Novi Sad • Cluj • Manchester

Guest Editor
Rafhael Milanezi de Andrade
Department of Mechanical
Engineering
Universidade Federal
do Espirito Santo
Vitoria
Brazil

Editorial Office
MDPI AG
Grosspeteranlage 5
4052 Basel, Switzerland

This is a reprint of the Special Issue, published open access by the journal *Biomimetics* (ISSN 2313-7673), freely accessible at: www.mdpi.com/journal/biomimetics/special_issues/862S8X2ESL.

For citation purposes, cite each article independently as indicated on the article page online and using the guide below:

Lastname, A.A.; Lastname, B.B. Article Title. *Journal Name* **Year**, *Volume Number*, Page Range.

ISBN 978-3-7258-3282-8 (Hbk)
ISBN 978-3-7258-3281-1 (PDF)
https://doi.org/10.3390/books978-3-7258-3281-1

Contents

About the Editor vii

Preface ix

Noémie Fortin-Bédard, Alice Pellichero, Stéphanie Leplaideur, Marie-Caroline Delebecque, Caroline Charette and Willy Allègre et al.
Acceptability of Overground Wearable Powered Exoskeletons for People with Spinal Cord Injury: A Multicenter Qualitative Study
Reprinted from: *Biomimetics* **2025**, *10*, 36, https://doi.org/10.3390/biomimetics10010036 . . . . . **1**

Pranav Madhav Kuber and Ehsan Rashedi
Training and Familiarization with Industrial Exoskeletons: A Review of Considerations, Protocols, and Approaches for Effective Implementation
Reprinted from: *Biomimetics* **2024**, *9*, 520, https://doi.org/10.3390/biomimetics9090520 . . . . . **17**

Ryota Kimura, Takahiro Sato, Yuji Kasukawa, Daisuke Kudo, Takehiro Iwami and Naohisa Miyakoshi
Automatic Assist Level Adjustment Function of a Gait Exercise Rehabilitation Robot with Functional Electrical Stimulation for Spinal Cord Injury: Insights from Clinical Trials
Reprinted from: *Biomimetics* **2024**, *9*, 621, https://doi.org/10.3390/biomimetics9100621 . . . . . **37**

Valentina Lanzani, Cristina Brambilla and Alessandro Scano
Kinematic–Muscular Synergies Describe Human Locomotion with a Set of Functional Synergies
Reprinted from: *Biomimetics* **2024**, *9*, 619, https://doi.org/10.3390/biomimetics9100619 . . . . . **45**

Alexandre B. S. da Silva, Gabriel E. P. Mendes, Eduardo S. Bragato, Guilherme L. Novelli, Marina Monjardim and Rafhael M. Andrade
Finger Prosthesis Driven by DEA Pairs as Agonist–Antagonist Artificial Muscles
Reprinted from: *Biomimetics* **2024**, *9*, 110, https://doi.org/10.3390/biomimetics9020110 . . . . . **62**

Xuanyi Zhou, Jianhua Zhang, Bangchu Yang, Xiaolong Ma, Hao Fu and Shibo Cai et al.
A Semi-Autonomous Hierarchical Control Framework for Prosthetic Hands Inspired by Dual Streams of Human
Reprinted from: *Biomimetics* **2024**, *9*, 62, https://doi.org/10.3390/biomimetics9010062 . . . . . . **76**

Jianwei Cui and Bingyan Yan
Research on Multimodal Control Method for Prosthetic Hands Based on Visuo-Tactile and Arm Motion Measurement
Reprinted from: *Biomimetics* **2024**, *9*, 775, https://doi.org/10.3390/biomimetics9120775 . . . . . **90**

Javier Andrés-Esperanza, José L. Iserte-Vilar and Víctor Roda-Casanova
Design and Optimization of a Custom-Made Six-Bar Exoskeleton for Pulp Pinch Grasp Rehabilitation in Stroke Patients
Reprinted from: *Biomimetics* **2024**, *9*, 616, https://doi.org/10.3390/biomimetics9100616 . . . . . **114**

Bettina Wollesen, Julia Gräf, Sander De Bock, Eligia Alfio, María Alejandra Díaz and Kevin De Pauw
Gender Differences in Performing an Overhead Drilling Task Using an Exoskeleton—A Cross-Sectional Study
Reprinted from: *Biomimetics* **2024**, *9*, 601, https://doi.org/10.3390/biomimetics9100601 . . . . . **138**

Suzanne J. Filius, Bas J. van der Burgh and Jaap Harlaar
The Design of the Dummy Arm: A Verification Tool for Arm Exoskeleton Development
Reprinted from: *Biomimetics* **2024**, *9*, 579, https://doi.org/10.3390/biomimetics9100579 . . . . . **151**

Pavan Senthil, Om Vishanagra, John Sparkman, Peter Smith and Albert Manero
Design and Assessment of Bird-Inspired 3D-Printed Models to Evaluate Grasp Mechanics
Reprinted from: *Biomimetics* **2024**, *9*, 195, https://doi.org/10.3390/biomimetics9040195 . . . . . **162**

About the Editor

Rafhael Milanezi de Andrade

Rafhael Milanezi de Andrade is an associate professor of Mechanical Engineering at the Universidade Federal do Espirito Santo Brazil (UFES) and a permanent professor in the Graduate Programs in Mechanical Engineering and Biotechnology at UFES. He received the B.S. and M.S. in Mechanical Engineering from UFES in 2009 and 2013, respectively, and the Ph.D. degree in Mechanical Engineering from Universidade Federal de Minas Gerais Brazil (UFMG) in 2018. Since 2019, he has collaborated as a research fellow at Harvard Medical School with a postdoctoral internship awarded by Fulbright (Fulbright Junior Faculty Member Award) working at the Motion Analysis Lab (MAL)—Harvard Medical School/Spaulding Rehabilitation Hospital in robotic rehabilitation projects. He is a senior member of the IEEE, belonging to the IEEE Robotics and Automation Society and the IEEE Engineering in Medicine and Biology Society. He is the coordinator of the Laboratory of Robotics and Biomechanics at UFES. He has experience in mechanical engineering, biomechanics, bioengineering, and biomechatronics.

Preface

The rapid advancements in bionic technology and biomechatronics have brought significant breakthroughs in enhancing human capabilities and improving the quality of life for individuals with disabilities. The integration of robotics, mechatronics, artificial intelligence, soft robotics, neuroscience, and advanced electronics has paved the way for a new generation of robotic prostheses and exoskeletons. These wearable robotic systems aim to restore or augment motor functions, offering new possibilities for rehabilitation and daily assistance.

However, the development of such advanced assistive devices presents numerous challenges. Wearable robots must be lightweight yet powerful enough to support or replace lost limb functions. They must also ensure seamless physical and cognitive interaction with users, adapting intuitively to their needs while maintaining safety and efficiency. The multidisciplinary nature of this field necessitates collaboration among experts in engineering, medicine, neuroscience, and human–computer interaction to address the complexities of human–robot synergy effectively.

This Reprint compiles cutting-edge research exploring various aspects of bionic technology, including the design and optimization of robotic exoskeletons and prostheses, novel control methods for intuitive user interaction, and innovative approaches to rehabilitation and functional restoration. The selected contributions present experimental studies, theoretical advancements, and practical implementations that highlight the latest progress in the field.

By gathering insights from researchers worldwide, this Reprint aims to foster further advancements in wearable robotics and bionic technology. We hope this serves as a valuable resource for scientists, engineers, and healthcare professionals committed to pushing the boundaries of assistive technology, ultimately improving mobility, autonomy, and quality of life for those in need.

We extend our gratitude to all contributors and researchers whose dedication and expertise have made this compilation possible. May this work inspire future innovations in the ever-evolving field of bionic technology.

Rafhael Milanezi de Andrade
Guest Editor

biomimetics

MDPI

Article

Acceptability of Overground Wearable Powered Exoskeletons for People with Spinal Cord Injury: A Multicenter Qualitative Study

Noémie Fortin-Bédard [1,2], Alice Pellichero [3], Stéphanie Leplaideur [3,4,5], Marie-Caroline Delebecque [3], Caroline Charette [1,2], Willy Allègre [3], Alyson Champagne [1,6], Caroline Rahn [1], Andréanne K. Blanchette [1,2], Laurent Bouyer [1,2], Jacques Kerdraon [3], Marie-Eve Lamontagne [1,2] and François Routhier [1,2,*]

1 Center for Interdisciplinary Research in Rehabilitation and Social Integration, Centre Intégré Universitaire de Santé et de Services Sociaux de la Capitale-Nationale, Quebec, QC G1C 3S2, Canada; noemie.fortin-bedard.1@ulaval.ca (N.F.-B.); caroline.charette.2@ulaval.ca (C.C.); alyson.champagne.1@ulaval.ca (A.C.); caroline.rahn.ciussscn@ssss.gouv.qc.ca (C.R.); andreanne.blanchette@fmed.ulaval.ca (A.K.B.); laurent.bouyer@rea.ulaval.ca (L.B.); marie-eve.lamontagne@fmed.ulaval.ca (M.-E.L.)

2 School of Rehabilitation Sciences, Université Laval, Quebec, QC G1V 0A6, Canada

3 Kerpape Rehabilitation Center, 56275 Ploemeur, France; alice.pellichero@coworkhit.com (A.P.); stephanie.leplaideur@vyv3.fr (S.L.); marie-caroline.delebecque@vyv3.fr (M.-C.D.); willy.allegre@vyv3.fr (W.A.); jacques.kerdraon@vyv3.fr (J.K.)

4 Physical and Rehabilitation Medicine Department, University Hospital of Rennes, 35000 Rennes, France

5 EMPENN-ERL, Inria Center, CNRS, Inserm, University of Rennes, 35065 Rennes, France

6 School of Psychology, Université Laval, Quebec, QC G1V 0A6, Canada

* Correspondence: francois.routhier@rea.ulaval.ca

Abstract: Background: Exoskeletons are used in rehabilitation centers for people with spinal cord injuries (SCI) due to the potential benefits they offer for locomotor rehabilitation. The acceptability of exoskeletons is crucial to promote rehabilitation and to ensure a successful implementation of this technology. The objective was to explore the acceptability of overground wearable powered exoskeleton used in rehabilitation among people with SCI. **Methods**: Fourteen individuals with SCI (9 men, mean [SD] age 47 years [14.8], a majority with traumatic and thoracic lesion (T6–T12)) who had utilized an exoskeleton in Canada or in France during their rehabilitation participated in a semi-structured interview. A thematic analysis using the theoretical framework of acceptability was carried out. **Results**: Participants were motivated to use an exoskeleton during their rehabilitation. They reported several perceived benefits to its use, including better walking pattern, increased endurance, and greater muscle mass. They also experienced mild pain, notable concentration demands, and fatigue. Most participants reported that using exoskeletons in their rehabilitation process was appropriate and relevant to them. **Conclusions**: Exoskeletons are generally well accepted by participants in this study. Adjustments in their use, such as conducting training sessions in obstacle-free environment and technological improvements to address the device's restrictive characteristics, heaviness, and massiveness are however still needed.

Keywords: spinal cord injury; exoskeleton; rehabilitation; acceptability; experience; locomotion

check for updates

Academic Editors: Rafhael Milanezi de Andrade and Dapeng Wei

Received: 17 October 2024
Revised: 16 December 2024
Accepted: 27 December 2024
Published: 8 January 2025

Citation: Fortin-Bédard, N.; Pellichero, A.; Leplaideur, S.; Delebecque, M.-C.; Charette, C.; Allègre, W.; Champagne, A.; Rahn, C.; Blanchette, A.K.; Bouyer, L.; et al. Acceptability of Overground Wearable Powered Exoskeletons for People with Spinal Cord Injury: A Multicenter Qualitative Study. *Biomimetics* **2025**, *10*, 36. https://doi.org/10.3390/biomimetics10010036

1. Introduction

Over the past few years, there has been a noticeable increase in the number of people living with disabilities, especially due to the aging population [1]. Nine million people worldwide were living with a spinal cord injury (SCI) in 2019 [2]. A SCI may result in changes depending on the level and severity of the injury, such as sensory and motor impairments [3]. These impairments often cause important issues with mobility which remains a vital concern for people living with SCI and their close relatives [4,5]. In this regard, walking has been identified as one of the rehabilitation priorities for people with SCI [6,7]. Altogether, these findings highlight the need for innovative technologies supporting gait rehabilitation. As a result, lower limb powered exoskeletons were greatly developed and are now being increasingly implemented.

Exoskeletons have been recently used in facilities offering rehabilitation services for people with physical disabilities, including those with SCI. Although evidence varies on the nature and extent of results, and among studies, recent studies support the safety of exoskeleton use and suggests that this technology may offer potential benefits in the rehabilitation of people with SCIs [8–10]. Systematic reviews of preliminary studies found that exoskeletons allowed people with SCI to walk at a modest speed [11] and engage in movements that could potentially yield health benefits [10]. More recently, a randomized trial reported an improvement in walking independence scores after participants with incomplete SCIs completed 15 one-hour training sessions [9]. However, evidence of effectiveness in improving independent gait speed has not been found among SCI participants with residual walking ability [12]. In the acute phase of post-injury recovery, people with SCIs reported many psychological benefits of using an exoskeleton [13].

Despite the promising effects of innovative approaches, high rates of abandonment by users in rehabilitation technology are reported notably due to failure in the implementation process [14,15]. Like other rehabilitation technologies, careful considerations should be given to the acceptability of exoskeletons by users with SCIs to promote their successful implementation and sustained use in clinical settings. In that regard, Sekhon et al., 2017 have defined the acceptability of a health intervention as "a multi-faceted construct that reflects the extent to which people delivering or receiving a healthcare intervention consider it to be appropriate, based on anticipated or experienced cognitive and emotional responses to the intervention" [16]. Low acceptability can be a major barrier to the use of a technology and limits its widespread use [17]. Users might simply refuse or restrain activities that require the use of the technology, opting instead for more familiar activities (e.g., stretching or exercise) that align better with their usual understanding of rehabilitation.

Previous studies have investigated, sometimes as a secondary objective, the user experience of exoskeletons from the point of view of people with SCIs. Various components of acceptability, such as high expectations toward this technology [18,19], considerable physical effort [13], and limited community use [20] after rehabilitation [20] have been reported. The presence or absence of these components may influence the overall acceptability of this technology. Despite its relevance for scaling up, no study has focused primarily on qualitatively exploring acceptability from the perspective of individuals with SCIs [21], nor using a theoretical framework of acceptability to analyze the results. The satisfaction and perceived acceptability of patients with SCIs are critical factors in fostering their active engagement in the rehabilitation process and ensuring the successful implementation of this technology in settings that offer rehabilitation care [16,22]. This knowledge could lead to better-targeted implementation strategies. Thus, the objective of this study was to explore the acceptability of overground wearable powered exoskeletons used in rehabilitation among people with SCIs.

2. Materials and Methods

2.1. Design and Participants

A qualitative study with a descriptive interpretative approach [23] using semi-structured interviews was conducted [24]. A qualitative design was used to provide an in-depth exploration of the acceptability of the SCI users [25]. The Consolidated Criteria for Reporting Qualitative Research (COREQ) checklist was used [26].

A multicenter study was conducted since participants were recruited from two rehabilitation facilities, namely the Centre intégré universitaire de santé et de services sociaux de la Capitale-Nationale (CIUSSS-CN) in Quebec, Canada, and the Centre Mutualiste de Réeducation et de Réadaptation Fonctionnelles de Kerpape (CMRRFK) in Ploemeur, France. These two rehabilitation facilities offer rehabilitation care for people with SCIs, which encompasses rehabilitation exoskeleton treatment among other modalities. To be eligible to participate in the interview, people with SCIs had to have used an exoskeleton during their rehabilitation at the CMRRFK or the CIUSSS-CN. People who were unable to consent or complete interviews were not included. In addition, each patient was screened by a referring physician before inclusion in the study to ensure safety and relevance of the use of an exoskeleton. This study was approved by local Research Ethics Boards from the Université de Lille (2023-701-S117) and from the CIUSSS-CN (MP-13-2020-2002). Procedures were conducted in accordance with relevant laws and institutional guidelines. Participants provided their informed consent.

2.2. Description of Exoskeletons and Context of Use

The CIUSSS-CN uses the Indego® powered lower limb exoskeleton with functional electrical stimulation developed by Parker Hannifin Corporation® (Cleveland, United States) and currently supported by Ekso Bionics® (San Rafael, United States). The Indego® is an exoskeleton used for people with SCIs (levels C7 to L5) and for people with hemiplegia [27]. However, at the time of this study, the use of this technology was restricted to a research project at the CIUSSS-CN for future clinical implementation. Hence, participants were required to adhere to a predetermined protocol outlined within the ongoing research project for the conduct of the rehabilitation sessions [28]. Table 1 illustrates the overall step-by-step process of using the exoskeleton at the CIUSSS-CN according to Charette et al., 2024 [28]. To have access to the locomotor training program with the exoskeleton at the CIUSSS-CN, participants had to: (1) be aged between 18 and 70 years old; (2) have an incomplete SCI; (3) be in the subacute stage of recovery (<1 year post-injury); (4) be able to stand and/or walk therapeutically; (5) have a minimum/sufficient upper limb strength to use a rolling walker; (6) have a height between 5′1″ and 6′1″; (7) have a weight <200 lbs (90 kg); (8) have a length of femurs between 37 and 49 cm; (9) have a seated hip width <42 cm; (10) have a medical approval to participate in the project following consultation with the multidisciplinary team; and (11) have a standing tolerance >15 min.

The CMRRFK uses the hand-free and self-balancing Atalante® exoskeleton developed by Wandercraft (Paris, France). The use of the exoskeleton is implemented into conventional clinical therapy sessions offered to people with SCIs by physiotherapists. Locomotor training with the exoskeleton is performed under medical prescription. To have access to the locomotor training program with the exoskeleton at the CMRRFK, participants had to (1) be at least 18 years old; (2) have a paraplegia; (3) tolerate daily verticalization; (4) have a height between 160 cm and 190 cm; (5) weight <90 kg; (6) have a hip width less than or equal to 460 mm in a sitting position; and (7) have thigh length between 380 and 460 mm. Additionally, leg length is also a criterion for inclusion, but this criteria varies according to user's range of motion in foot dorsiflexion.

The exclusion criteria for the use of the Indego® and Atalante® exoskeletons are provided in Supplementary Materials; Table S1 and Table S2, respectively.

Table 1. Step-by-step process of using the exoskeleton at the CIUSSS-CN.

Sessions	Duration	Procedures
(1) Initial session (Exoskeleton adjustment)	60 min for fitting the exoskeleton depending on the user	(1.1) Adjustment of the exoskeleton for the user (sitting position). (1.2) Teaching and practicing sit-to-stand and stand-to-sit transfers (2–3 repetitions). (1.3) Re-adjustment of the exoskeleton in standing or sitting position, if necessary. (1.4) For some patients, walking between parallel bars was initiated.
(2) Familiarization period	2–4 sessions, 60 min each, depending on the user's capacity	(2.1) Practice of sit-to-stand and stand-to sit transfers. (2.2) Walking between parallel bars, including learning to initiate gait, continue walking, and stop using the exoskeleton. (2.3) Once the user was deemed safe using the exoskeleton between parallel bars, walking with a walker in a corridor of the rehabilitation center was attempted for approximately 20 m. (2.4) For patient eligible for functional electrical stimulation (FES), the final session of the familiarization period was dedicated to adjusting FES parameters (amplitude, pulse width and frequency) for each stimulated muscle group.
(3) Locomotor training program	12 sessions, 60–90 min depending on the user	(3.1) At the beginning of each session, electrodes for FES were applied, and the exoskeleton was adjusted for the user (12–20 min) (3.2) Adjustment of the exoskeleton's walking parameters (e.g., step length and height, walking speed, motorized assistance for both hip and knee joints) based on the user's capacity. (3.3) The session duration and walking time were individualized based on the user's capacity, with the primary goal of optimizing the number of steps taken during each session. (3.4) After four sessions, the training program was progressed by either reducing motorized assistance, transitioning from a walker to forearm crutches, or further increasing the number of steps taken during each session. FES parameters were also adjusted after four sessions, if necessary.

2.3. Data Collection

The recruitment of participants differed between the two rehabilitation centers, but the data collection process was similar. The recruitment began on 20 August 2022 at the CIUSSS-CN and on 1 June 2023 at the CMRRFK. At the CIUSSS-CN, the study was introduced to individuals with SCIs who had used a rehabilitation exoskeleton by a member of the research team (C.C). If they were interested in taking part in the project, a Master's student, having previous experience and knowledge in conducting interviews (N.F.-B.), contacted participants by telephone or by email. The sociodemographic data of participants at the CIUSSS-CN were collected as part of the ongoing research project. At the CMRRFK, the study was presented to clinicians by a member of the research team (N.F.-B.). People with SCIs were then offered to participate in the present project by the clinicians. Sociodemographic data were collected during the interview session for participants at the CMRRFK (e.g., gender, age, exoskeleton used, time using an exoskeleton).

Semi-structured interviews were conducted in person for participants still residing in the rehabilitation center and via telephone or videoconference using Zoom [29] for participants who had transitioned home or to other facilities offering rehabilitation services. Interviews were conducted by two authors (N.F.-B. and A.C) who had no prior relationship with the participants. Interviews were conducted in French or in English between 7 July 2021 and 26 July 2023 at the CIUSSS-CN and between 5 June 2023 and 22 June 2023 at the CMRRFK. All interviews were audio recorded for transcription.

An interview guide was developed by the research team members who all together have knowledge of exoskeletons and the context in which the technology is used. The interview guide (in Supplementary Material) was developed with consideration to different determinants of behavioral change according to the theoretical domains framework (TDF) (e.g., knowledge, abilities, motivations, and emotions toward the technology) [30] and inspired by various contextual elements that were deemed as potentially influencing the adoption and use of the exoskeleton in a larger implementation study [28]. An additional question was added to explore the satisfaction with the exoskeleton. The interview guide included a total of 30 questions and 7 prompts.

The interview guide was not pretested considering the limited number of individuals with SCIs who had used the exoskeleton, but it was validated by an expert at the CIUSSS-CN (C.R.) Additionally, the interview guide underwent validation with a physiatrist (S.L.) working with individuals with SCIs at the CMRRFK, ensuring the questions were culturally relevant and appropriately adapted to the context.

2.4. Data Analysis

Interviews were analyzed through thematic analysis [31] with a mixed approach (deductive and inductive) [32] using the conceptual theoretical framework of acceptability (TFA) developed by Sekhon et al. [16] to explore the acceptability of the exoskeleton [16] (see Figure 1). An initial coding tree covering all the seven domains of the TFA (i.e., affective attitude, burden, perceived effectiveness, ethicality, intervention coherence, opportunity costs, and self-efficacy) [16] was used to analyze the interviews [16]. After becoming familiar with the interviews, two co-authors (N.F-B and A.P) with complementary experience (social worker and occupational therapist) independently analyzed two interviews as part of a standardization process. Then, they met to compare their interpretation of the coding tree. The first author (N.F-B) proceeded to code the remaining interviews, and meetings were scheduled to discuss the interpretation of the content. A third person, a professor-researcher (M-E.L.), was involved to provide guidance on the analysis of the interviews. The quotes included in this article have been translated from French to English using a translation software (i.e., DeepL software; version 24.10.11297902).

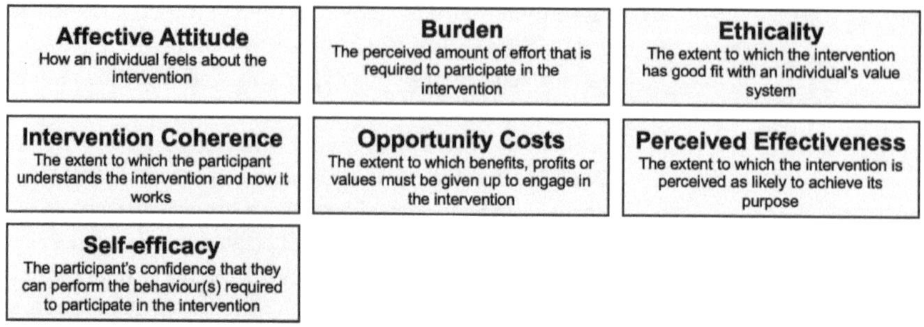

Figure 1. Domains of the theoretical framework of acceptability (TFA) [16].

3. Results

Nine participants who completed the exoskeleton training program at the CIUSSS-CN, and five participants at the CMRRFK agreed to take part in the interview, for a total of 14 participants. The mean (SD) age was 46.9 (14.8) years old; most participants were men (64%) and had a traumatic SCI (79%) (Table 2). Interview duration ranged from 19 to 63 min.

Table 2. Characteristics of the participants.

Characteristics	n (%)
Age, years (mean [SD])	46.9 [14.8]
Gender	
Women	5 (36)
Men	9 (64)
Country	
France	5 (36)
Canada	9 (64)
Time since the injury of all participants, months (mean [SD]) [1]	9.9 [7.6]
Time since the injury of Canada's participants, months (mean [SD]) [2]	6.5 [3.54]
Time since the injury of France's participants, months (mean [SD]) [3]	43.8 [9.45]
Mechanism of injury	
Non-traumatic	3 (21)
Traumatic	11 (79)
Self-reported level of injury	
High-Cervical Nerves (C1–C4)	1 (7.1)
Low-Cervical Nerves (C5–C8)	2 (14.3)
Thoracic Nerves (T1–T5)	1 (7.1)
Thoracic Nerves (T6–T12)	8 (57.1)
Lumbar Nerves (L1–L5)	2 (14.3)
Sacral Nerves (S1–S5)	0

[1] Missing information for three participants. [2] Missing information for two participants. [3] Missing information for one participant.

Results relevant to the acceptability of exoskeletons are presented according to overall main themes including all seven domains. Figure 2 presents the summary of the main findings.

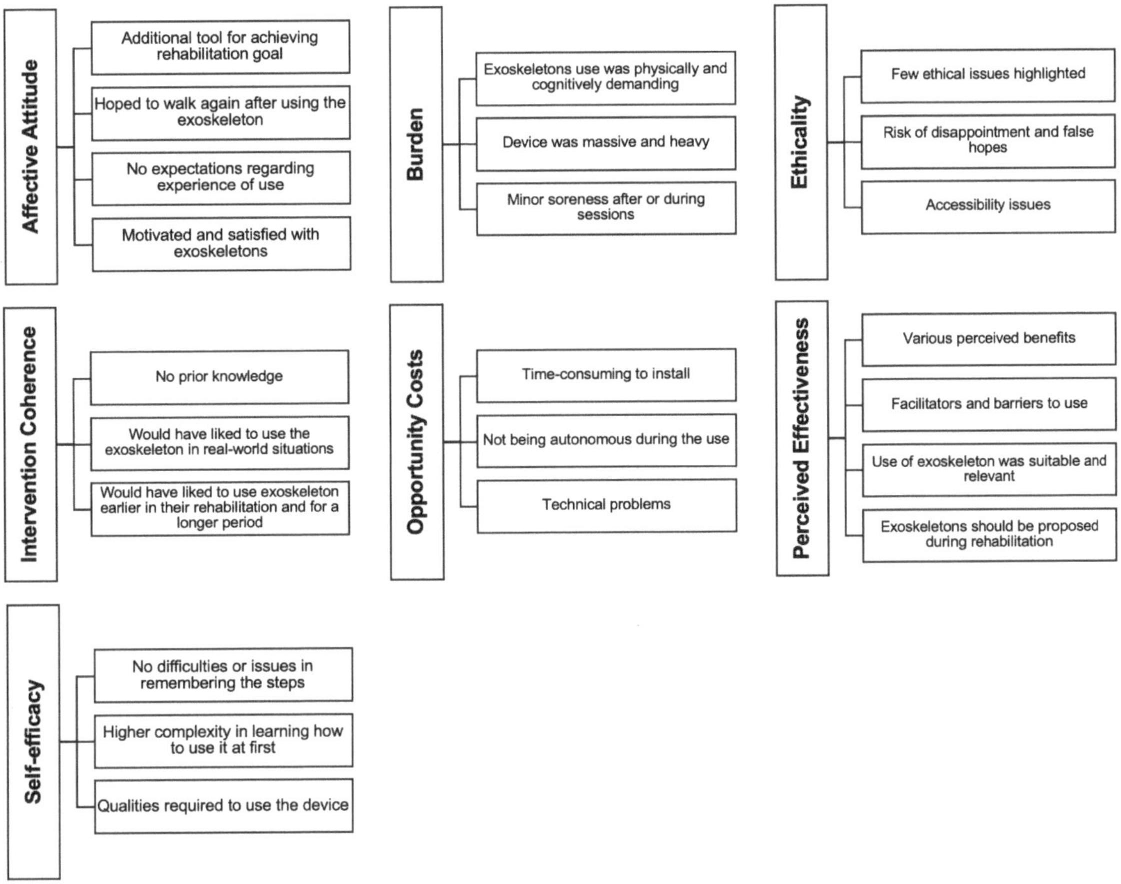

Figure 2. Summary of the main findings.

3.1. General Positive Affective Attitude

Most participants had no initial apprehension about using the exoskeleton in their rehabilitation. The exoskeleton was considered an additional tool for achieving their rehabilitation goal. In this regard, half of the participants hoped to walk again after using the exoskeleton:

> "That's what I was hoping for and still hope for that it would help me walk again. [. . .] I was ready to take part in anything. That was my hope, to shock my body and then stand up most of the time. Then, the exoskeleton was part of it. We did big sessions. I wanted to take a lot of steps too, to give myself every chance."
>
> (Participant in Quebec City, Canada)

Some participants mentioned having no expectations regarding experience or outcomes to avoid disappointment. A few participants were primarily interested in better understanding the device. Three participants reported that their expectations had been met:

> "Because it put me back on my feet, because it gave me a little hope because it gave me hope. And it [the exoskeleton] really exceeded my expectations, I was able to have fun with it."
>
> (Participant in Ploemeur, France)

Almost all participants indicated that they were extremely motivated to use an exoskeleton. While walking again was identified as their main goal and motivation, others had an objective to achieve a more natural walking pattern. Their motivation stemmed from a willingness to explore every available tool that could assist in their rehabilitation, as well as to contribute to the advancement of knowledge about the exoskeleton for the benefit of others. In this regard, three participants expressed feeling privileged to use an exoskeleton during their rehabilitation:

> "I was very motivated. I was told that I was lucky to be able to participate in this [to use the device], and that it wasn't for everyone. So, well, I saw it as another chance, another tool, to help me achieve walking as normally as possible."
>
> (Participant in Quebec City, Canada)

One participant, who felt that the exoskeleton had not been used sufficiently to observe significant improvement, expressed moderate satisfaction. However, almost all participants reported being satisfied or very satisfied with the exoskeleton, even if they did not all achieve their desired outcomes:

> "[I] am very grateful, but [I] don't have the end result we'd like. We'd all like the miracle."
>
> (Participant in Quebec City, Canada)

A few participants reported that their relatives were excited, happy, and impressed to see the potential benefits of exoskeleton use on their loved ones with an SCI. However, participants had to explain the actual effects to their relatives:

> "For them [relative], it's magical to be able to see us again through videos of us walking vertically, especially since they mostly see us in a wheelchair. However, from their perspective, they see it more as an indication that, since we're in it, we must be going to walk again. It's always a bit more complicated to explain to them that it's not necessarily the case, but it's a good tool for rehabilitation."
>
> (Participant in Ploemeur, France)

In summary, participants were generally motivated and satisfied with using an exoskeleton during their rehabilitation, and the majority had no apprehensions or expectations.

3.2. Requirements for Physical and Cognitive Engagement

Almost all participants reported that using an exoskeleton was physically and cognitively demanding. On the one hand, it required a high level of attention to follow instructions given by the therapist during sessions:

> "You have to be 100% focused on every step, on every breath, leaning forward, which is unnatural. You must be 100% focused. And especially at the beginning; I couldn't talk and walk at the same time."
>
> (Participant in Quebec City, Canada)

On the other hand, the use of the exoskeleton was physically demanding, especially at the beginning of use, even if the participants were in good physical shape:

> "It's a lot of energy, a lot of energy for a little progress."
>
> (Participant in Quebec City, Canada)

In this regard, a few individuals reported that the device was massive and heavy. A participant also expressed that the device looked like a robot. Additionally, four participants reported experiencing minor soreness during or after sessions, including back pain, discomfort around the lesion, tendon rubbing, spasms, and minor grazes. Nevertheless, participants said that the therapists had been alerted and that the pain was minor:

> "I was more satisfied to have done it than the pain it would bring. The pain was small. [...] All the happiness of having done it made up for the pain it brought."
>
> (Participant in Quebec City, Canada)

Finally, one participant living outside the area where the exoskeleton sessions took place reported significant costs, particularly for travel and accommodation. Certain expenses should have been covered to enable more participants to use the exoskeleton to advance knowledge of this device:

> "Candidates from outside [...] they should absolutely have an allowance for the hotel, an allowance for meals, then a travel allowance. Because if not, it [knowledge about exoskeletons] won't advance as much."
>
> (Participant in Quebec City, Canada)

All these results highlighted, in sum, the physical and cognitive demands of using an exoskeleton, as well as the massiveness and heaviness of the device.

3.3. Emerging Ethical Considerations

Few participants discussed ethical issues regarding the use of exoskeletons. One participant reported disappointment due to the temporary effects of the use of the device:

> "On the downside, there's the disappointment, of course. When you finish, you go back into a wheelchair. That's how it is, it's part of the game; you must accept it. Accepting it is already a chance to be able to try it. I was aware of that."
>
> (Participant in Ploemeur, France)

Another participant expressed the possibility of false hope regarding the outcomes of the use of the exoskeleton, highlighting the importance of carefully selecting participants:

> "These are the tests that are conducted [before use] which will determine whether you can go in or not. I think putting someone in it, especially someone on whom we are almost certain will not walk again, and giving them false hope, may not necessarily be a good idea; hence the tests conducted at the beginning."
>
> (Participant in Ploemeur, France)

Finally, a participant reported that access to the use of an exoskeleton in the rehabilitation process of people with SCI should be more inclusive:

> "Like me, I was pretty standard, no health issues, height, size, it fits, everything is fine. Granted, there are those who may not be as fortunate and might possibly need it, but in a future, maybe not too distant, we can understand that it might be enjoyable for those individuals to have access to a similar experience"
>
> (Participant in Quebec City, Canada)

In sum, few ethical considerations were discussed, but the risk of disappointment, the possibility of false hope, and accessibility issues were reported by participants.

3.4. Variable Intervention Coherence

Most participants did not have prior knowledge about exoskeletons before being approached by therapists to use the device. Even if the majority of participants received sufficient information from their therapists, others reported insufficient information about the benefits and risks. A participant suggested presenting short videos to future participants before use to increase their knowledge:

> "Perhaps there could have been some small videos featuring 2–3 people without showing their faces, maybe doing the workout sequence. And, maybe showing when they put on the device. [...] It would probably be interesting to be able to see a short video before signing the consent. I don't think it scares people."
>
> (Participant in Quebec City, Canada)

One participant discovered exoskeletons through the media:

> "Like everyone else. I saw it a little in the media, but without thinking that one day I would have the opportunity to try it as a patient at the center."
>
> (Participant in Ploemeur, France)

Participants reported that individuals with an SCI having strength, endurance, and determination and who fit the inclusion criteria should have the opportunity to use an exoskeleton during rehabilitation. Some participants would have liked to use the device earlier in their rehabilitation and to continue using it for a longer period. However, one participant said that the frequency of sessions was adequate, given the effort required to participate:

> "I had two sessions a week, and then that was fine. I'm not sure I would have taken three, certainly not four, because it's very demanding."
>
> (Participant in Quebec City, Canada)

Finally, three participants reported that they would have liked to use the exoskeleton in real-world situations but that the usage of the exoskeleton is currently mainly in the context of rehabilitation:

> "What does it give me to take 2500 steps with two crutches if I can't go grocery shopping? I can't go for a walk. Because I walk on hard surfaces, so..."
>
> (Participant in Quebec City, Canada)

Overall, these results highlighted the lack of prior knowledge about exoskeletons and that participants would have liked to use the device earlier and longer in their rehabilitation.

3.5. *Remaining Opportunity Costs*

The exoskeleton was deemed time-consuming to install, especially during the initial fitting sessions. One participant reported that this initial preparation takes therapy time:

> "It's the implementation because we have an hour of physiotherapy, and as a result, it takes three-quarters of an hour to set up. But that's the way it is. You must accept it, so you're a guinea pig for 45 min for ten minutes of walking. You have to accept it."
>
> (Participant in Ploemeur, France)

A participant also reported not being autonomous during the use of the exoskeleton, unlike when his wheelchair is used, resulting in a temporary loss of autonomy:

> "I could never walk as fast as the exoskeleton allows me to, but with my canes, I walk more slowly but I can walk, I can climb stairs, which I can't do with an exoskeleton, it would limit me. I can go down a ramp, I can turn, I can... and then when I want to shave, and I stand up because I am able to stand, it won't take me 15 min to put on the device."
>
> (Participant in Quebec City, Canada)

Finally, while most participants reported no technical problems and adequate equipment, four participants reported that the device freezes and jams, as well as problems with the waist belt, resulting in loss of time and inefficient use of therapeutic expertise.

In sum, these findings highlighted the additional time needed for setting up exoskeletons and the occurrence of technical and equipment-related issues.

3.6. Positive Perception of Effectiveness

The participants' perceptions of the effectiveness of the exoskeleton during their rehabilitation varied widely. A few participants reported benefits such as better walking pattern, increased endurance, increased muscle mass, greater leg strength, less spasticity, and improved ability to walk and walk faster:

> "I liked feeling like I did before my accident."
>
> (Participant in Quebec City, Canada)

However, one participant reported not feeling enough benefit from using this technology. In addition, few participants noted the challenge to distinguish the effects of the exoskeleton from those of other conventional rehabilitation therapies to achieve a clear picture of the benefits:

> "I was still willing to do anything to try to improve my condition. So, I don't know if it improved anything, but I think it's a combination of all the little progress I've had, but it's connected to physiotherapy, the exoskeleton, and all that. That's for sure better than lying in bed."
>
> (Participant in Quebec City, Canada)

Participants identified several barriers that could limit the exoskeleton's effectiveness in rehabilitation settings, including physical constraints in the training environment, such as narrow corridors, crowded spaces, and uneven floors. However, the support received from the therapists and their knowledge, such as giving advice, being attentive to participants' needs, and establishing a sense of trust, was reported to be an important facilitator of exoskeleton use:

> "Like being in symbiosis with the person behind you to support you. So, you need to communicate together, and it helps a lot."
>
> (Participant in Quebec City, Canada)

Overall, most participants perceived that the use of the exoskeleton during their rehabilitation was suitable and relevant for them. When questioned, almost all participants said that the use of an exoskeleton should be offered to improve the rehabilitation of people with SCI:

> "I think it can be a good complement. [. . .] It's a somewhat less medical approach to things. As I was saying, through games or through activities like this, where ultimately, one can regain a certain movement without even realizing it, to strengthen muscles, even at the level of the torso, arms, and all that."
>
> (Participant in Ploemeur, France)

In conclusion, participants recognized the overall benefits of exoskeleton use, although attributing these benefits specifically to the device remains challenging. They also identified barriers and facilitators that could influence the effectiveness of this technology.

3.7. Confident Self-Efficacy

Despite the physical and cognitive demands of using the device, most participants reported no difficulties or issues in remembering the steps involved in using it. Of note, a few participants reported higher complexity in learning how to use it at first, but that the usability improved over time:

> "At first, [it's] quite challenging. [. . .] Because it's not just a machine, it's not just a robot. You must understand it. Its functioning is designed to be as close as possible to our natural movements, and you must understand that it's a robot that's doing it. So, you also need to. . . it accompanies you, and you must accompany it at the same time. We must become one. That's why I talk about a second skin, because it's a bit like that."
>
> (Participant in Ploemeur, France)

Participants identified key qualities for using the device, including mental focus, realistic expectations, and motivation. They emphasized the importance of determination, ambition, and persistence:

> "You must be ready to stand up, to be supported by a machine. [. . .] When you've been lying down, or at least sitting for long months and suddenly, it's a machine that allows us to [walk]—there's no need to be apprehensive about that, I think."
>
> (Participant in Ploemeur, France)

In summary, participants identified necessary skills for using exoskeletons and reported no difficulties in remembering the steps for using the device.

4. Discussion

The objective of this study was to explore, among individuals with SCIs, the acceptability of locomotor exoskeletons as used in two rehabilitation centers located in France and Canada. Participants generally expressed positive acceptability regarding the exoskeleton used. Almost all participants reported a high level of satisfaction with the use of the exoskeleton in their rehabilitation. This satisfaction remained high despite the exoskeleton not meeting all of the participants' expectations and a few experiencing discomforts. Most participants perceived the use of the exoskeleton as effective, experiencing little to no difficulty in its use, despite its physical and cognitive demands. While few participants reported ethical issues, some highlighted considerations regarding access to exoskeletons and possibilities of unrealistic expectations toward the use of the technology. To favorize the successful implementation of exoskeletons in facilities offering rehabilitation services for people with SCI, it is crucial to have before considering other features sufficient information on the acceptability of this technology by users [33]. The results of this multicenter study, based on TFA, provide in-depth insights into the diverse experiences of individuals with SCI regarding the perceived acceptability of this technology. These results add additional knowledge to the previous studies conducted about the feasibility and effectiveness of exoskeletons used in rehabilitation. Overall, these results are consistent with a systematic literature review reporting a favorable acceptability of the use of exoskeletons among people with SCIs [21]. This literature review identified the need to assess the acceptability of this technology more systematically. The use of TFA thus enables this study to respond to this knowledge limitation.

The personal characteristics of participants, including individual culture and previous knowledge, might be an important factor that influences the acceptability and use of technologies. In the present study, most participants were unfamiliar with the exoskeleton before its use, although they had been informed about the risks and benefits and their few questions have been answered. A previous study found that most people with SCI learned about exoskeletons through traditional media channels (e.g., documentaries or news on television) [19]. The extent to which suboptimal knowledge about the effects of exoskeletons influences expectations should be examined in future studies. In this regard, clinicians involved in the study of Heinemann et al., 2018 reported that high expectations in terms of perceived benefits may be a risk for users [34]. Considering that potential benefits of exoskeleton use extend beyond physiological effects and can have a great psychological influence, these high expectations may be particularly common [35]. Consequently, future studies should focus on the development of an intervention to facilitate the adoption of realistic expectations by people using a rehabilitation exoskeleton to avoid disappointment toward technology, and, in turn, a low acceptability. This intervention could take the form of an educational activity aimed at enhancing the knowledge of both current and potential users.

Moreover, users with an SCI may feel privileged to be selected to use the exoskeleton. This feeling of privilege may bias the reported acceptability toward the exoskeleton in a favorable direction and thus potentially influence the findings. Indeed, the limited number of exoskeletons currently available in rehabilitation centers as well as the stringent eligibility criteria (e.g., morphology, level of injury) due to the characteristics of the device are major accessibility issues [36]. This context of limited technology availability highlights the relevance of the global principles of diversity, equity, and inclusion in the use and development of innovative technologies such as exoskeletons. Consequently, as proposed in the present study, developers, managers, decision makers, and others should initiate a collective reflection to ensure that exoskeletons can be used by a greater number of people with SCIs or any other conditions leading to limitations in walking in the future.

The context of use of the exoskeleton [37], whether for clinical or research purpose, can also influence user acceptability. For example, in a research context, participants and clinicians must follow predetermined protocols. These protocols restrict the adaptability of exoskeletons and, to some extent, fail to consider users' expectations, goals, and capacities. This lack of adaptability can greatly influence the acceptance of the technology, as the personalization of interventions provided is an important factor of acceptability [38]. In addition, the intrinsic features of a specific exoskeleton model that influence perceived effectiveness may also impact user acceptability [16]. For example, the Atalante exoskeleton allows users to have their arms free during its use which, in turn, allows the combination of upper-body activities with walking rehabilitation. Therefore, it is essential to analyze the goals of users when implementing a rehabilitation exoskeleton.

Limitations of our study must be considered. First, the context of use in two different rehabilitation centers with potential cultural and organizational differences, may influence the results obtained. Future studies should comprehensively assess how the context of use influences the acceptability of exoskeletons among people with SCI. Second, the time since the occurrence of the SCI was much shorter for people with SCIs using an exoskeleton at the CIUSSS-CN center than at the CMRRFK center. This may influence the perceived acceptability regardless of the exoskeleton per se. The different walking prognosis among the participants could also influence the acceptability of the technology. However, we did not discern any major differences in the participants' opinions. In addition, people with SCI at the CIUSSS-CN had to follow a research protocol for the use of the exoskeleton. The research protocol restricted the use of the exoskeleton to predefined sessions compared with more flexible sessions at the CMRRFK, where users had the possibility of performing different programs and even sporting activities. The number of usage sessions by participants could also influence their perspective. Indeed, Sekhon et al. describe well the influence of time on the acceptability of technology [16]. In addition, the interview guide was based on the TDF, and, consequently, users were not specifically asked about certain areas of the TFA (e.g., ethics). Finally, CIUSSS-CN participants were invited to take part in interviews only if they completed the twelve-session training program of the research protocol. Thus, the few participants who dropped out of the program were not invited to participate in the interviews and their perceived acceptability of the exoskeleton may differ from those included in the study. Hypothetically, these individuals could report different experiences influencing their acceptability of the technology.

5. Conclusions

The exoskeleton is a promising rehabilitation technology that was generally accepted by the participants in this study. Indeed, the present study highlighted that users with SCI were satisfied with using an exoskeleton during their rehabilitation, even if the use was physically and cognitively demanding. Several perceived benefits to its use, such as better walking pattern, increased endurance and greater muscle mass, and mild pain were reported. The use of an exoskeleton during rehabilitation process was appropriate and relevant to them for most of the participants. However, adjustments in its use, such as conducting training sessions in obstacle-free environment and technological improvements to address the device's restrictive characteristics, heaviness, and massiveness are, however, still needed. In future studies, a deeper exploration of the ethical considerations will be important, along with efforts to increase the inclusivity of exoskeleton characteristics and accessibility for a broader spectrum of individuals with walking limitations.

Supplementary Materials: The following supporting information can be downloaded at: https://www.mdpi.com/article/10.3390/biomimetics10010036/s1, Table S1: Exclusion criteria for the use of the Indego® exoskeleton; Table S2: Exclusion criteria for the use of the Atalante® exoskeleton; Interview questions.

Author Contributions: Conceptualization, N.F.-B., F.R., and M.-E.L.; methodology, N.F.-B., S.L., F.R., and M.-E.L.; formal analysis, N.F.-B., A.P., and M.-E.L.; investigation, N.F.-B. and A.C.; writing—original draft preparation, N.F.-B.; writing—review and editing, N.F.-B., A.P., S.L., M.-C.D., C.C., W.A., A.C., C.R., A.K.B., L.B., J.K., F.R., and M.-E.L.; supervision, S.L., M.-E.L., and F.R. All authors have read and agreed to the published version of the manuscript.

Funding: This work was partially carried out within the Handicap Innovation Territoire project, supported by the French Government as part of France 2030—Territoires d'innovation program for innovative territories and received funding from the Praxis Spinal Cord Institute (Research Grant No: G2020-33) and the Social Sciences and Humanities Research Council (Research Grant No: 892-2021-2033). Noémie Fortin-Bédard was supported through a scholarship from the Réseau provincial de recherche en adaptation-réadaptation (REPAR) and holds a master training award from the CIHR and from the Quebec Health Research Funds (FRQS) (#311138) in partnership with the Unité de soutien au système de santé apprenant (SSA) Québec. Salary support for François Routhier were provided by FRQS Senior Research Scholar program (#296761).

Institutional Review Board Statement: The study was conducted in accordance with the Declaration of Helsinki and approved by local Research Ethics Boards from the Université de Lille (2023-701-S117) and from the CIUSSS-CN (MP-13-2020-2002).

Data Availability Statement: Due to ethical reasons (i.e., the potential risk of identifying participants), the qualitative data are not available.

Acknowledgments: The authors thank the clinical teams of the myelopathy program of CIUSSS-CN center and from the spinal cord injury programs of CMRRFK for their contribution to the completion of this study. Conducting interviews in France by Noémie Fortin-Bédard was made possible with travel grants received from the Canadian Institutes of Health Research (CIHR), Mitacs Globalink program, and the Bureau International of Université Laval.

Conflicts of Interest: The authors declare no conflicts of interest.

References

1. Statistiques Canada. Canadian Survey on Disability, 2017 to 2022. Available online: https://www150.statcan.gc.ca/n1/daily-quotidien/231201/dq231201b-eng.htm (accessed on 7 June 2023).
2. Liu, Y.; Yang, X.; He, Z.; Li, J.; Li, Y.; Wu, Y.; Manyande, A.; Feng, M.; Xiang, H. Spinal cord injury: Global burden from 1990 to 2019 and projections up to 2030 using Bayesian age-period-cohort analysis. *Front. Neurol.* **2023**, *14*, 1304153. [CrossRef] [PubMed]
3. World Health Organization. Spinal Cord Injury 2013. Available online: https://www.who.int/news-room/fact-sheets/detail/spinal-cord-injury (accessed on 7 June 2023).
4. Juguera Rodriguez, L.; Pardo Rios, M.; Leal Costa, C.; Castillo Hermoso, M.; Perez Alonso, N.; Diaz Agea, J.L. Relatives of people with spinal cord injury: A qualitative study of caregivers' metamorphosis. *Spinal Cord* **2018**, *56*, 548–559. [CrossRef] [PubMed]
5. Estores, I.M. The consumer's perspective and the professional literature: What do persons with spinal cord injury want? *J. Rehabil. Res. Dev.* **2003**, *40* (Suppl. S1), 93–98. [CrossRef]
6. Ditunno, P.L.; Patrick, M.; Stineman, M.; Ditunno, J.F. Who wants to walk? Preferences for recovery after SCI: A longitudinal and cross-sectional study. *Spinal Cord* **2008**, *46*, 500–506. [CrossRef]
7. Anderson, K.D. Targeting recovery: Priorities of the spinal cord-injured population. *J. Neurotrauma* **2004**, *21*, 1371–1383. [CrossRef]
8. Stampacchia, G.; Olivieri, M.; Rustici, A.; D'avino, C.; Gerini, A.; Mazzoleni, S. Gait rehabilitation in persons with spinal cord injury using innovative technologies: An observational study. *Spinal Cord* **2020**, *58*, 988–997. [CrossRef] [PubMed]
9. Gil-Agudo, Á.; Megía-García, Á.; Pons, J.L.; Sinovas-Alonso, I.; Comino-Suárez, N.; Lozano-Berrio, V.; Del-Ama, A.J. Exoskeleton-based training improves walking independence in incomplete spinal cord injury patients: Results from a randomized controlled trial. *J. NeuroEng. Rehabil.* **2023**, *20*, 36. [CrossRef]
10. Miller, L.; Zimmermann, A.K.; Herbert, W.G. Clinical effectiveness and safety of powered exoskeleton-assisted walking in patients with spinal cord injury: Systematic review with meta-analysis. *Med. Devices Evid. Res.* **2016**, *9*, 455–466. [CrossRef] [PubMed]
11. Louie, D.R.; Eng, J.J.; Lam, T. Gait speed using powered robotic exoskeletons after spinal cord injury: A systematic review and correlational study. *J. NeuroEng. Rehabil.* **2015**, *12*, 82. [CrossRef] [PubMed]
12. Edwards, D.J.; Forrest, G.; Cortes, M.; Weightman, M.M.; Sadowsky, C.; Chang, S.-H.; Furman, K.; Bialek, A.; Prokup, S.; Carlow, J.; et al. Walking improvement in chronic incomplete spinal cord injury with exoskeleton robotic training (WISE): A randomized controlled trial. *Spinal Cord* **2022**, *60*, 522–532. [CrossRef]
13. Charbonneau, R.; Loyola-Sanchez, A.; McIntosh, K.; MacKean, G.; Ho, C. Exoskeleton use in acute rehabilitation post spinal cord injury: A qualitative study exploring patients' experiences. *J. Spinal Cord Med.* **2022**, *45*, 848–856. [CrossRef] [PubMed]
14. Chau, T.; Moghimi, S.; Popovic, M.R. Knowledge translation in rehabilitation engineering research and development: A knowledge ecosystem framework. *Arch. Phys. Med. Rehabil.* **2013**, *94*, S9–S19. [CrossRef]
15. Waldron, D.; Layton, N. Hard and soft assistive technologies: Defining roles for clinicians. *Aust. Occup. Ther. J.* **2008**, *55*, 61–64. [CrossRef] [PubMed]
16. Sekhon, M.; Cartwright, M.; Francis, J.J. Acceptability of healthcare interventions: An overview of reviews and development of a theoretical framework. *BMC Health Serv. Res.* **2017**, *17*, 88. [CrossRef]
17. Ramsey, A.; Lord, S.; Torrey, J.; Marsch, L.; Lardiere, M. Paving the Way to Successful Implementation: Identifying Key Barriers to Use of Technology-Based Therapeutic Tools for Behavioral Health Care. *J. Behav. Health Serv. Res.* **2016**, *43*, 54–70. [CrossRef] [PubMed]
18. Ben Mortenson, W.; Pysklywec, A.; Chau, L.; Prescott, M.; Townson, A. Therapists' experience of training and implementing an exoskeleton in a rehabilitation centre. *Disabil. Rehabil.* **2022**, *44*, 1060–1066. [CrossRef]
19. van Silfhout, L.; Hosman, A.J.F.; van de Meent, H.; Bartels, R.H.M.A.; Edwards, M.J.R. Design recommendations for exoskeletons: Perspectives of individuals with spinal cord injury. *J. Spinal Cord Med.* **2023**, *46*, 256–261. [CrossRef]
20. van Dijsseldonk, R.B.; van Nes, I.J.; Geurts, A.C.; Keijsers, N.L. Exoskeleton home and community use in people with complete spinal cord injury. *Sci. Rep.* **2020**, *10*, 15600. [CrossRef]
21. Fortin-Bedard, N.; Dery, J.; Simon, M.; Blanchette, A.K.; Bouyer, L.; Gagnon, M.; Routhier, F.; Lamontagne, M.E. Acceptability of Rehabilitation Exoskeleton from the Perspective of Users with Spinal Cord Injury and Healthcare Professionals: A Mixed Methods Systematic Review. *medRxiv* **2024**. medRxiv:2024.09.18.24313846.
22. Keith, R.A. Patient satisfaction and rehabilitation services. *Arch. Phys. Med. Rehabil.* **1998**, *79*, 1122–1128. [CrossRef] [PubMed]
23. Elliott, R.; Timulak, L. Descriptive and interpretive approaches to qualitative research. In *A Handbook of Research Methods for Clinical and Health Psychology*; Oxford University Press: New York, NY, USA, 2005; pp. 147–159.
24. Greenfield, B.; Jensen, G.M. Phenomenology: A powerful tool for patient-centered rehabilitation. *Phys. Ther. Rev.* **2012**, *17*, 417–424. [CrossRef]
25. Curry, L.A.; Nembhard, I.M.; Bradley, E.H. Qualitative and mixed methods provide unique contributions to outcomes research. *Circulation* **2009**, *119*, 1442–1452. [CrossRef]
26. Tong, A.; Sainsbury, P.; Craig, J. Consolidated criteria for reporting qualitative research (COREQ): A 32-item checklist for interviews and focus groups. *Int. J. Qual. Health Care* **2007**, *19*, 349–357. [CrossRef] [PubMed]

27. Ekso Bionics. Introduction to Ekso Indego® Therapy 2023. Available online: https://eksobionics.com/static/4d9b53aafcee35933e3df49873c26b3a/Ekso-Indego-Therapy-Exoskeleton-Website.pdf (accessed on 12 February 2024).
28. Charrette, C.; Faure, C.; Mailloux, O.; Simon, M.; Routhier, F.; Bouyer, L.; Blanchette, A.K. Feasibility and preliminary effects of a locomotor training program using an overground powered exoskeleton combined with functional electrical stimulation in clinical practice for individuals with a subacute spinal cord injury. Center for Interdisciplinary Research in Rehabilitation and Social Integration, Centre Intégré Universitaire de Santé et de Services Sociaux de la Capitale-Nationale, Quebec, QC, Canada. 2024; *In preparation*.
29. Zoom Video Communications Inc. Zoom 2024. Available online: https://zoom.us/fr. (accessed on 7 March 2024).
30. Michie, S.; Johnston, M.; Abraham, C.; Lawton, R.; Parker, D.; Walker, A. Making psychological theory useful for implementing evidence based practice: A consensus approach. *Qual. Saf. Heal. Care* **2005**, *14*, 26–33. [CrossRef]
31. Braun, V.; Clarke, V. Thematic analysis. In *APA Handbook of Research Methods in Psychology, Vol 2: Research designs: Quantitative, Qualitative, Neuropsychological, and Biological. APA Handbooks in Psychology®*; Psychological Association: Washington, DC, USA, 2012; pp. 57–71.
32. Fereday, J.; Muir-Cochrane, E. Demonstrating Rigor Using Thematic Analysis: A Hybrid Approach of Inductive and Deductive Coding and Theme Development. *Int. J. Qual. Methods* **2006**, *5*, 80–92. [CrossRef]
33. Taherdoost, H. A review of technology acceptance and adoption models and theories. *Procedia Manuf.* **2018**, *22*, 960–967. [CrossRef]
34. Heinemann, A.W.; Jayaraman, A.; Mummidisetty, C.K.; Spraggins, J.; Pinto, D.; Charlifue, S.P.; Tefertiller, C.P.; Taylor, H.B.; Chang, S.-H.; Stampas, A.; et al. Experience of Robotic Exoskeleton Use at Four Spinal Cord Injury Model Systems Centers. *J. Neurol. Phys. Ther.* **2018**, *42*, 256–267. [CrossRef]
35. Maribo, T.; Jensen, C.M.; Madsen, L.S.; Handberg, C. Experiences with and perspectives on goal setting in spinal cord injury rehabilitation: A systematic review of qualitative studies. *Spinal Cord* **2020**, *58*, 949–958. [CrossRef]
36. Søraa, R.A.; Fosch-Villaronga, E. Exoskeletons for all: The interplay between exoskeletons, inclusion, gender, and intersectionality. *Paladyn J. Behav. Robot.* **2020**, *11*, 217–227. [CrossRef]
37. Charette, C.; Déry, J.; Blanchette, A.K.; Faure, C.; Routhier, F.; Bouyer, L.J.; Lamontagne, M.-E. A systematic review of the determinants of implementation of a locomotor training program using a powered exoskeleton for individuals with a spinal cord injury. *Clin. Rehabil.* **2023**, *37*, 1119–1138. [CrossRef] [PubMed]
38. Bennett, S.D.; Shafran, R. Adaptation, personalization and capacity in mental health treatments: A balancing act? *Curr. Opin. Psychiatry* **2023**, *36*, 28–33. [CrossRef]

biomimetics

Review

Training and Familiarization with Industrial Exoskeletons: A Review of Considerations, Protocols, and Approaches for Effective Implementation

Pranav Madhav Kuber and Ehsan Rashedi *

Rochester Institute of Technology, 1 Lomb Memorial Dr, Rochester, NY 14623, USA; pmk2015@rit.edu
* Correspondence: exreie@rit.edu

Abstract: Effective training programs are essential for safely integrating exoskeletons (EXOs) in industrial workplaces. Since the effects of wearable systems depend highly upon their proper use, lack of training of end-users may cause adverse effects on users. We reviewed articles that incorporated training and familiarization protocols to train novices on proper operation/use of EXOs. Findings showed variation in training methods that were implemented to train study participants in EXO evaluation studies. Studies also indicate that multiple (up to four) sessions may be needed for novice EXO wearers to match movement patterns of experts, and training can offer benefits in enhancing motor learning in novices. Biomechanical assessments and ergonomic evaluations can be helpful in developing EXO-specific training protocols by determining training parameters (duration/number of sessions and task difficulty). Future directions include development of personalized training approaches by assessing user behavior/performance through integration of emerging sensing technologies. Application of simulators and use of data-driven approaches for customizing training protocols to individuals, tasks, and EXO design are provided along with a comprehensive training framework. Discussed elements in this article can be helpful to exoskeleton researchers in familiarizing novice users to EXOs prior to evaluation, and to practitioners in developing protocols for training workforce.

Keywords: wearable assistive devices; exoskeletons; motor learning; workforce training; human factors; instructional design; technology acceptance

Citation: Kuber, P.M.; Rashedi, E. Training and Familiarization with Industrial Exoskeletons: A Review of Considerations, Protocols, and Approaches for Effective Implementation. *Biomimetics* **2024**, *9*, 520. https://doi.org/10.3390/biomimetics9090520

Academic Editor: Rafhael Milanezi de Andrade

Received: 15 July 2024
Revised: 25 August 2024
Accepted: 27 August 2024
Published: 30 August 2024

1. Introduction

The integration of exoskeletons (EXOs) in industry holds promise, revolutionizing the way we work by enhancing the capabilities of the human workforce. Various types of such wearable assistive devices have been developed and are commonly categorized based on their application area as military, rehabilitation, or industry. Industrial variants of these systems are typically classified based on the method of their actuation, as activity (powered), or passive (mechanical, or pseudo-mechanical) systems. Passive industrial EXOs (i-EXOs), as shown in Figure 1, are more likely to be adopted by the workforce as these devices are compact, lightweight, and are easy to use. Each of these systems is designed to augment specific body regions (Shoulder/Back/Leg) or aid during specific tasks (like overhead assembly, lifting/bending, squatting) [1,2]. Use of i-EXOs offer several benefits, such as reducing the risk of musculoskeletal injuries, enhancing worker performance and endurance, and improving overall work quality, as displayed by detailed laboratory and field evaluations [3–6]. Despite these advantages, the adoption and implementation of exoskeletons in industry has been relatively slow. This may be attributed to various factors, including the need for customization to individual users and tasks, concerns regarding cost-effectiveness, and limited awareness and understanding of the devices among employers and workers. Furthermore, regulatory frameworks and standards for exoskeletons in the workplace are still evolving, posing challenges to their widespread implementation [7,8].

In addition to laboratory and field-based studies that examine the impacts of exoskeletons, it is imperative to prioritize the transfer of knowledge regarding their proper use to end-users. This approach can be crucial in ensuring that results from controlled evaluations are effectively translated towards real-world applications.

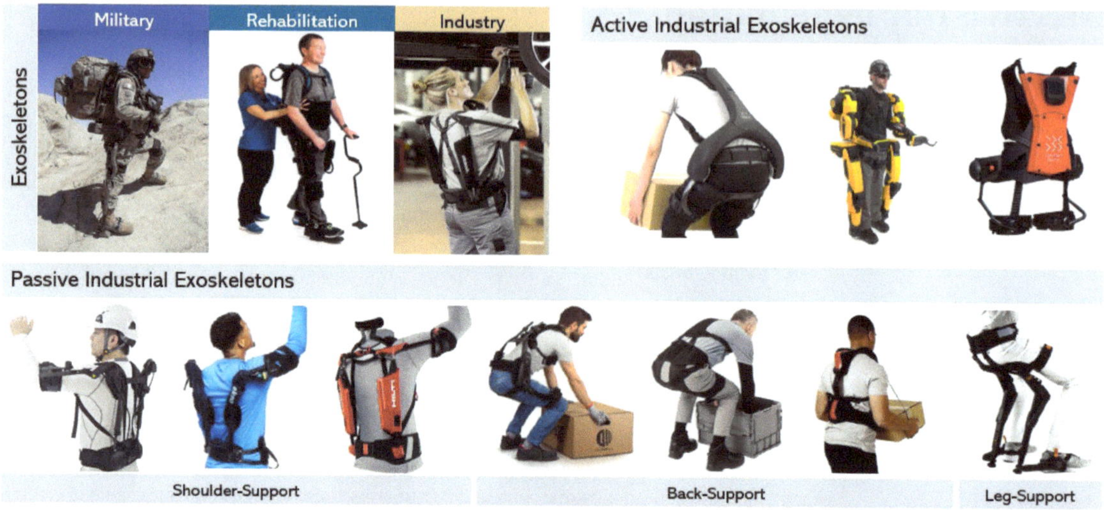

Figure 1. Application areas of exoskeletons (**top-left**), active (**top-right**) and passive (**bottom**) industrial exoskeletons classified into the shoulder, back, and leg support exoskeletons (adapted from [9]).

Training plays a key role in integrating i-EXOs in industrial environments. If proper training is not provided to exoskeleton users, they may apply the devices in use-cases wherein they are not beneficial or may use them in improper manner (e.g., with a lack of fit by not adjusting device to anthropometric dimensions or using an incorrect assistance level). This may even lead to potential biomechanical and perceived effects that can impact their safety, performance, and overall well-being in the long run (Figure 2). Without adequate training, users may struggle with suboptimal posture and movement patterns while wearing the exoskeleton. Adaptation to EXO-assisted body movements requires effective motor learning. Specifically, improper alignment or excessive reliance on the exoskeleton's assistance cause misalignment, increased joint loading, and adverse effects on wearers [10]. Users may also experience discomfort, fatigue, and reduced range of motion [11]. Additionally, without training, users may lack the necessary understanding of the exoskeleton's capabilities and limitations, which can result in overestimation or underestimation of their own physical abilities, potentially leading to accidents or inefficiencies in task performance. From a perceived standpoint, the absence of training can result in frustration, anxiety, and decreased user confidence in operating the exoskeleton, inhibiting their willingness to fully utilize potential benefits of i-EXOs [12]. Proper training is therefore vital to ensure biomechanical alignment, reduce the risk of injuries, and enhance the user's perceived comfort, confidence, and overall acceptance of exoskeleton technology.

The field of instructional design involves a systematic approach of developing effective and engaging training programs for product users [13–15]. Several methods are commonly employed in this process. Firstly, a thorough analysis of the target audience and their specific needs is conducted to identify learning objectives and create a learner profile [14]. This information forms the foundation for designing the training content and determining the most suitable delivery methods. Next, instructional designers often employ a variety of techniques such as storyboarding, concept mapping, and task analysis to organize the training materials and sequence the content logically [14,15]. Additionally, the use of formative and summative assessments can help in evaluating effectiveness of the training [13]. Throughout the process, collaboration with subject matter experts, pilot testing, and iterative design approaches can contribute to the development of a comprehensive and impactful training program for the device wearers [14].

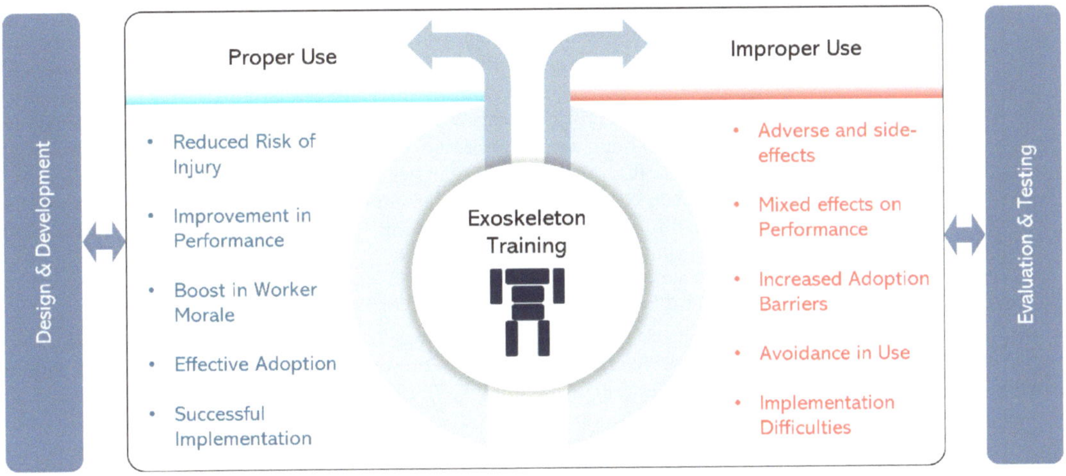

Figure 2. Illustration depicting the exoskeleton training being at the core of integration of the devices in industrial environments and its effects categorized according to proper/improper use of the devices.

Despite considerable research on evaluating effects of exoskeletons in the past 5 years, with numerous research [16–21] and review articles being published [1–3,10,22–24], there remains a noticeable lack of emphasis on the topic of training in scientific literature. This gap highlights a need to study and understand the importance of proper training for exoskeleton users. Given the intricacies involved in operating, adjusting, and maintaining exoskeletons, an instruction manual, although thoroughly prepared, may not suffice. Even after providing demonstrations, replicating movement patterns of experts may be difficult for novices. Comprehensive and in-depth training is essential to ensure user competence, minimize the risk of injury, optimize performance, and harness the potential benefits of these devices. There is a need to understand training protocols and develop best practices/guidelines to facilitate the safe and efficient integration of exoskeletons in occupational settings. The purpose of this review article was to understand currently used methods for training/familiarizing wearers of industrial EXOs. Objectives also included identifying key considerations for designing a comprehensive training program specifically for industrial EXOs. In addition, we also reviewed the effects of training and familiarization protocols on motor learning. Subsequent sections also list potential application of data-driven approaches and ergonomic evaluations in providing effective training to the workforce. Finally, a generalized training framework for users of industrial EXOs is provided along with potential applications of emerging technologies to augment training.

2. Review Methods

This narrative review involved searching relevant scientific literature on training/familiarization as well as motor learning and adaptations when using EXOs. While we mainly focused on studies related to design/development and evaluation of industrial EXOs where training was provided to study subjects, those with rehabilitation and mobility applications were also considered to understand effects of different familiarization protocols. Search methods involved conducting literature search in Web of Science, Google

Scholar, and Scopus databases for journal articles, conference proceedings, and thesis published after 2010. The search was performed with the following keywords: "exoskeleton", "evaluation", "assessment", "development", "ergonomics", "biomechanics", and "wearable assistive device" in either/OR strategy. The number of published research/review articles specifically focused on "training" or having "training" within the title of such articles was very low, and thus we expanded our search to articles involving "training" and "exoskeleton" for rehabilitation purposes. We manually screened full text of each of the obtained articles and synthesized details relating to participant training (within the methods and procedures sections in most cases). After presenting a list of articles in the form of a table, articles were also subjectively rated by the authors based on the depth and comprehensiveness of provided training to study participants into categories of low/medium/high. Subsequent searches were also performed to review fundamental concepts in the field of instructional design and to identify emerging technologies that can be applied to improve EXO training programs. Findings of our review are presented and discussed in detail in the following sections.

3. Training and Familiarization Methods for Proper Use of Exoskeletons

Providing training in new technology for the workforce has been found to have significant positive effects compared to not providing training. Research has shown that training programs enhance employees' knowledge, skills, and competencies, leading to improved job performance and productivity [25,26]. In addition, training plays a crucial role in promoting technological acceptance of technology among employees. Training helps individuals develop a better understanding of technology and its potential applications, increasing their confidence and reducing resistance to change [27]. Employees who receive adequate training are more likely to use the devices in the intended manner, enabling them to perceive the technology as useful. Moreover, training facilitates a smoother transition to the new technology, reducing errors, minimizing downtime, and improving overall operational efficiency [28]. On the other hand, a lack of training can lead to difficulties in adopting and effectively utilizing new technology, resulting in frustration, decreased job satisfaction, and increased resistance among employees. When employees do not receive adequate training, they may perceive the technology as complex, difficult to understand, and irrelevant to their job tasks, leading to lower levels of acceptance and utilization [29,30]. Besides perception, using wearable assistive devices requires adaptation to changes in body movement due to assistive/resistive forces. Subsequent sections focus on training related to operation and use of EXOs and provide an overview of approaches implemented to train study participants during exoskeleton evaluation studies, and effects of familiarization protocols on motor learning and biomechanical/perceived measures.

3.1. *Training Novices during Exoskeleton Evaluation Studies*

Evaluation-focused studies are conducted to assess the impact of i-EXOs on the human body and to estimate performance improvement/risk reduction. Such evaluations include recruitment of human subjects and simulation of industrial tasks. We reviewed articles involving design validation, laboratory, and field assessment of i-EXOs to understand currently used training methods. Findings from the review have been listed in Table 1 categorized based on the type of study (Validation/Lab/Field) and methods used for training (Duration and Tasks). Each of these studies were then categorized based on the identified level of depth, and comprehensiveness of the training session was categorized into three levels as High/Medium/Low level.

We reviewed studies that considered training sessions within their experimental protocol. Within the pool of articles, only one study conducted a laboratory evaluation to compare different familiarization methods, specifically reading instruction manual vs. observing a demonstration [31]. Participants of exoskeleton evaluation studies underwent a comprehensive training program on i-EXOs prior to the experiment [18,32], where participants were trained. In some cases, a separate session was dedicated for training participants on the use of the i-EXO [18,33,34], while in other studies, training was provided for a duration (either fixed, or until participants were comfortable/confident on using the device) prior to the experimental session. Among the reviewed articles, the studies wherein duration was fixed considered a time in the range of 2 min to 2 h [32,34–36]. In one field evaluation, participants were asked to wear and practice using the device for several days [37]. Upon coming for evaluation, participants were first educated with detailed instructions on safe use of the device [38] and a demonstration of its working [34]. In one study, an instructional video was shown to participants on proper way of donning/doffing the device [39]. The training tasks provided to participants often started with donning the i-EXO and adjusting the device to ensure proper fit. Participants were instructed to mostly practice the experimental tasks. Once proper fit was confirmed by the investigator, participants were asked to walk around until they were comfortable/familiar with the device [35,40]. This was followed by practicing the pre-determined task in each experimental condition [18]. During training, participants were also provided with the experience of max/min levels of assistance and were asked to select their preferred support level. Meanwhile, in other cases [40,41] the training included instructions on donning and doffing the exoskeleton, adjusting the assistance level, and maintaining proper body mechanics during tasks. Although not explicitly stated, the reviewed studies collectively show that incorporating training programs before experimental evaluations of i-EXOs can be beneficial for investigators as study participants need to acquire necessary motor adaptations to effectively perform intended body movements while being assisted with the devices. On the other hand, if sufficient training is not provided, study participants may perform body movements in a manner that the assistance from the device is not used to reduce the wearer's effort during the task, or in worse cases, wearers may struggle or fight against the assistive torque, leading to increased demands.

A study that compared the effects of training (demonstration vs. reading an instruction manual) demonstrated that providing study participants with a live demonstration increased performance (and reduced errors) and users acceptance, as well as led to lower global and local perceived effort [31]. In fact, easiness of learning almost doubled when study participants were provided a tutorial vs. reading a brochure. While further evaluations are needed to assess the impact of training on the safety and performance, some can be expected based on current understanding of the design of exoskeletons [10]. Designing effective exoskeleton training for industrial workers requires consideration of various variables that can influence their performance and well-being. These include mainly the efficient delivery, depth of topics covered, comprehensiveness, and duration/timing, as well as the type of task. It is expected that a longer training duration generally leads to better biomechanical outcomes, as users have more time to adapt to the exoskeleton and optimize their movement patterns. The specific nature of the task is also important, as different tasks may require varying levels of physical exertion and repetitive movements [10,42]. A lack of proper exoskeleton training may lead to detrimental effects. Without adequate training, users may not fully understand how to operate and utilize the exoskeletons correctly. A recent study that implemented that unified theory of acceptance and use of technology (UTAUT) model to EXOs and showed explanations for 75% of variation in intention to use the devices [43]. Specifically, the findings denoted that effort expectancy (how easy it seems to use an exoskeleton) played an important role in model prediction.

Table 1. List of evaluation-based studies of industrial exoskeletons (i-EXO) that trained novice study participants prior to conducting experimental evaluations of the devices. Each of the listed studies has been rated according to the level of training provided to the study participants (L: Low, M: Medium, H: High).

Article Type	Exoskeleton	Sample Size	Training Duration	Training Tasks	Training Level (L/M/H)
Lab [18]	Fawcett Exsovest with ZeroG, EksoWorks Vest, FORTIS (Mechanical Arm, Back, Full-Body	12 (6♂, 6♀)	2 h session (5 min each condition), 2 min practice for each condition in actual session	Proper fit, Practice on each experimental condition, Selecting preferred assistance level.	H
Lab [44]	VT Lowe (Back)	12♂	~30 min	15–25 practice lifts with boxes of 0–25% body weight.	H
Lab [32]	VT Lowe (Back)	15 (13♂, 2♀)	~30 min	15–25 practice lifts with boxes of 0–25% body weight.	H
Lab [40]	Laevo (Back)	18♂	-	Proper fit, movement with walking and squatting.	L
Lab [41]	SPEXOR (Back)	26♂	~2 min	Movement to get habitual with device.	L
Lab [45]	Laevo (Back)	36♂	-	Proper fit, familiarization with performing tasks.	L
Lab [46]	Exo4Work (Back)	16♂	2 h	Two familiarization trials, task practiced seven times in each trial.	H
Lab [47]	StrongArm V22 (Back)	6 (4♂, 2♀)	15 min	Wearing and getting familiar with task.	L
Field [37]	Levitate Airframe (Shoulder)	6	Several days	Proper fit, familiarization with the work.	H
Lab [48]	Laevo (Back)	13♂	-	Wearing and getting familiar with task.	M
Lab [49]	SuitX BackX (Back)	10	-	Wearing and getting familiar with task.	M
Lab [34]	HeroWear (Back)	20 (15♂, 5♀)	1.5 h session	Demonstration, donning, unstructured movement with different levels of assistance.	H
Lab [38]	Noonee chairless chair (Leg)	17 (-)	-	Detailed instruction in the use (safe use) of the exoskeleton was practiced.	L
Lab [50]	Prototype (Shoulder)	8 (4♂, 4♀)	-	Practice the task and demonstrate the understanding.	M
Lab [51]	StrongArm Flx Ergoskeleton (Back)	20 (3♂, 17♀)	-	Instructions on performing the tasks.	M
Lab [52]	BackX (Back)	8	-	Wearing the exoskeleton few days before data collection.	L
Lab [5]	EksoVest prototype	12 (6♂, 6♀)	40 min	Proper fit, practice tasks, and encouraged to determine preferred postures.	H
Lab [53]	CEX (Leg)	20♂	-	Informed about wearing the exoskeleton.	L
Lab [35]	SPEXOR (Back)	14 (7♂, 7♀)	5 min	Proper fit, movement to get familiar with the exoskeleton.	L
Lab [54]	Active Pelvis Orthosis	5♂	~20 min	Familiarization with exoskeleton, tuning of assistance level based on comfort levels.	M
Lab [33]	Noonee chairless chair (Leg)	46♂	1st visit to lab (one session—30 min)	Getting familiar with wearing and handling the device, practice the tasks with/without exoskeleton.	M
Lab [55]	BackX Model AC (Back)	18 (9♂, 9♀)	3 h	Proper fit, familiarization with each condition, finding preferred support and postures.	H

Table 1. *Cont.*

Article Type	Exoskeleton	Sample Size	Training Duration	Training Tasks	Training Level (L/M/H)
Lab [56]	Laevo (Back)	9♂	10 min	Donning, movement, and familiarization with exoskeleton followed by performing tasks.	M
Lab [57]	Laevo (Back)	10	-	Introduction to exoskeleton, donning/doffing device until confident to perform task.	M
Field [20]	Noonee chairless chair (Leg)	4 (3♂, 1♀)	30 min (day before experiment)	Training with safety instructions on use of exoskeleton.	M
Lab [58]	Paexo Back (Back)	10 (5♂, 5♀)	20 min	Instructions, adaptation, and training on functions of exoskeleton and practice trial (for 5 min).	H
Lab [39]	BackX (Back)	30 (20♂, 10♀)	-	Watching instructional videos for the proper procedures to don and wear exoskeleton.	L
Validation [59]	E-Leg (Leg)	10♂	15 min warm-up phase before experiment	Practice (wearing, squatting).	L
Validation [60]	Custom prototype (Leg)	-	-	Orientation session, walking and squatting while holding a support rail. Fit was adjusted simultaneously.	M
Validation [61]	Custom Prototype (Leg)	10♂	Before experiment (unspecified)	Informed of the detailed test procedure and exoskeleton settings.	L
Lab [31]	SkelEx	36 (18♂, 18♀)	5 min for reading/demo, 3 min for testing participants on ability to install/adjust	Instruction manual vs. demonstration.	H

3.2. Familiarization Protocols and Their Effects on Motor Learning

A subsequent review was conducted to study familiarization protocols and metrics used to quantify motor learning across training sessions. The outcomes of the review can be viewed in Table 2. To summarize, the effects of familiarization protocols on benefits provided by EXOs were found to be mixed and depended on the design, purpose, task, and users. For instance, one study found that novices could not achieve motor strategies of experts even after three sessions [62]. A recent study incorporated a detailed familiarization protocol that consisted of loaded marches, don/doff, adjustment/assembly, warm- up, and muscle-activation exercises [63]. These tasks were carried out in three phases over a duration of ~3 weeks with the assistance of an evaluator. Interestingly, the findings showed that familiarization processes did not provide benefits of reducing metabolic cost and muscular benefits. However, an earlier study by the same group [64] showed that familiarization affected differently for different designs of the same EXO, and the adjustable design affected motor learning. The duration of required training for an ankle EXO was found to be more than the duration used for training in exoskeleton studies, and training type had a significant effect [65]. Another study conducted an assessment to determine the duration and number of training sessions required to get familiar with a soft back-assist EXO by evaluating changes in biomechanical parameters (such as postural stability and muscle activity) and recommended need for at least four sessions of 1 h duration [66].

As opposed to i-EXOs, those in the healthcare sector are often used for re-training lost body functions in patients, where therapists guide and assist. A prior study conducted a qualitative assessment by comparing two groups of therapists, where one was formally trained and the other was exposed to clinical practice [12]. Findings from the study highlighted a steep learning curve and perceived difficulties in implementation of the device. Benefits of using a familiarization protocol were found in a study evaluating perception and usability of an EXO for walking in study participants that were provided training [67].

A detailed step-by-step method consisting of 11 steps for training patients on proper use of the devices has been proposed in a prior study [68], and similar protocols could be developed for training users of their industrial variants. Studies also demonstrate the use of biomechanical metrics as indicators of familiarization. For example, gait parameters of stride duration, mediolateral deviation, and polygon of support area were found to be good familiarization indicators for gait-training EXOs. Another study developed a method to replicate body movements of therapists in patients and promote correct motor learning [69].

Table 2. List of studies that implemented familiarization protocols to study their effects on motor learning when performing tasks while assisted with an exoskeleton (EXO).

Article	EXO (Primary Purpose)	Study Goals	Sample Size	Familiarization Protocol	Outcomes
[64]	UPRISE Gen 3.0, 4.0 (Military)	To assess differences between customizable (Gen 3.0) and adjustable (Gen 4.0) designs of EXO.	Dataset 1: 3♂ soldiers, Dataset 2: 3 healthy participants	Gen 3.0: 3 h daily sessions for 9 days distributed over 2 weeks. Gen 4.0: 14 days of training (around 1 h 30 min) distributed over a period of 4 weeks.	After familiarization, users reduced their MCW when using the customized but not adjustable EXO. Adjustment of the EXO affected user and motor learning.
[63]	UPRISE Gen 4.0 (Military)	To understand changes in metabolic cost of walking after a familiarization process.	13 (12♂, 1♀)	~14 days of tasks (~1–1.5 h) with an evaluator. Distributed practice and gradual progression of loads and difficulty across three phases. Tasks included loaded marches, don/doff, adjustment/assembly, warm-up, and muscle-activation exercises.	Familiarization process did not provide benefits of reducing the metabolic cost of walking and muscle activation.
[62]	Guardian XO 2019 prototype (Industrial)	Adaptation of novices to EXOs and comparison with experts.	11♂ (6 novices, 5 experts)	Novices: initial familiarization and three subsequent gait sessions. Experts: one gait session.	Novices demonstrated progress but could not achieve similar motor strategies as experts even after three sessions.
[68]	ExoAtlet (Rehabilitation)	Propose steps for training patients on using the EXO.	1♂	Validation: Electroencephalography (EEG) topographic maps, pressure insoles, and discomfort.	Training was beneficial in clinical settings.
[70]	TWIN (Rehabilitation)	To provide four biomechanical metrics as familiarization indicators.	5	Five walking bouts. Evaluation measures: gait parameters, support area, and muscle activity.	Stride duration, mediolateral deviation from a straight path, and polygon of support area were found to be good familiarization indicators.
[71]	Soft hip EXO (Military)	To study metabolic adaptations over training sessions.	8♂ military cadets	Five training sessions within 20 days. Task consisted of loaded walking (with ~20 kg).	Percentage benefits of EXO in net metabolic cost improved across sessions from $-6.2 \pm 3.9\%$ (session one) to $-10.3 \pm 4.7\%$ (session five).
[72]	Prototype (Rehabilitation)	To conduct wearability evaluation of EXO.	15 healthy individuals (7♂, 8♀), 2♂ stroke survivors	Trials with donning/doffing the device. Evaluation included donning/doffing time and usability.	Participants were able to independently don and doff the device after four practice trials.
[31]	SkelEx (Industrial)	To assess effects of media used for familiarization.	36 (18♂, 18♀)	Instruction manual vs. demonstration. 5 min for reading/demo, 3 min for testing participants on ability to install/adjust.	Live tutorial led to higher task performance and user acceptance and reduced global and local perceived effort as well as errors.

Table 2. *Cont.*

Article	EXO (Primary Purpose)	Study Goals	Sample Size	Familiarization Protocol	Outcomes
[67]	TWIN EXO (Rehabilitation)	To study effects of familiarization protocol.	9 (6♂, 4♀)	Proposed protocol included: preparation, tutorial, exoskeleton session, and ending. assessment that included System Usability Scale, NASA Raw Task Load Index, and surveys.	The protocol was found to be beneficial for improving familiarity.
[66]	CORFOR soft back EXO (Industrial)	To characterize the familiarization procedure and determine the time for stabilizing biomechanical parameters.	18♂	Six familiarization sessions of 1 h. Measurements included kinematics, posture stability, perception, muscle activity, and performance during lifting tasks (8 kg box).	Recommend four sessions of 1 h to stabilize parameters.
[65]	Bi-lateral ankle EXO (Mobility)	To understand the effects of training and type of training on metabolic cost.	15 (10♂, 5♀)	Three training groups with variation of Low, Medium, and High in device behavior.	Training required more exposure than typical studies of 109 min of assisted walking. Low variation group needed 2× duration of moderate group. High group never reached expertise. Metabolic cost reduced by 39%, and training contributed to half of the benefits.

4. Limitations of Current Training Methods and Implementation Challenges

Review findings indicate a wide variation in the training procedures (duration and tasks) for training study participants during EXO evaluation studies. For instance, duration ranged from as low as a few mins per experimental task up to several days depending on the complexity of the experiment, aims of the experiment, the design of the EXO being evaluated, and the application area. While in some cases participants were trained in donning/doffing, proper fit, and safety procedures, others mostly focused on specific experimental tasks (Table 1). As observed from Table 2, effects of the type, duration, and comprehensiveness of the familiarization protocols were observed on physiological parameters during biomechanical assessments. Differences in training protocols may have affected the level of adaptation across study participants, as well as between studies. This implies that EXO evaluation studies require standardized training protocols so that the outcomes are more generalizable across studies. It should also be noted that reviewed studies (Table 1) involved novice participants that were recruited from the university population, and there could be several challenges when training industrial workers. For example, depending on the education level of workers, learning about using the device may first require providing fundamental knowledge about the function of the devices. Similarly, demonstrations and familiarization protocols may require instructor interventions as well as assessments to test whether these devices are being properly used.

Designing proper training programs for users of i-EXOs poses several challenges. One significant challenge is the customization of training to accommodate individual differences in anthropometry, physical abilities, and prior experience with exoskeletons [23]. These factors must be thoroughly assessed to tailor the training program effectively. Another challenge is achieving optimal fit and alignment of the exoskeleton to the user's body. Exoskeletons come in various designs, and anatomical variability among users can make it challenging to ensure a proper fit. Although alteration in design may be controllable, especially during implementation phases, ensuring proper fit prior to use may help in reducing such side effects. Besides relying on subjective measures (perspectives from users) about the fit, objective methods could be utilized. To elaborate, along with simpler verification based on body size (anthropometric dimensions) and provided instruction manuals (that often provide adjustability steps based on body/segment lengths), body shape considerations are also needed. We recommend that instructors in exoskeleton training focus on structural components of the device, their adjustment, and their relation to discomfort caused by high-pressure spots at interfaces/attachments. After a low-fidelity testing (subjective rating of comfort), a detailed assessment may be conducted using objective tools such as pressure mat systems to assess the presence of such discomfort regions [73–75]. Similar studies have also provided insights on user comfort level based on body demands and range of motion [76,77]. Inputs from such studies (such as setting an optimal assistance level to minimize discomfort and side effects) could play a crucial role in the development of instructional materials for specific tasks/conditions in which the i-EXO in consideration may be helpful. The following section provides a more detailed account on the role of ergonomic evaluations in informing design of training programs.

5. Role of Ergonomic Evaluations in Exoskeleton Training Design

The field of ergonomics involves the study of human interactions with products, processes, and the environment to optimize system safety and performance. Often, ergonomic evaluations include obtaining qualitative (like comfort and usability) and quantitative (such as kinematics, kinetics, muscular demands, and metabolic rate) measures to estimate the impacts of interventions (i.e., EXO) on the human body. Earlier controlled assessments have provided a detailed account of tasks where EXOs can be beneficial and those where EXOs are known to increase demands (side- and adverse-effects). A recent review [10] has stated that the majority of side effects (discomfort and reduced usability being the most prominent) are related to incorrect fit, uncomfortable materials (e.g., straps, belts, plates), or misalignment between i-EXO and human joints. Such conditions may cause variation in distribution of assistive forces/moments, slippage of attachments during work, or increased joint loading in other body regions. Proper donning/doffing and use of EXOs can be evaluated by testing motor learning patterns of novice users and comparing biomechanical measures to those of experts. Training can incorporate biomechanical assessments to study forces, movements, and loads on the human body (for instance, calculating low-back compression force at the L5-S1 joint). This can help in obtaining insights into optimal movement patterns (comfortable range of motion of upper limbs) and muscle-activation strategies (increase/decrease in muscular demands within specific muscle groups). Previous studies that evaluated temporal effects of using i-EXOs denote that the devices may be helpful in reducing fatigue [6,34,78], yet longitudinal and field studies have shown mixed outcomes [3].

Few efforts were found on understanding the effects of adaptation/learning during early phases of using EXOs as depicted in Table 2. One study [31] involved a lab-based study that compared the effects of familiarization protocols of providing a live demonstration vs. reading an instruction manual. As expected, the findings showed that providing a demonstration was beneficial in improving performance as well as usability and reducing physical demand. In another study, differences in walking patterns between six novices vs. five expert users were compared after they wore a whole-body active EXO in a recent article [79]. The outcomes showed significant differences in spatial (25% less step/stride lengths, and ~8° less knee/hip flexion among novices) parameters during walking. These differences decreased after three training sessions showing the benefits of training EXO users. Meanwhile, another study showed that during early adaptation phases, users may require additional cognitive resources during movement while wearing an EXO, as interpreted from increased visual reaction time [80]. A more detailed study was also conducted to study adaption of naïve users to an ankle-assist EXO by offering different levels of variation in training (Low, Medium, High) [65]. With moderate training, large benefits were displayed for customized assistance mode as shown by ~39% reduction in metabolic rates compared to walking with the EXO turned off. Furthermore, the study also showed that types of training can significantly affect training duration. In another study about evaluating motor adaptation of an ankle EXO controlled using muscle activity, recruitment of soleus muscle decreased by ~35% after practice [81]. Moreover, study participants retained adapted locomotor pattern, as assessed by similar muscle activity, kinematic, and kinetic forms between different sessions. These studies not only highlight the benefits of providing training to EXO users, but also show the role of ergonomic/biomechanical evaluations in developing training programs.

6. Comprehensive Framework for Training Exoskeleton Users

The successful integration of i-EXOs into the workplace requires a comprehensive and effective training program. For instance, a prior study mentioned a familiarization protocol consisting of the following steps: Demystification, Technics, Potential, Limits, Donning/Adjusting/doffing, and Free experience (without industrial constraints) [31]. The study provided a 7-level familiarization chart, and suggested that subjects need to reach a level of four prior to using i-EXOs [31]. Fundamental theories and models in the field of instructional design were reviewed to understand methodologies used to develop training programs. According to the principles of effective instruction, learning is promoted when (a) it is problem-centered, (b) prior knowledge is activated, (c) new knowledge is demonstrated, (d) applied, and (e) integrated in the real world [13]. These principles can be related to designing training protocols for specific applications. Based on the instructional design practices, we propose a training framework abbreviated as IBDEI (Identify, Brief, Demonstrate, Evaluate, Implement) for training the workforce on i-EXOs [13,25,29,82]. This five-step approach (Figure 3) encompasses the identification of user needs, delivery of fundamental knowledge, hands-on demonstrations, evaluation of performance, and real-world implementation.

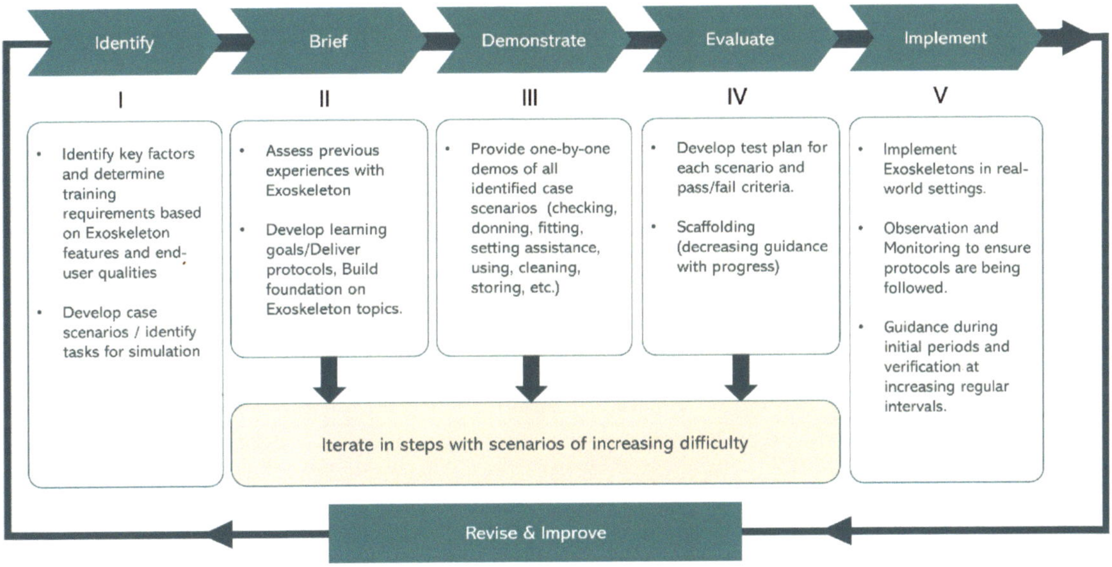

Figure 3. Illustration depicting the IBDEI (Identify, Brief, Demonstrate, Evaluate, Implement) exoskeleton training framework for guiding the workforce towards effective and safe use of industrial exoskeletons (i-EXOs).

6.1. Identifying Key Factors and Training Requirements

Planning a training initially involves understanding the target user group and their specific needs [15]. In order to identify important factors in designing an exoskeleton-specific training, we reviewed studies on exoskeletons focusing on the design of the devices, their usability, wearability determinants, and adoption barriers [12,72,83–85]. i-EXOs are designed to be worn by the working population. Literature demonstrates the application of these devices in industrial tasks such as assembly [17], construction [57], and healthcare [56] where repetitive and sustaining postures (such as flexed trunk during bending or holding hands above head during overhead work) are common. Furthermore, end-user specific factors such as physical abilities, job tasks, and prior experience with exoskeletons should be assessed prior to developing a training. Although the end-users of i-EXOs, being a manual workforce, could be assumed to represent specific characteristics (such as education level), such characteristics may vary depending upon the industry sector. For example, user requirements in manufacturing (male dominant workforce, majority of overhead and/or lifting tasks, rugged environments) vs. healthcare (female-dominant workforce, high-paced work, sterile environments) may be fundamentally different (including the use of i-EXOs). Thus, the comprehensive instruction on the correct operation, adjustment, and maintenance of the i-EXO can vary based on the use case scenario of end-users.

Improper fit and lack of adjustment could cause detrimental effects on the wearer's body [86]. Thus, training should cover proper fitting procedures, adjustment of device joint stiffness and assistance levels, troubleshooting common issues, and emphasize education on ergonomics and body mechanics to promote proper movement patterns, which depend highly upon the users and type of task during their routine work, as well as the work culture (level of high demands and exposure time). Although i-EXOs aim to reduce the risk of musculoskeletal injuries, if incorrectly worn or used, the assistance provided may impede natural body motion or harm the otherwise healthy joints of users [10,11,87]. Similarly, safety protocols and emergency procedures should also be emphasized, especially during rare cases if the structure of the i-EXO breaks or the mechanism malfunctions, leading to unintended forces on body joints and body movement [8]. Considerations are also required for covering instances and action plans for accidents occurring during use of i-EXOs. Due to a wide variation in the type and design of i-EXOs, developing a regime that fits every case may be infeasible; hence, developing a systematic approach to training design is needed. In summary, these eight key factors as depicted in Figure 4, namely, end-user qualities, workforce trends, maintenance, safety, work culture, end-user needs, environment, and work type could be crucial in developing training for exoskeleton users.

To understand the individual user needs and expectations, tools such as questionnaires and one-to-one semi-structured interviews may be used [13,26,29]. Individual problem statements can then be designed to target each of the challenging aspects required in effectively and safely performing the tasks using i-EXOs. Problem statements can be generated based on some of the identified key factors, and Table 3 depicts a list of such potential statements that may be used for developing case scenarios to train i-EXO users.

Table 3. List of potential problem statements to be used for developing training protocols.

No.	Problem Statements for Developing Problem-Centered Instruction Protocols
1.	How to determine and set required assistance levels for actuators of the exoskeleton?
2.	How to adjust the exoskeleton to varying body sizes/shapes?
3.	How to detect and test if the actuators are working properly?
4.	What is the step-by-step process of donning, and doffing the exoskeleton?
5.	How to engage/disengage actuator assistance of the exoskeleton?
6.	How to perform tasks by using assistance provided by the exoskeleton?
7.	How to determine whether the exoskeleton requires maintainence, or repairs?
8.	What is the procedure for cleaning the exoskeleton without damaging?
9.	What is the procedure to safety remove the exoskeleton in the event of a breakdown?

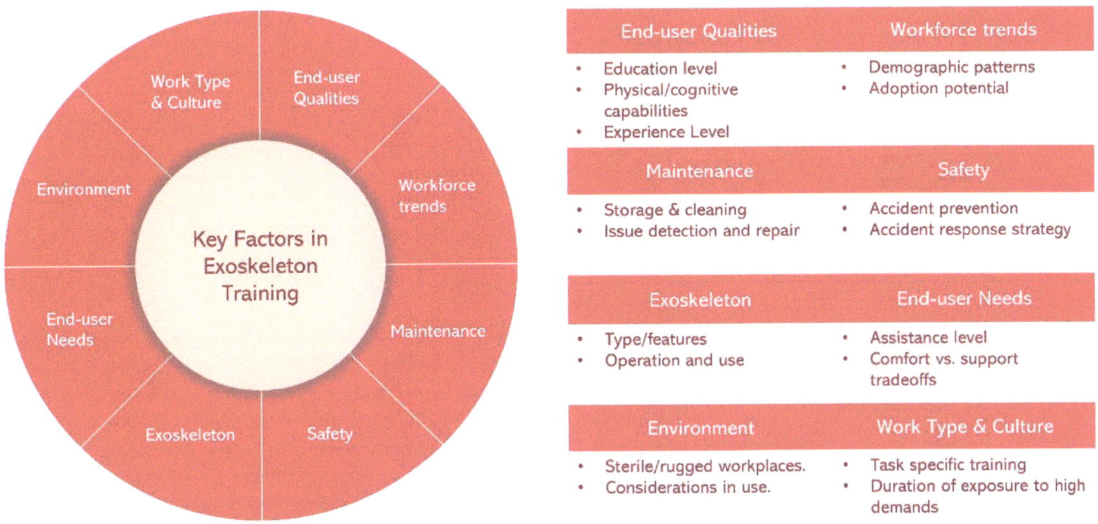

Figure 4. A schematic showing an overview of key factors to be considered in developing proper training for exoskeleton users.

6.2. Briefing, Demonstration and Evaluation

The second step, "Brief" step focuses on delivering fundamental knowledge and educating users on the basics of exoskeleton operation, adjustment, and maintenance. This may include providing information on the correct fitting procedures, adjustment of joint stiffness and assistance levels, as well as troubleshooting common issues of i-EXOs. Particularly, fundamental operating principles of the device can be delivered during this phase (such as checking mechanism/actuator functionality and assistance levels prior to donning). Training materials and instructional strategies may also include mixed media (video/audio descriptions of taught content) and must be pre-designed to effectively deliver the essential knowledge to users [26].

In the demonstration step, learners should be provided with one-by-one demonstrations of task scenarios that are based on the identified key factors (Figure 3), such as fitting the exoskeleton to the user's body, setting appropriate assistance levels, and executing the task safely and efficiently. Such demos can be conducted by a trained instructor or by individuals with previous experience in using i-EXOs. Demos should guide learners through the entire procedure of donning, setting assistance, using, and doffing. Furthermore, simulations of activities that may not be evaluated can be demonstrated under controlled circumstances (such as response strategy in the event of mechanism malfunction during use).

Evaluation is a crucial step in the training framework, where a test plan is developed to assess the user's performance and understanding of the training content [88,89] as per the discussed considerations for the role of ergonomic/biomechanical evaluations in ensuring training effectiveness. This evaluation process may include hands-on assessments, practical exercises, and simulated tasks. Gamification alternatives can also be explored to improve motivation and learning experience [90]. Specifically, learners may be asked to perform the demonstrated activities under vigilance from an expert/instructor with appropriate interventions to correct learners' performance. The training can be designed with scaffolding, which involves gradually reducing guidance as users progress through increasing difficulty levels. Due to wide variations in commercial EXO products (active/passive, shoulder/back/leg, rigid/soft) as well as industrial activities, assessments can be designed on a case-by-case basis depending on the type of i-EXO and specific task under consideration.

The three phases of briefing, demonstrating, and evaluation, treated as a module, can be iterated for the number of identified problem statements (e.g., optimizing assistance for individuals, visual inspection of mechanism, etc.). Each such module can be defined based on increasing level of difficulty (with the easiest ones completed first). After completion of each module, appropriate tests must be provided to ensure that learners can independently solve each of the defined problem statements. More complex tasks may be assessed at this stage with the help of biomechanical data-acquisition methods (such as measuring metabolic rate differences or muscle activation [6,78,91–93]) to ensure that the devices are beneficial. Lastly, feedback from assessments and user observations helps identify areas for improvement in subsequent training modules.

6.3. Implementation in the Real World

The final step in the framework focuses on implementing the trained skills in real-world settings. Users may be observed/provided with continued guidance to ensure proper adherence to protocols and safe operation of the exoskeleton. Observations and user feedback at this stage may also be used to further enhance and refine the training program for future iterations [29]. With the proposed framework, training programs for i-EXOs can be designed to address specific factors and challenges, ultimately enhancing user competency and facilitating their safe and effective integration into the workplace.

7. Data-Driven and Emerging Technologies for Augmenting Exoskeleton Training

Performing tasks safely and effectively with an i-EXO depends on the ability of the individual to learn the needed changes in motor skills when wearing the device. Emerging technologies such as Virtual Reality (VR) or Augmented Reality (AR) can be utilized for simulating realistic tasks [94]. For example, VR simulators have been found to be effective in improving training on surgical robotic systems [95]. Prior efforts demonstrate the use of a Virtual Reality (VR)-based framework to test, experience, and train users based on a set of feedback modalities [96]. The simulation platform enabled users to select parameters for the simulation, which provided feedback in the form of visual, vibro-tactile, and force-feedback to the users. In addition, outcomes on motion and metabolic exposures were also provided to users, along with trajectories which may assist in reducing adaptation time while also providing real-time feedback. In another study, a haptic-based sensation transfer system was developed to migrate haptic and kinematic feeling of experts of i-EXOs to novices [97]. Similarly, a simulator-style device representing different types of i-EXOs, and their actuation capabilities, can be developed, which can be used to train the workforce in the real world. Such a simulator can incorporate features from a diverse set of i-EXOs in the market along with adjustable parameters that can enable training the workforce on different types of devices. Along with helping users to get trained on different types of devices, such a system can also be helpful in identifying devices in the market that are better suited to the user characteristics/preferences and the task.

Sensors and data-driven approaches can be utilized to estimate real-time physical demands by recording measures like muscle activity, trunk motion, and stability as the trainee performs the tasks, as demonstrated in previous evaluation-based studies [98]. More importantly, this can be utilized to identify and predict instances (postures or movements) and demands for selected tasks, device parameters, and user characteristics that can lead to detrimental effects on the wearer's body. In addition, real-time fatigue-level prediction and forecasting models can be implemented to predict temporal effects, as described in our previous work [99]. Providing this feedback to the trainer can drive interventions, such as correcting body postures/movements or even the selection of appropriate device parameters. With recent technological developments in digitization, intelligent tutoring systems can be developed and implemented to provide personalized training to users [100]. For instance, data from trainees related to performance or physiological measures can be collected during training sessions and compared with previously established standards (e.g., with experts) to customize delivery and pace of the training sessions. A training platform utilizing integrated mechanisms, sensing, Artificial Intelligence (AI), and VR/AR was proposed in a recent study where performance evaluation was conducted throughout as workers performed tasks [101]. Another advantage of using VR/AR-based systems is that they can provide real-time visualization of training progress, performance, as well as provide guidance and instructions. To summarize, AI-assisted training can be beneficial by providing customized training based on individual experiences/skill levels and improving overall outcomes.

8. Study Limitations

Due to the rapid growth in exoskeleton research and development, the implementation requirements of the devices have not been thoroughly explored. Specifically, topics related to training of i-EXO users in actual workplaces have not yet been explored in academic literature. Although this article considered articles that considered training novice users of i-EXOs, many such studies included training/familiarization protocols to ensure motor adaptations prior to experimental assessments. As training itself was not an objective, many details regarding the training were observed to have been omitted. Only a few studies with primary aims of evaluating effects of familiarization protocols were found in the literature. Future research in the field of training for i-EXO users holds significant potential for enhancing user safety, performance, and overall experience. One important direction for future investigation is the development of personalized training approaches that consider individual user characteristics and needs, such as anthropometry, physical abilities, and prior experience. Additionally, further research is needed to explore the long-term effects of exoskeleton-provided training (vs. not providing detailed training) on user performance, musculoskeletal health, and work-related outcomes. Investigating the optimal timing and duration of training programs, as well as the potential benefits of ongoing training and refresher courses, can provide valuable insights for designing comprehensive and sustainable training protocols. Furthermore, the integration of emerging technologies (such as the use of motion capture systems), in training programs warrants exploration to enhance engagement, simulate realistic work scenarios, and facilitate skill acquisition. In this study, a framework has been proposed for training i-EXO users. Although the method is derived from foundational concepts in instructional design, the proposed framework needs to be tested by conducting detailed evaluations to ensure the retainability of the learnt topics. Future research efforts in these areas can contribute to the advancement of training practices and optimize the utilization of i-EXOs across diverse work settings.

9. Conclusions

Educating workers on the proper use of assistive devices plays a vital role in ensuring the safe and effective integration of i-EXOs into the workplace. Review findings indicated variations in training methods (duration and familiarization tasks) across evaluation-based studies. Training was provided for donning/doffing, ensuring proper fit, body movement, experimental tasks, setting assistance levels, and safe use. Means of instruction included instruction manuals, videos, and demonstrations by instructors. Studies that assessed the effects of familiarization protocols showed that novice users may require up to four sessions to achieve motor strategies of experts. On the other hand, evaluation-based studies often considered training sessions of much shorter durations, indicating the need to test motor adaptations of study participants prior to testing. Incorporating inputs from ergonomics and biomechanics can enhance the design of training and adaptation to using the device, ultimately ensuring proper movement patterns and safe operation. Considerations such as understanding the target user group, providing comprehensive instruction on operation and maintenance, emphasizing ergonomics and body mechanics, and utilizing a combination of instructional methods contribute to the success of training programs. Training sessions can be assisted with ergonomic/biomechanical evaluations (such as comparisons between novice/experts) to gain insights on motor learning patterns of novices and could be potentially helpful in determining the optimal duration and instructional depth across different types of EXOs and tasks. Future studies can integrate technologies such as simulators, VR/AR, and sensor-based data-driven approaches, including the use of machine learning and AI-assisted guidance for improving training outcomes.

Author Contributions: Conceptualization, P.M.K. and E.R.; methodology, P.M.K. and E.R.; software, P.M.K.; validation, P.M.K.; formal analysis, P.M.K.; investigation, P.M.K.; resources, E.R.; writing—original draft preparation, P.M.K.; writing—review and editing, P.M.K.; visualization, P.M.K.; supervision, E.R.; project administration, P.M.K. All authors have read and agreed to the published version of the manuscript.

Funding: This research received no external funding.

Acknowledgments: The authors would like to thank Hrushikesh Godbole for sharing his insights on the approaches commonly implemented for training and education of the workforce.

Conflicts of Interest: The authors declare no conflict of interest.

References

1. De Bock, S.; Ghillebert, J.; Govaerts, R.; Tassignon, B.; Rodriguez-Guerrero, C.; Crea, S.; Veneman, J.; Geeroms, J.; Meeusen, R.; De Pauw, K. Benchmarking Occupational Exoskeletons: An Evidence Mapping Systematic Review. *Appl. Ergon.* **2022**, *98*, 103582. [CrossRef] [PubMed]
2. Crea, S.; Beckerle, P.; De Looze, M.; De Pauw, K.; Grazi, L.; Kermavnar, T.; Masood, J.; O'Sullivan, L.W.; Pacifico, I.; Rodriguez-Guerrero, C.; et al. Occupational Exoskeletons: A Roadmap toward Large-Scale Adoption. Methodology and Challenges of Bringing Exoskeletons to Workplaces. *Wearable Technol.* **2021**, *2*, e11–e22. [CrossRef] [PubMed]
3. Kuber, P.M.; Abdollahi, M.; Alemi, M.M.; Rashedi, E. A Systematic Review on Evaluation Strategies for Field Assessment of Upper-Body Industrial Exoskeletons: Current Practices and Future Trends. *Ann. Biomed. Eng.* **2022**, *50*, 1203–1231. [CrossRef] [PubMed]
4. Butler, T. Exoskeleton Technology. *Ergonomics* **2016**, *61*, 32–36.
5. Kim, S.; Nussbaum, M.A.; Mokhlespour Esfahani, M.I.; Alemi, M.M.; Jia, B.; Rashedi, E. Assessing the Influence of a Passive, Upper Extremity Exoskeletal Vest for Tasks Requiring Arm Elevation: Part II—"Unexpected" Effects on Shoulder Motion, Balance, and Spine Loading. *Appl. Ergon.* **2018**, *70*, 323–330. [CrossRef]
6. De Bock, S.; Ghillebert, J.; Govaerts, R.; Elprama, S.A.; Marusic, U.; Serrien, B.; Jacobs, A.; Geeroms, J.; Meeusen, R.; De Pauw, K. Passive Shoulder Exoskeletons: More Effective in the Lab Than in the Field? *IEEE Trans. Neural Syst. Rehabil. Eng.* **2021**, *29*, 173–183. [CrossRef]
7. Nussbaum, M.A.; Lowe, B.D.; de Looze, M.; Harris-Adamson, C.; Smets, M. An Introduction to the Special Issue on Occupational Exoskeletons. *IISE Trans. Occup. Ergon. Hum. Factors* **2019**, *7*, 153–162. [CrossRef]
8. O'Sullivan, L.; Nugent, R.; van der Vorm, J. Standards for the Safety of Exoskeletons Used by Industrial Workers Performing Manual Handling Activities: A Contribution from the Robo-Mate Project to Their Future Development. *Procedia Manuf.* **2015**, *3*, 1418–1425. [CrossRef]
9. Exoskeletonreport.com. Types and Classifications of Exoskeletons. Available online: https://exoskeletonreport.com/2015/08/types-and-classifications-of-exoskeletons/ (accessed on 8 August 2024).
10. Kranenborg, S.E.; Greve, C.; Reneman, M.F.; Roossien, C.C. Side-Effects and Adverse Events of a Shoulder- and Back-Support Exoskeleton in Workers: A Systematic Review. *Appl. Ergon.* **2023**, *111*, 104042. [CrossRef]
11. Kim, S.; Madinei, S.; Alemi, M.M.; Srinivasan, D.; Nussbaum, M.A. Assessing the Potential for "Undesired" Effects of Passive Back-Support Exoskeleton Use during a Simulated Manual Assembly Task: Muscle Activity, Posture, Balance, Discomfort, and Usability. *Appl. Ergon.* **2020**, *89*, 103194. [CrossRef]
12. Mortenson, W.B.; Pysklywec, A.; Chau, L.; Prescott, M.; Townson, A. Therapists' Experience of Training and Implementing an Exoskeleton in a Rehabilitation Centre. *Disabil. Rehabil.* **2020**, *44*, 1060–1066. [CrossRef]
13. Merrill, M.D. First Principles of Instruction. *Educ. Technol. Res. Dev.* **2002**, *50*, 43–59. [CrossRef]
14. Morrison, G.R.; Ross, S.J.; Morrison, J.R.; Kalman, H.K. *Designing Effective Instruction*; John Wiley & Sons: Hoboken, NJ, USA, 2019; ISBN 1119465931.
15. Smith, P.L.; Ragan, T.J. *Instructional Design*; John Wiley & Sons: Hoboken, NJ, USA, 2004; ISBN 0471393533.
16. Rashedi, E.; Kim, S.; Nussbaum, M.A.; Agnew, M.J. Ergonomic Evaluation of a Wearable Assistive Device for Overhead Work. *Ergonomics* **2014**, *57*, 1864–1874. [CrossRef]
17. Kim, S.; Nussbaum, M.A.; Mokhlespour Esfahani, M.I.; Alemi, M.M.; Alabdulkarim, S.; Rashedi, E. Assessing the Influence of a Passive, Upper Extremity Exoskeletal Vest for Tasks Requiring Arm Elevation: Part I—"Expected" Effects on Discomfort, Shoulder Muscle Activity, and Work Task Performance. *Appl. Ergon.* **2018**, *70*, 315–322. [CrossRef] [PubMed]
18. Alabdulkarim, S.; Kim, S.; Nussbaum, M.A. Effects of Exoskeleton Design and Precision Requirements on Physical Demands and Quality in a Simulated Overhead Drilling Task. *Appl. Ergon.* **2019**, *80*, 136–145. [CrossRef] [PubMed]

19. Picchiotti, M.T.; Weston, E.B.; Knapik, G.G.; Dufour, J.S.; Marras, W.S. Impact of Two Postural Assist Exoskeletons on Biomechanical Loading of the Lumbar Spine. *Appl. Ergon.* **2019**, *75*, 1–7. [CrossRef] [PubMed]
20. Onofrejova, D.; Balazikova, M.; Glatz, J.; Kotianova, Z.; Vaskovicova, K. Ergonomic Assessment of Physical Load in Slovak Industry Using Wearable Technologies. *Appl. Sci.* **2022**, *12*, 3607. [CrossRef]
21. del Cid, D.A.; Larranaga, D.; Leitao, M.; Mosher, R.L.; Berzenski, S.R.; Gandhi, V.; Drew, S.A. Exploratory Factor Analysis and Validity of the Virtual Reality Symptom Questionnaire and Computer Use Survey. *Ergonomics* **2021**, *64*, 69–77. [CrossRef]
22. Kermavnar, T.; de Vries, A.W.; de Looze, M.P.; O'Sullivan, L.W. Effects of Industrial Back-Support Exoskeletons on Body Loading and User Experience: An Updated Systematic Review. *Ergonomics* **2021**, *64*, 685–711. [CrossRef]
23. Stirling, L.; Kelty-Stephen, D.; Fineman, R.; Jones, M.L.H.; Daniel Park, B.K.; Reed, M.P.; Parham, J.; Choi, H.J. Static, Dynamic, and Cognitive Fit of Exosystems for the Human Operator. *Hum. Factors* **2020**, *62*, 424–440. [CrossRef]
24. de Looze, M.P.; Bosch, T.; Krause, F.; Stadler, K.S.; O'Sullivan, L.W. Exoskeletons for Industrial Application and Their Potential Effects on Physical Work Load. *Ergonomics* **2016**, *59*, 671–681. [CrossRef]
25. Wentland, D. The Strategic Training of Employees Model: Balancing Organizational Constraints and Training Content. *SAM Adv. Manag. J.* **2003**, *68*, 56.
26. Brown, J. Training Needs Assessment: A Must for Developing an Effective Training Program. *Public Pers. Manag.* **2002**, *31*, 569–578. [CrossRef]
27. Gallivan, M.J.; Spitler, V.K.; Koufaris, M. Does Information Technology Training Really Matter? A Social Information Processing Analysis of Coworkers' Influence on IT Usage in the Workplace. *J. Manag. Inf. Syst.* **2005**, *22*, 153–192. [CrossRef]
28. Wang, G.G.; Wilcox, D. Training Evaluation: Knowing More Than Is Practiced. *Adv. Dev. Hum. Resour.* **2006**, *8*, 528–539. [CrossRef]
29. Martin, H.J. Improving Training Impact through Effective Follow-up: Techniques and Their Application. *J. Manag. Dev.* **2010**, *29*, 520–534. [CrossRef]
30. Aiman-smith, L.; Carolina, N. Implementing New Manufacturing Technology: The Related Effects of Technology Characteristics and User Learning Activities. *Acad. Manag. J.* **2002**, *45*, 421–430. [CrossRef]
31. Moyon, A.; Petiot, J.-F.; Poirson, E. Investigating the Effects of Passive Exoskeletons and Familiarization Protocols on Arms-Elevated Tasks. In Proceedings of the Human Factors and Ergonomics Society Europe Chapter 2019 Annual Conference, Nantes, France, 2–4 October 2019; Volume 4959, pp. 1–17.
32. Alemi, M.M.; Simon, A.A.; Geissinger, J.; Asbeck, A.T. Modeling the Metabolic Reductions of a Passive Back-Support Exoskeleton. *J. Appl. Physiol.* **2022**, *132*, 737–760. [CrossRef]
33. Luger, T.; Seibt, R.; Cobb, T.J.; Rieger, M.A.; Steinhilber, B. Influence of a Passive Lower-Limb Exoskeleton during Simulated Industrial Work Tasks on Physical Load, Upper Body Posture, Postural Control and Discomfort. *Appl. Ergon.* **2019**, *80*, 152–160. [CrossRef]
34. Goršič, M.; Song, Y.; Dai, B.; Novak, D. Evaluation of the HeroWear Apex Back-Assist Exosuit during Multiple Brief Tasks. *J. Biomech.* **2021**, *126*, 110620. [CrossRef]
35. Kozinc, Ž.; Baltrusch, S.; Houdijk, H.; Šarabon, N. Short-Term Effects of a Passive Spinal Exoskeleton on Functional Performance, Discomfort and User Satisfaction in Patients with Low Back Pain. *J. Occup. Rehabil.* **2021**, *31*, 142–152. [CrossRef]
36. Baltrusch, S.J.; Houdijk, H.; van Dieën, J.H.; van Bennekom, C.A.M.; de Kruif, A.J. Perspectives of End Users on the Potential Use of Trunk Exoskeletons for People With Low-Back Pain: A Focus Group Study. *Hum. Factors* **2020**, *62*, 365–376. [CrossRef]
37. Gillette, J.C.; Stephenson, M.L. Electromyographic Assessment of a Shoulder Support Exoskeleton during On-Site Job Tasks. *IISE Trans. Occup. Ergon. Hum. Factors* **2019**, *7*, 302–310. [CrossRef]
38. Groos, S.; Fuchs, M.; Kluth, K. *Determination of the Subjective Strain Experiences during Assembly Activities Using the Exoskeleton "Chairless Chair"*; Springer International Publishing: New York, NY, USA, 2020; Volume 962, ISBN 9783030204662.
39. So, B.C.L.; Cheung, H.H.; Liu, S.L.; Tang, C.I.; Tsoi, T.Y.; Wu, C.H. The Effects of a Passive Exoskeleton on Trunk Muscle Activity and Perceived Exertion for Experienced Auxiliary Medical Service Providers in Cardiopulmonary Resuscitation Chest Compression. *Int. J. Ind. Ergon.* **2020**, *76*, 102906. [CrossRef]
40. Baltrusch, S.J.; van Dieën, J.H.; van Bennekom, C.A.M.; Houdijk, H. The Effect of a Passive Trunk Exoskeleton on Functional Performance in Healthy Individuals. *Appl. Ergon.* **2018**, *72*, 94–106. [CrossRef]
41. Baltrusch, S.J.; Van Dieen, J.H.; Van Bennekom, C.A.M.; Houdijk, H. Testing an Exoskeleton That Helps Workers with Low-Back Pain: Less Discomfort with the Passive Spexor Trunk Device. *IEEE Robot. Autom. Mag.* **2020**, *27*, 66–76. [CrossRef]
42. Kozinc, Ž.; Baltrusch, S.; Houdijk, H.; Šarabon, N. Reliability of a Battery of Tests for Functional Evaluation of Trunk Exoskeletons. *Appl. Ergon.* **2020**, *86*, 103117. [CrossRef] [PubMed]
43. Elprama, S.A.; Vannieuwenhuyze, J.T.A.; De Bock, S.; Vanderborght, B.; De Pauw, K.; Meeusen, R.; Jacobs, A. Social Processes: What Determines Industrial Workers' Intention to Use Exoskeletons? *Hum. Factors* **2020**, *62*, 337–350. [CrossRef] [PubMed]
44. Alemi, M.M.; Geissinger, J.; Simon, A.A.; Chang, S.E.; Asbeck, A.T. A Passive Exoskeleton Reduces Peak and Mean EMG during Symmetric and Asymmetric Lifting. *J. Electromyogr. Kinesiol.* **2019**, *47*, 25–34. [CrossRef] [PubMed]
45. Bär, M.; Luger, T.; Seibt, R.; Rieger, M.A.; Steinhilber, B. Using a Passive Back Exoskeleton during a Simulated Sorting Task: Influence on Muscle Activity, Posture, and Heart Rate. *Hum. Factors* **2022**, *66*, 40–55. [CrossRef] [PubMed]
46. Bock, S.D.; Ampe, T.; Rossini, M.; Tassignon, B.; Lefeber, D.; Rodriguez-guerrero, C.; Roelands, B.; Geeroms, J.; Meeusen, R.; Pauw, K. De Passive Shoulder Exoskeleton Support Partially Mitigates Fatigue-Induced Effects in Overhead Work. *Appl. Ergon.* **2023**, *106*, 103903. [CrossRef] [PubMed]

47. Debusk, H.; Babski-reeves, K.; Chander, H. Preliminary Analysis of StrongArm® Ergoskeleton on Knee and Hip Kinematics and User Comfort. In Proceedings of the Human Factors and Ergonomics Society Annual Meeting, Rome, Italy, 28–30 September 2017; pp. 1346–1350.
48. Giustetto, A.; Vieira Dos Anjos, F.; Gallo, F.; Monferino, R.; Cerone, G.L.; Di Pardo, M.; Gazzoni, M.; Micheletti Cremasco, M. Investigating the Effect of a Passive Trunk Exoskeleton on Local Discomfort, Perceived Effort and Spatial Distribution of Back Muscles Activity. *Ergonomics* **2021**, *64*, 1379–1392. [CrossRef] [PubMed]
49. Gonsalves, N.J.; Ogunseiju, O.R.; Akanmu, A.A.; Nnaji, C.A. Assessment of a Passive Wearable Robot for Reducing Low Back DIsorders during Rebar Work. *J. Inf. Technol. Constr.* **2021**, *26*, 936–952. [CrossRef]
50. Huysamen, K.; Bosch, T.; de Looze, M.; Stadler, K.S.; Graf, E.; O'Sullivan, L.W. Evaluation of a Passive Exoskeleton for Static Upper Limb Activities. *Appl. Ergon.* **2018**, *70*, 148–155. [CrossRef] [PubMed]
51. Hwang, J.; Kumar Yerriboina, V.N.; Ari, H.; Kim, J.H. Effects of Passive Back-Support Exoskeletons on Physical Demands and Usability during Patient Transfer Tasks. *Appl. Ergon.* **2021**, *93*, 103373. [CrossRef]
52. Kazerooni, H.; Tung, W.; Pillai, M. Evaluation of Trunk-Supporting Exoskeleton. In Proceedings of the Human Factors and Ergonomics Society Annual Meeting, Seattle, WA, USA, 28 October–1 November 2019; Volume 63, pp. 1080–1083. [CrossRef]
53. Kong, Y.K.; Park, C.W.; Cho, M.U.; Kim, S.Y.; Kim, M.J.; Hyun, D.J.; Bae, K.; Choi, J.K.; Ko, S.M.; Choi, K.H. Guidelines for Working Heights of the Lower-Limb Exoskeleton (Cex) Based on Ergonomic Evaluations. *Int. J. Environ. Res. Public Health* **2021**, *18*, 5199. [CrossRef]
54. Lanotte, F.; Grazi, L.; Chen, B.; Vitiello, N.; Crea, S. A Low-Back Exoskeleton Can Reduce the Erector Spinae Muscles Activity during Freestyle Symmetrical Load Lifting Tasks. In Proceedings of the 2018 7th IEEE International Conference on Biomedical Robotics and Biomechatronics (Biorob), Enschede, The Netherlands, 26–29 August 2018; pp. 701–706. [CrossRef]
55. Madinei, S.; Alemi, M.M.; Kim, S.; Srinivasan, D.; Nussbaum, M.A. Biomechanical Evaluation of Passive Back-Support Exoskeletons in a Precision Manual Assembly Task: "Expected" Effects on Trunk Muscle Activity, Perceived Exertion, and Task Performance. *Hum. Factors* **2020**, *62*, 441–457. [CrossRef]
56. Maurice, P.; Enault-Cuny, F.; Ivaldi, S. Influence of a Passive Back Support Exoskeleton on Simulated Patient Bed Bathing: Results of an Exploratory Study. *Ergonomics* **2022**, *66*, 859–873. [CrossRef]
57. Ogunseiju, O.; Gonsalves, N.; Akanmu, A.; Nnaji, C. Subjective Evaluation of Passive Back-Support Exoskeleton for Flooring Work. In Proceedings of the ASC 2021—57th Annual Associated Schools of Construction International Conference, Chico, CA, USA, 5–8 April 2021; Volume 2, pp. 10–11.
58. Schmalz, T.; Colienne, A.; Bywater, E.; Fritzsche, L.; Gärtner, C.; Bellmann, M.; Reimer, S.; Ernst, M. A Passive Back-Support Exoskeleton for Manual Materials Handling: Reduction of Low Back Loading and Metabolic Effort during Repetitive Lifting. *IISE Trans. Occup. Ergon. Hum. Factors* **2022**, *10*, 7–20. [CrossRef]
59. Tu, Y.; Zhu, A.; Song, J.; Zhang, X.; Cao, G. Design and Experimental Evaluation of a Lower-Limb Exoskeleton for Assisting Workers With Motorized Tuning of Squat Heights. *IEEE Trans. Neural Syst. Rehabil. Eng.* **2022**, *30*, 184–193. [CrossRef]
60. Wehner, M.; Rempel, D.; Kazerooni, H. Lower Extremity Exoskeleton Reduces Back Forces in Lifting. In Proceedings of the ASME 2009 Dynamic Systems and Control Conference, Hollywood, CA, USA, 12–14 October 2009; pp. 961–968.
61. Yan, Z.; Han, B.; Du, Z.; Huang, T.; Bai, O.; Peng, A. Development and Testing of a Wearable Passive Lower-Limb Support Exoskeleton to Support Industrial Workers. *Biocybern. Biomed. Eng.* **2021**, *41*, 221–238. [CrossRef]
62. Park, H.; Kim, S.; Nussbaum, M.A.; Srinivasan, D. A Pilot Study Investigating Motor Adaptations When Learning to Walk with a Whole-Body Powered Exoskeleton. *J. Electromyogr. Kinesiol.* **2023**, *69*, 102755. [CrossRef] [PubMed]
63. Diamond-Ouellette, G.; Le Quang, M.; Karakolis, T.; Bouyer, L.J.; Best, K.L. Exploring the Influence of Structured Familiarization to an Adjustable, Passive Load-Bearing Exoskeleton on Oxygen Consumption and Lower Limb Muscle Activation during Walking. *IEEE Trans. Neural Syst. Rehabil. Eng.* **2024**, *32*, 2441–2449. [CrossRef] [PubMed]
64. Diamond-Ouellette, G.; Bouyer, L.J.; Best, K.L. Influence of Customization on Familiarization of Two Passive Exoskeletons on Metabolic Cost of Walking. In Proceedings of the 2023 RESNA Conference on Move to the BeAT of Innovation, New Orleans, LA, USA, 24–26 July 2023; pp. 1–5.
65. Poggensee, K.L.; Collins, S.H. How Adaptation, Training, and Customization Contribute to Benefits from Exoskeleton Assistance. *Sci. Robot.* **2021**, *6*, eabf1078. [CrossRef]
66. Favennec, A.; Frère, J.; Mornieux, G. Changes in Human Motor Behavior during the Familiarization with a Soft Back-Support Occupational Exoskeleton. *Appl. Sci.* **2024**, *14*, 1160. [CrossRef]
67. Lau, J.C.L.; Mombaur, K. Preliminary Study on a Novel Protocol for Improving Familiarity with a Lower-Limb Robotic Exoskeleton in Able-Bodied, First-Time Users. *Front. Robot. AI* **2022**, *8*, 785251. [CrossRef]
68. Pais-Vieira, C.; Allahdad, M.; Neves-Amado, J.; Perrotta, A.; Morya, E.; Moioli, R.; Shapkova, E.; Pais-Vieira, M. Method for Positioning and Rehabilitation Training with the ExoAtlet® Powered Exoskeleton. *MethodsX* **2020**, *7*, 100849. [CrossRef] [PubMed]
69. Luciani, B.; Roveda, L.; Braghin, F.; Pedrocchi, A.; Gandolla, M. Trajectory Learning by Therapists' Demonstrations for an Upper Limb Rehabilitation Exoskeleton. *IEEE Robot. Autom. Lett.* **2023**, *8*, 4561–4568. [CrossRef]
70. Marinou, G.; Sloot, L.; Mombaur, K. Towards Efficient Lower-Limb Exoskeleton Evaluation: Defining Biomechanical Metrics to Quantify Assisted Gait Familiarization. In Proceedings of the 2022 9th IEEE RAS/EMBS International Conference for Biomedical Robotics and Biomechatronics (BioRob), Seoul, Republic of Korea, 21–24 August 2022; pp. 1–8. [CrossRef]

71. Panizzolo, F.A.; Freisinger, G.M.; Karavas, N.; Eckert-Erdheim, A.M.; Siviy, C.; Long, A.; Zifchock, R.A.; LaFiandra, M.E.; Walsh, C.J. Metabolic Cost Adaptations during Training with a Soft Exosuit Assisting the Hip Joint. *Sci. Rep.* **2019**, *9*, 9779. [CrossRef]
72. Lambelet, C.; Temiraliuly, D.; Siegenthaler, M.; Wirth, M.; Woolley, D.G.; Lambercy, O.; Gassert, R.; Wenderoth, N. Characterization and Wearability Evaluation of a Fully Portable Wrist Exoskeleton for Unsupervised Training after Stroke. *J. Neuroeng. Rehabil.* **2020**, *17*, 132. [CrossRef]
73. Chander, D.S.; Cavatorta, M.P. Modelling Interaction Forces at a Curved Physical Human-Exoskeleton Interface. *Adv. Transdiscipl. Eng.* **2020**, *11*, 217–226. [CrossRef]
74. Lee, W.; Yang, X.; Jung, D.; Park, S.; Kim, H.; You, H. Ergonomic Evaluation of Pilot Oxygen Mask Designs. *Appl. Ergon.* **2018**, *67*, 133–141. [CrossRef] [PubMed]
75. Shah, P.; Luximon, Y.; Luximon, A. Measurement of Soft Tissue Deformation at Discomfort and Pain Threshold in Different Regions of the Head. *Ergonomics* **2022**, *65*, 1286–1301. [CrossRef] [PubMed]
76. Kölsch, M.; Beall, A.C.; Turk, M. The Postural Comfort Zone for Reaching Gestures. In Proceedings of the Human Factors and Ergonomics Society Annual Meeting, Denver, CO, USA, 13–17 October 2003; Volume 47, pp. 787–791. [CrossRef]
77. Yang, J.; Abdel-Malek, K. Human Reach Envelope and Zone Differentiation for Ergonomic Design. *Hum. Factors Ergon. Manuf.* **2009**, *19*, 15–34. [CrossRef]
78. Poon, N.; van Engelhoven, L.; Kazerooni, H.; Harris, C. Evaluation of a Trunk Supporting Exoskeleton for Reducing Muscle Fatigue. In Proceedings of the Human Factors and Ergonomics Society Annual Meeting, Seattle, WA, USA, 28 October 2019–1 November 2019; Volume 63, pp. 980–983. [CrossRef]
79. Park, H.; Lee, Y.; Kim, S.; Nussbaum, M.A.; Srinivasan, D. Gait Kinematics When Learning to Use a Whole-Body Powered Exoskeleton. In Proceedings of the Human Factors and Ergonomics Society Annual Meeting, Baltimore, MD, USA, 3–8 October 2021; Volume 65, pp. 1367–1368. [CrossRef]
80. Bequette, B.; Norton, A.; Jones, E.; Stirling, L. Physical and Cognitive Load Effects Due to a Powered Lower-Body Exoskeleton. *Hum. Factors* **2020**, *62*, 411–423. [CrossRef]
81. Gordon, K.E.; Ferris, D.P. Learning to Walk with a Robotic Ankle Exoskeleton. *J. Biomech.* **2007**, *40*, 2636–2644. [CrossRef]
82. Griffin, R.P. Means and Ends: Effective Training Evaluation. *Ind. Commer. Train.* **2010**, *42*, 220–225. [CrossRef]
83. Cha, J.S.; Monfared, S.; Ecker, K.; Lee, D.; Stefanidis, D.; Nussbaum, M.A.; Yu, D. Identifying Barriers and Facilitators of Exoskeleton Implementation In The Operating Room. In Proceedings of the Human Factors and Ergonomics Society Annual Meeting, Seattle, WA, USA, 28 October–1 November 2019; Volume 63, p. 1113. [CrossRef]
84. Charette, C.; Déry, J.; Blanchette, A.K.; Faure, C.; Routhier, F.; Bouyer, L.J.; Lamontagne, M.E. A Systematic Review of the Determinants of Implementation of a Locomotor Training Program Using a Powered Exoskeleton for Individuals with a Spinal Cord Injury. *Clin. Rehabil.* **2023**, *37*, 1119–1138. [CrossRef]
85. Shore, L.; Power, V.; Hartigan, B.; Schülein, S.; Graf, E.; de Eyto, A.; O'Sullivan, L. Exoscore: A Design Tool to Evaluate Factors Associated With Technology Acceptance of Soft Lower Limb Exosuits by Older Adults. *Hum. Factors* **2020**, *62*, 391–410. [CrossRef]
86. Kuber, P.M.; Rashedi, E. Product Ergonomics in Industrial Exoskeletons: Potential Enhancements for Workforce Safety and Efficiency. *Theor. Issues Ergon. Sci.* **2020**, *22*, 729–752. [CrossRef]
87. Schwerha, D.J.; McNamara, N.; Nussbaum, M.A.; Kim, S. Adoption Potential of Occupational Exoskeletons in Diverse Enterprises Engaged in Manufacturing Tasks. *Int. J. Ind. Ergon.* **2021**, *82*, 103103. [CrossRef]
88. Foxon, M. Evaluation of Training and Development Programs: A Review of the Literature. *Australas. J. Educ. Technol.* **1989**, *5*. [CrossRef]
89. Berge, Z.L. Why It Is so Hard to Evaluate Training in the Workplace. *Ind. Commer. Train.* **2008**, *40*, 390–395. [CrossRef]
90. Grünewald, H.; Kneip, P.; Kozica, A. The Use of Gamification in Workplace Learning to Encourage Employee Motivation and Engagement. In *The Wiley Handbook of Global Workplace Learning*; John Wiley & Sons: Hoboken, NJ, USA, 2019; pp. 557–575. [CrossRef]
91. Iranzo, S.; Piedrabuena, A.; Iordanov, D.; Martinez-Iranzo, U.; Belda-Lois, J.M. Ergonomics Assessment of Passive Upper-Limb Exoskeletons in an Automotive Assembly Plant. *Appl. Ergon.* **2020**, *87*, 103120. [CrossRef] [PubMed]
92. Wang, H.M.; Le, D.K.L.; Lin, W.C. Evaluation of a Passive Upper-Limb Exoskeleton Applied to Assist Farming Activities in Fruit Orchards. *Appl. Sci.* **2021**, *11*, 757. [CrossRef]
93. Baltrusch, S.J.; Houdijk, H.; Van Dieën, J.H.; Bruijn, S.M.; Koopman, A.S. The Effect of a Passive Trunk Exoskeleton on Metabolic Costs during Lifting and Walking. *Ergonomics* **2019**, *62*, 903–916. [CrossRef]
94. Lerner, M.A.; Ayalew, M.; Peine, W.J.; Sundaram, C.P. Does Training on a Virtual Reality Robotic Simulator Improve Performance on the Da Vinci® Surgical System? *J. Endourol.* **2010**, *24*, 467–472. [CrossRef] [PubMed]
95. Liu, M.; Curet, M. A Review of Training Research and Virtual Reality Simulators for the Da Vinci Surgical System. *Teach. Learn. Med.* **2015**, *27*, 12–26. [CrossRef]
96. Burks, G.; Lee, Y.; Kim, S.; Ye, Y.; Beiter, B.; Herron, C.; Shi, Y.; Leonessa, A.; Du, J.; Srinivasan, D. A Framework for Virtual Reality-Based Motor Skills Training for the Use of Exoskeletons. In Proceedings of the Human Factors and Ergonomics Society Annual Meeting, Baltimore, MD, USA, 3–8 October 2021; Volume 65, pp. 277–278. [CrossRef]
97. Ye, Y.; Shi, Y.; Srinivasan, D.; Du, J. Sensation Transfer for Immersive Exoskeleton Motor Training: Implications of Haptics and Viewpoints. *Autom. Constr.* **2022**, *141*, 104411. [CrossRef]

98. Kuber, P.M.; Rashedi, E. Investigating Spatiotemporal Effects of Back-Support Exoskeletons Using Unloaded Cyclic Trunk Flexion—Extension Task Paradigm. *Appl. Sci.* **2024**, *14*, 5564. [CrossRef]
99. Kuber, P.M.; Godbole, H.; Rashedi, E. Detecting Fatigue during Exoskeleton-Assisted Trunk Flexion Tasks: A Machine Learning Approach. *Appl. Sci.* **2024**, *14*, 3563. [CrossRef]
100. Woolf, B.; Ghosh, A.; Lan, A.; Zilberstein, S.; Juravizh, T.; Cohen, A.; Geho, O. *AI-Enabled Training in Manufacturing Workforce Development*; NSF's Convergence Accelerator Pilot; University of Massachusetts: Amherst, MA, USA, 2020.
101. Wang, W.; Wu, X.; Wang, P.; Maybury, M.; Lu, A. Towards AI-Assisted Smart Training Platform for Future Manufacturing Workforce. In Proceedings of the AAAI 2020 Spring Symposium, AI in Manufacturing, Stanford, CA, USA, 23–25 March 2020.

biomimetics

Article

Automatic Assist Level Adjustment Function of a Gait Exercise Rehabilitation Robot with Functional Electrical Stimulation for Spinal Cord Injury: Insights from Clinical Trials

Ryota Kimura [1,*], Takahiro Sato [1], Yuji Kasukawa [2], Daisuke Kudo [2], Takehiro Iwami [3] and Naohisa Miyakoshi [1]

1 Department of Orthopedic Surgery, Akita University Graduate School of Medicine, Akita 010-8543, Japan; twrofwsh0930@outlook.com (T.S.); miyakosh@doc.med.akita-u.ac.jp (N.M.)
2 Department of Rehabilitation, Akita University Hospital, Akita 010-8543, Japan; kasukawa@doc.med.akita-u.ac.jp (Y.K.); dkudo@doc.med.akita-u.ac.jp (D.K.)
3 Department of Systems Design Engineering, Faculty of Engineering Science, Akita University Graduate School of Engineering Science, Akita 010-8502, Japan; iwami@gipc.akita-u.ac.jp
* Correspondence: rkimura@med.akita-u.ac.jp; Tel.: +81-18-884-6148

Abstract: This study aimed to identify whether the combined use of functional electrical stimulation (FES) reduces the motor torque of a gait exercise rehabilitation robot in spinal cord injury (SCI) and to verify the effectiveness of the developed automatic assist level adjustment in people with paraplegia. Acute and chronic SCI patients (1 case each) performed 10 min of gait exercises with and without FES using a rehabilitation robot. Reinforcement learning was used to adjust the assist level automatically. The maximum torque values and assist levels for each of the ten walking cycles when walking became steady were averaged and compared with and without FES. The motor's output torque and the assist level were measured as outcomes. The assist level adjustment allowed both the motor torque and assist level to decrease gradually to a steady state. The motor torque and the assist levels were significantly lower with the FES than without the FES under steady conditions in both cases. No adverse events were reported. The combined use of FES attenuated the motor torque of a gait exercise rehabilitation robot for SCI. Automatic assistive level adjustment is also useful for spinal cord injuries.

Keywords: spinal cord injury; functional electrical stimulation; rehabilitation robot; reinforcement learning; assistive level

Citation: Kimura, R.; Sato, T.; Kasukawa, Y.; Kudo, D.; Iwami, T.; Miyakoshi, N. Automatic Assist Level Adjustment Function of a Gait Exercise Rehabilitation Robot with Functional Electrical Stimulation for Spinal Cord Injury: Insights from Clinical Trials. *Biomimetics* **2024**, *9*, 621. https://doi.org/10.3390/biomimetics9100621

Academic Editor: Rafhael Milanezi de Andrade

Received: 8 September 2024
Revised: 6 October 2024
Accepted: 12 October 2024
Published: 13 October 2024

1. Introduction

Worldwide, approximately 15.4 million people have been living with spinal cord injury (SCI). SCI is a major cause of long-term disability, accounting for over 4.5 million years of life lived with disability in 2021 [1]. Many of these injuries are reported to result from high-energy trauma, such as motor vehicle accidents and falls. In contrast, the number of cases of incomplete SCI due to low-energy trauma from falls on level surfaces in older adults is increasing in Japan [2]. In particular, in the case of incomplete paralysis, appropriate assistance is needed in rehabilitation, in accordance with motor learning theory, depending on the recovery process of motor paralysis. A basic premise of motor learning is that of high-volume repetition and task-oriented training. Treatments based on this premise have become a major focus of research on the recovery of motor function in central nervous system disorders such as spinal cord injury and stroke [3,4]. For high-frequency training, robotic rehabilitation can provide more extensive training while compensating for the patient's lost functions and reducing the burden on supporting medical staff.

Neuromodulation techniques, such as functional electrical stimulation (FES), are important tools in restorative neurology [5]. FES can noninvasively activate paralyzed muscles or muscle groups through electrodes placed on the skin. Depending on whether the upper or lower motor neurons are damaged, stimulation directly activates the motor nerves or muscle fibers. Nerve stimulation relies on intact peripheral nerves and neural signal processing in the intact portion of the spinal cord below the lesion [6]. Kralj et al. [7] developed an FES method and system to facilitate ambulation in individuals with spinal cord injury. Furthermore, FES has been used in combination with ankle–foot orthosis [8] and hip–knee–ankle–foot orthosis [9]. There have also been several reports of the addition of FES into gait training rehabilitation robots [10,11], and it has been reported to reduce the electric motor torque of an exoskeletal assistive walking robot [12]. A systematic review and network meta-analysis has also indicated that FES was the most effective treatment for improving walking velocity and distance in incomplete spinal cord injury [13]. Furthermore, combining FES with a robot resulted in exercise with less muscle fatigue than FES alone [14], making hybrid FES–robot training a viable option for prolonged exercise. As a result, the use of a gait rehabilitation robot with FES improves the range of motion of the joints, the muscle strength and the ability to walk [15,16]. Robot-assisted training with FES appears to support the recovery of residual function after SCI and has been observed to lead to improvements in motor function and strength in the lower extremities [17]. Previously, we developed a gait training rehabilitation robot with FES [18] and confirmed that the robot's torque was reduced by using FES in pseudo paraplegics [19]. In addition, a function for automatically adjusting the level of assistance using reinforcement learning has been developed, and its effectiveness in healthy subjects has been confirmed [20].

It is unknown whether using FES in combination with a gait training rehabilitation robot reduces motor torque in people with paraplegia with spinal cord injury. Moreover, there are currently no devices that automatically adjust the level of assistance. This study aimed to verify whether concurrent use of FES reduces the motor torque of a gait exercise rehabilitation robot in SCI and the effectiveness of the developed automatic assist level adjustment in people with paraplegia.

2. Materials and Methods

Patients with paraparesis were recruited from our institution. Paraparesis was defined as those with SCI because of trauma. The inclusion criteria were as follows: (1) inability to walk unaided; (2) recognition of the significance of this research and participation in the research of their own free will; and (3) written informed consent after explanation in the informed consent document. The exclusion criteria were as follows: (1) inability to follow the therapist's instructions, (2) conditions in which exercise load leads to deterioration of physical condition, (3) severe joint contractures or deformities, and (4) other movement limitations for any reason.

Patients performed 10 min of gait exercises with and without FES using a rehabilitation robot. The robot's exoskeleton was designed based on the hip–knee–ankle–foot orthosis for paraplegia. The trunk, thigh, and lower leg were secured with belts. An ankle-foot orthosis (RAPS, Tomei Brace, Seto, Japan) was used for the ankle joint, allowing for angle adjustment based on the patient's spasticity. The actuators driving the hip and knee joints were equipped with encoders, enabling the acquisition of joint angle data. Although the entire exoskeleton weighed 40 kg, it did not affect the patient during use owing to the weight being offset by a counterweight. The orthosis could be adjusted in length to fit each patient's thigh and lower leg. The patients were lifted, and their weight was supported by a rehabilitation lift (SP-1000, Moritoh, Ichinomiya, Japan); following this, they were walked under this condition on a treadmill (8.1T, JohnsonHealth Tech Japan, Tokyo, Japan). Bilateral quadriceps and hamstrings were stimulated using FES (Dynamid, DM2500, Minato Medical Science, Osaka, Japan) (Figure 1). The quadriceps, primarily the rectus femoris, were stimulated from mid-swing to mid-stance during the robotic gait cycle. The stimulation point was located at the motor point, identified by palpation of the anterior superior iliac spine and lateral femoral condyle. The hamstrings, mainly on the lateral side, were stimulated from pre-swing to mid-stance, and the stimulation point was located at the motor point identified by palpation of the sciatic tuberosity and the head of the fibula. The exoskeleton system was pre-programmed using gait data from the joint angles of a healthy individual. The system performs walking motions by changing the positions of the hip and knee joints according to the gait data. The stimulus intensity was set to the lowest stimulus (15–20 mA, 25 Hz) that produced joint movement, and the stimulus timing was synchronized to the gait cycle [19].

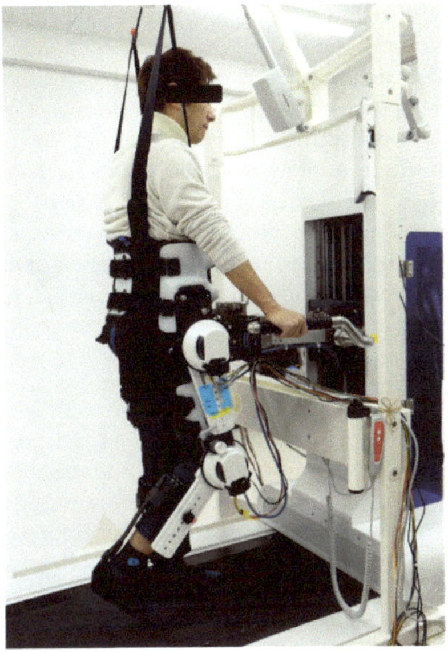

Figure 1. Gait exercise rehabilitation robot. The robot has an exoskeleton, rehabilitation lift, treadmill, and functional electrical stimulation (FES).

In this device, the motor drive was controlled by a computer to reproduce the walking motion. The motor's output torque (Nm) was proportional to the stiffness parameter, and the range was divided into 50 parts, defined as the assist level. The higher the assistance level, the greater the amount of assistance. Force control was used to control the motor, and compliance control was used to vary the amount of assistance. The control Equation (1) used for compliance control is shown below.

$$\tau = K{\cdot}\theta + C{\cdot}\dot{\theta} + I{\cdot}\ddot{\theta} \tag{1}$$

In the equation, K (Nm/deg) represents the stiffness; C (Nm/deg·s) represents the viscosity; I (Nm/deg $\cdot$ s^2) represents the inertia; θ (deg) represents the deviation between the target joint angle and measured angle; $\dot{\theta}$ (deg/s) and $\ddot{\theta}$ (deg/s^2) represent first and second derivative values of θ, respectively; and τ (Nm) is the motor torque. The equation follows the equation of motion, and the first term indicates that the motor output increases as the difference between the target joint angle and measured joint angle increases. In the first term, the output torque is proportional to the stiffness parameter K, i.e., when the stiffness parameter represented by K is large, the output torque increases; conversely, when K is small, the output torque decreases. In this study, the ranges of stiffness, viscosity, and inertia parameters were determined experimentally based on a previous study [16] and the range of stiffness (K) was divided into 50 parts and defined as the assist level. The assistance level provided feedback to the patient through data displayed on the computer monitor placed in front of the robot.

Reinforcement learning was used to adjust the assist level automatically. The method used was Q-learning, a type of off-policy temporal difference learning, and the ε-greedy method was used to determine the policy [21]. In this study, the reinforcement learning environment was defined as "the device and the entire subject wearing it". The subject was trained to select actions by choosing between three options: increasing, maintaining, or decreasing the level of assistance. The subject was rewarded with angular error and motor torque for each walking cycle. The ε-greedy method introduced a search rate ε $(0 \leq \varepsilon \leq 1)$ into the decision process and prevented the system from falling into a local solution through exploration by selecting a random action with probability ε regardless of the action value and an action corresponding to the maximum action value with probability $1 - \varepsilon$. In the initial stages of learning, we increased the proportion of exploration by setting ε to a large value and collecting knowledge. Then, in the advanced stages of learning, we set ε to a small value so that the robot can select the optimal action by using the collected knowledge. The Q-learning algorithm is an off-policy type and the aim was not to optimize the policy, but to optimize the action state value function Q, which indicates the effectiveness of actions, and to construct a decision-making standard that can select the optimal action for the environment. The update equation for updating the Q value $Q(s_t, a_t)$ when action a_t is selected in state s_t is shown in Equation (2).

$$Q(s_t, a_t) \leftarrow Q(s_t, a_t) + \alpha(\gamma + \gamma \cdot maxQ(s_{t+1}, a_{t+1}) - Q(s_t, a_t)) \tag{2}$$

In Equation (2), s represents the state, a represents the action selected in state s, $\alpha(0 \leq \alpha \leq 1)$ represents the learning coefficient, r represents the reward obtained as a result of the action, $\gamma(0 \leq \gamma \leq 1)$ represents the discount rate, and $maxQ(s_{t+1}, a_{t+1})$ represents the maximum Q value of the actions that can be selected in the next action. The action value function Q for selecting action a in a given state s is expressed as $Q(s, a)$. The Equation (2) means that if the value of the reward r, obtained as a result of the action, is positive, then the value of $Q(s, a)$ is increased, and if the reward is negative, the value of $Q(s, a)$ is decreased. By repeating the process of taking actions and updating the Q value using the Equation (2), it becomes possible to proceed with learning about the Q value. In addition, in the Equation (2), the state s, action a, and reward r represent variables that are brought about by interaction with the environment, so there was no need to adjust the parameters. The parameters that had be adjusted here were the learning coefficient α and the discount rate γ. The actions that the agent could take were classified as increasing, maintaining, or decreasing the assist level, and the corresponding Q values were set for each. Because the ε-greedy method was used here to determine the policy, the action corresponding to the largest Q value among each of these was selected, while a random action was selected with probability ε, and the Q value was updated according to the reward obtained from the environment as a result. By repeating this process, the Q values were updated to minimize the reward r obtained, thereby reducing the discrepancy between the joint angle and motor torque during the walking cycle. This allowed the assist level to be adjusted to an optimal level for the patient.

The walking speed was set at 0.8 km/h. The maximum torque values and assist levels for each of the 10 walking cycles when walking became a steady condition were averaged and compared with and without FES. All statistical analyses were conducted using EZR (Saitama Medical Center, Jichi Medical University, Saitama, Japan) [22]. The motor torque and assist level were compared using the paired t-test, with statistical significance set at $p < 0.05$.

This study was approved by our institution's ethics committee (approval number: CRB2180005). All of the individuals voluntarily participated in the study and provided written informed consent.

3. Results

The subjects were a patient with acute thoracic spinal cord injury (33-year-old man), 2 weeks after injury, American Spinal Injury Association (ASIA) Impairment Scale (AIS) C, neurological level of injury (NLI) T12 without spasticity, and a patient with chronic thoracic spinal cord injury (34-year-old man), 2 years after injury, AIS C, and NLI T11 with spasticity (modified Ashworth scale: grade 2).

Assist level adjustment allowed both the motor torque and assist level to decrease gradually to a steady state. Each value reached a steady state between 60 and 120 s. The motor torque was significantly lower with the FES than without the FES under steady conditions in both cases (Table 1). Furthermore, the assist levels were significantly lower with FES than without FES in both cases (Table 2).

The electrical stimulation delivered via the FES did not cause any adverse effects, such as pain, and did not lead to any adverse events associated with robotic gait exercises.

Table 1. Motor torque.

(Nm)	Case 1			Case 2		
	FES (−)	FES (+)	*p*	FES (−)	FES (+)	*p*
Hip Right	18.6 ± 1.5	16.6 ± 1.7	0.0237	17.9 ± 1.2	8.7 ± 1.2	<0.001
Hip Left	18.1 ± 1.4	16 ± 1.6	0.0181	14.7 ± 1.3	11.9 ± 1.7	<0.001
Knee Right	20.3 ± 1.8	13.1 ± 1.5	<0.001	15.4 ± 1.2	13.4 ± 1.6	0.0047
Knee Left	19.2 ± 1.4	17.1 ± 1.9	0.0226	18.4 ± 1.5	12.2 ± 1.3	<0.001

Table 2. Assist levels.

(Nm)	Case 1			Case 2		
	FES (−)	FES (+)	*p*	FES (−)	FES (+)	*p*
Hip Right	28.5 ± 0.5	26.5 ± 0.5	<0.001	42.9 ± 0.3	30.9 ± 0.7	<0.001
Hip Left	24.1 ± 0.6	23.4 ± 0.7	0.0445	38.9 ± 0.3	30.9 ± 0.3	<0.001
Knee Right	30.8 ± 0.4	23.9 ± 0.3	<0.001	25.4 ± 0.5	20.6 ± 0.5	<0.001
Knee Left	29.5 ± 0.5	17.5 ± 0.5	<0.001	33.9 ± 0.3	21.4 ± 0.5	<0.001

4. Discussion

The combined use of FES attenuated the motor torque of the gait exercise rehabilitation robot for spinal cord injury. Furthermore, automatic adjustment of the assistance level using reinforcement learning proved to be effective in gait exercises for patients with spinal cord injuries, and the combined use of FES attenuated the assistance level. It was shown that the automatic assist level adjustment system could be used in conjunction with the torque generated by the FES. This suggests that the intrinsic muscle activity generated by FES reconstructed some of the torque required for walking (Figure 2). In conventional robotic gait training, the gait is reconstructed by combining the robot torque with the patient's muscle torque; the use of FES in conjunction may help attenuate the robot torque.

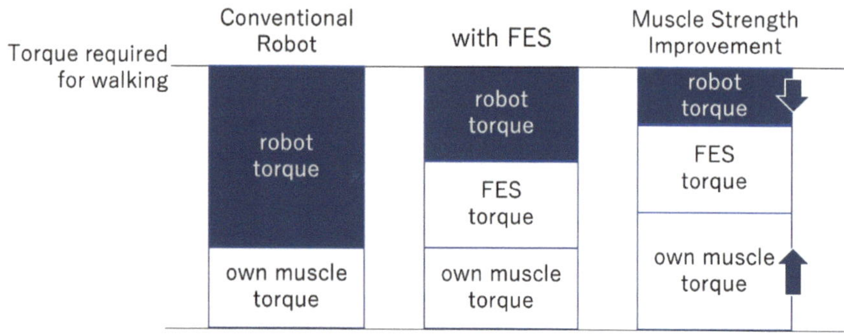

Figure 2. The motor torque. Conventional robots were constructed using the robot's torque and their own muscle's torque in order to prove the torque required for walking. Motor torque is attenuated by the combined use of functional electrical stimulation (FES). Furthermore, the required motor torque decreases as the patient's muscle torque improves.

In functional electrical stimulation therapy (FEST), three factors are crucial: the patient, FES, and the therapist [23]. A phase I randomized control trial (RCT) has revealed that locomotion function improved significantly more with FEST than a non-FEST controlled intervention [24]. The therapist could be replaced by a robot. Regarding the combination of FES and gait rehabilitation robots, we have reported that the smoothness of movement was not lost even when FES was used in combination [18], and that the robot torque was reduced by pseudo-paraplegia [19]. The present study builds on these previous reports and extends them by demonstrating that the combined use of a gait rehabilitation robot and FES reduces the motor torque of the robot in SCI. In order to substantiate the clinical efficacy of FEST when utilizing robots, it is necessary to conduct RCTs; for example, to compare FEST with FES monotherapy. Furthermore, the potential of combining robotics and FES in rehabilitating patients with disorders affecting the central nervous system remains to be fully validated. Future validation is needed because robotic rehabilitation is expected to be integrated with brain–computer interface (BCI) [24] in the future.

Although machine learning and reinforcement learning in exoskeletal rehabilitation robots have previously been studied and reported on [25–27], this is the first report of their combination with FES for SCI. The automatic and appropriate adjustment of task difficulty according to the degree of paralysis is an effective rehabilitation tool from the perspective of motor learning [28]. Furthermore, the patients did not experience any discomfort during gait, and no adverse events were observed. Further research is required to ascertain the clinical efficacy of this system's rehabilitative intervention.

The integration of FES into the robot resulted in a reduction in the torque of the motor. This could ultimately result in a reduction in power consumption and size. The large size of conventional gait rehabilitation robots has been a substantial barrier to their implementation in a broader range of settings. It is of paramount importance to reduce the size of the robot if the objective of achieving the generalization of robotic rehabilitation is to be met. The incorporation of artificial muscles may prove an efficacious solution to these challenges [29]. Given that the concurrent use of an exoskeleton-type robot and a treadmill represents a safe method for gait training rehabilitation, it is imperative to reduce the size of each to promote the overall effectiveness of robotic rehabilitation. As FES reduces the torque generated by the robot's motor, it seems reasonable to employ FES to reduce the overall size of the robot. Consequently, the treadmill size can be reduced, thereby increasing compatibility with the treadmills that are commonly used. This may, in turn, facilitate the implementation of gait training rehabilitation robots in a wide range of applications, including the prevention of disabilities in the aging population. From this study, clinicians should consider combining FES with a gait rehabilitation robot, including the clinical efficacy of FES in SCI.

This study has several limitations. First, it included only two cases, but data were available for both acute and chronic cases. The automatic assistance level adjustment system proved useful in both cases. However, further validation of the clinical effects of rehabilitation with a sufficient number of patients is needed. Second, the walking speed and FES settings were fixed; further verification of the variations caused by changes in speed and FES settings is needed.

5. Conclusions

The combined use of a gait rehabilitation robot and FES reduced the robot's motor torque in SCI, and automatic assist level adjustment through reinforcement learning was effective in people with paraplegia. The clinical outcomes need to be evaluated.

Author Contributions: Conceptualization, R.K.; methodology, R.K. and T.S.; software and hardware, T.I.; formal analysis, R.K.; investigation, T.S.; writing—original draft preparation, R.K.; writing—review and editing, Y.K., D.K. and N.M.; supervision, N.M.; project administration, R.K.; funding acquisition, R.K. All authors have read and agreed to the published version of the manuscript.

Funding: This study was partly supported by the JSPS KAKENHI (Grant No. 23K16600) and the Japan Orthopaedics and Traumatology Research Foundation (Grant No. 474).

Institutional Review Board Statement: The study was conducted in accordance with the Declaration of Helsinki, and approved by our institution's ethics committee (approval number: CRB2180005). All of the individuals voluntarily participated in the study and provided written informed consent.

Data Availability Statement: Data are contained within the article. Contact the author for additional data.

Acknowledgments: We thank Sumito Musaka, Atsuko Harata, Yasufumi Yamaji, Tetsuya Yamauchi, Daiki Miura, Kai Maeda, Kaname Sasaki, Cao Yu, and Kota Odanagi for developing the rehabilitation robot. We thank Kazutoshi Hatakeyama, Motoyuki Watanabe, and Tomohiro Suda, for rehabilitation assistance and ideas for equipment development.

Conflicts of Interest: The authors declare no conflicts of interest.

References

1. WHO. Spinal Cord Injury. Available online: https://www.who.int/news-room/fact-sheets/detail/spinal-cord-injury (accessed on 5 September 2024).
2. Miyakoshi, N.; Suda, K.; Kudo, D.; Sakai, H.; Nakagawa, Y.; Mikami, Y.; Suzuki, S.; Tokioka, T.; Tokuhiro, A.; Takei, H.; et al. A nationwide survey on the incidence and characteristics of traumatic spinal cord injury in Japan in 2018. *Spinal Cord* **2021**, *59*, 626–634. [CrossRef] [PubMed]
3. Hubli, M.; Dietz, V. The physiological basis of neurorehabilitation—Locomotor training after spinal cord injury. *J. Neuroeng. Rehabil.* **2013**, *10*, 5. [CrossRef] [PubMed]
4. Veerbeek, J.M.; van Wegen, E.; van Peppen, R.; van der Wees, P.J.; Hendriks, E.; Rietberg, M.; Kwakkel, G. What is the evidence for physical therapy poststroke? A systematic review and meta-analysis. *PLoS ONE* **2014**, *9*, e87987. [CrossRef]
5. Holsheimer, J. Concepts and methods in neuromodulation and functional electrical stimulation: An introduction. *Neuromodulation* **1998**, *1*, 57–61. [CrossRef]
6. Illis, L.S. Central nervous system regeneration does not occur. *Spinal Cord* **2012**, *50*, 259–263. [CrossRef]
7. Kralj, A.; Bajd, T.; Turk, R. Enhancement of gait restoration in spinal injured patients by functional electrical stimulation. *Clin. Orthop. Relat. Res.* **1988**, *233*, 34–43. [CrossRef]
8. Andrews, B.J.; Baxendale, R.H.; Barnett, R. Hybrid FES orthosis incorporating closed loop control and sensory feedback. *J. Biomed. Eng.* **1988**, *10*, 189–195. [CrossRef] [PubMed]
9. Solomonow, M.; Baratta, R.; Hirokawa, S.; Rightor, N.; Walker, W.; Beaudette, P.; Shoji, H.; D'Ambrosia, R. The RGO generation II: Muscle stimulation powered orthosis as a practical walking system for thoracic paraplegics. *Orthopedics* **1989**, *12*, 1309–1315. [CrossRef] [PubMed]
10. Bae, Y.H.; Ko, Y.J.; Chang, W.H.; Lee, J.H.; Lee, K.B.; Park, Y.J.; Ha, H.G.; Kim, Y.H. Effects of Robot-assisted Gait Training Combined with Functional Electrical Stimulation on Recovery of Locomotor Mobility in Chronic Stroke Patients: A Randomized Controlled Trial. *J. Phys. Ther. Sci.* **2014**, *26*, 1949–1953. [CrossRef]
11. Schicketmueller, A.; Rose, G.; Hofmann, M. Feasibility of a Sensor-Based Gait Event Detection Algorithm for Triggering Functional Electrical Stimulation during Robot-Assisted Gait Training. *Sensors* **2019**, *19*, 4804. [CrossRef]

12. Bao, X.; Molazadeh, V.; Dodson, A.; Dicianno, B.E.; Sharma, N. Using person-specific muscle fatigue characteristics to optimally allocate control in a hybrid exoskeleton—Preliminary results. *IEEE Trans. Med. Robot. Bionics* **2020**, *2*, 226–235. [CrossRef] [PubMed]
13. Patathong, T.; Klaewkasikum, K.; Woratanarat, P.; Rattanasiri, S.; Anothaisintawee, T.; Woratanarat, T.; Thakkinstian, A. The efficacy of gait rehabilitations for the treatment of incomplete spinal cord injury: A systematic review and network meta-analysis. *J. Orthop. Surg. Res.* **2023**, *18*, 60. [CrossRef]
14. Kirsch, N.A.; Bao, X.; Alibeji, N.A.; Dicianno, B.E.; Sharma, N. Model-based dynamic control allocation in a hybrid neuroprosthesis. *IEEE Trans. Neural Syst. Rehabil. Eng.* **2018**, *26*, 224–232. [CrossRef] [PubMed]
15. Zhang, D.; Ren, Y.; Gui, K.; Jia, J.; Xu, W. Cooperative control for a hybrid rehabilitation system combining functional electrical stimulation and robotic exoskeleton. *Front. Neurosci.* **2017**, *11*, 725. [CrossRef]
16. del-Ama, A.J.; Gil-Agudo, Á.; Pons, J.L.; Moreno, J.C. Hybrid FES–robot cooperative control of ambulatory gait rehabilitation exoskeleton. *J. Neuroeng. Rehabil.* **2014**, *11*, 27. [CrossRef]
17. Bersch, I.; Alberty, M.; Fridén, J. Robot-assisted training with functional electrical stimulation enhances lower extremity function after spinal cord injury. *Artif. Organs* **2022**, *46*, 2009–2014. [CrossRef] [PubMed]
18. Kimura, R.; Matsunaga, T.; Iwami, T.; Kudo, D.; Saitoh, K.; Hatakeyama, K.; Watanabe, M.; Takahashi, Y.; Miyakoshi, N.; Shimada, Y. Development of a Rehabilitation Robot Combined with Functional Electrical Stimulation Controlled by Non-disabled Lower Extremity in Hemiplegic Gait. *Prog. Rehabil. Med.* **2018**, *3*, 20180005. [CrossRef]
19. Inoue, J.; Kimura, R.; Shimada, Y.; Saito, K.; Kudo, D.; Hatakeyama, K.; Watanabe, M.; Maeda, K.; Iwami, T.; Matsunaga, T.; et al. Development of a gait rehabilitation robot using an exoskeleton and functional electrical stimulation: Validation in a pseudo-paraplegic model. *Prog. Rehabil. Med.* **2022**, *7*, 20220001. [CrossRef]
20. Maeda, K.; Iwami, T.; Kimura, R.; Shimada, Y. Development of an automatic adjustment system for the amount of assist using reinforcement leaning in gait rehabilitation robot for hemiplegic patients. *Jpn. J. Inst. Ind. Appl. Eng.* **2022**, *10*, 28–37. (In Japanese)
21. Sutton, R.S.; Barto, A.G. *Reinforcement Learning: An Introduction*, 2nd ed.; MIT Press: Cambridge, MA, USA, 2018.
22. Kanda, Y. Investigation of the freely available easy-to-use software 'EZR' for medical statistics. *Bone Marrow Transplant.* **2013**, *48*, 452–458. [CrossRef]
23. Marquez-Chin, C.; Popovic, M.R. Functional electrical stimulation therapy for restoration of motor function after spinal cord injury and stroke: A review. *Biomed. Eng. Online* **2020**, *19*, 34. [CrossRef] [PubMed]
24. Kapadia, N.; Masani, K.; Craven, C.; Giangregorio, L.M.; Hitzig, S.L.; Richards, K.; Popovic, M.R. A randomized trial of functional electrical stimulation for walking in incomplete spinal cord injury: Effects on walking competency. *J. Spinal Cord. Med.* **2014**, *37*, 511–524. [CrossRef] [PubMed]
25. Pizzolato, C.; Saxby, D.J.; Palipana, D.; Diamond, L.E.; Barrett, R.S.; Teng, Y.D.; Lloyd, D.G. Neuromusculoskeletal modeling-based prostheses for recovery after spinal cord injury. *Front. Neurorobot* **2019**, *13*, 97. [CrossRef] [PubMed]
26. Luo, S.; Androwis, G.; Adamovich, S.; Nunez, E.; Su, H.; Zhou, X. Robust walking control of a lower limb rehabilitation exoskeleton coupled with a musculoskeletal model via deep reinforcement learning. *J. Neuroeng. Rehabil.* **2023**, *20*, 34. [CrossRef]
27. Xu, J.; Xu, L.; Cheng, G.; Shi, J.; Liu, J.; Liang, X.; Fan, S. A robotic system with reinforcement learning for lower extremity hemiparesis rehabilitation. *Ind. Robot. Int. J. Robot. Res. Appl.* **2021**, *48*, 388–400. [CrossRef]
28. Spiess, M.R.; Steenbrink, F.; Esquenazi, A. Getting the best out of advanced rehabilitation technology for the lower limbs: Minding motor learning principles. *PM&R* **2018**, *10*, S165–S173. [CrossRef]
29. Gonzalez-Vazquez, A.; Garcia, L.; Kilby, J.; McNair, P. Soft wearable rehabilitation robots with artificial muscles based on smart materials: A review. *Adv. Intell. Syst.* **2023**, *5*, 2200159. [CrossRef]

biomimetics

Article

Kinematic–Muscular Synergies Describe Human Locomotion with a Set of Functional Synergies

Valentina Lanzani, Cristina Brambilla * and Alessandro Scano

Advanced Methods for Biomedical Signal and Image Processing Laboratory, Institute of Intelligent Industrial Systems and Technologies for Advanced Manufacturing (STIIMA), Italian Council of National Research (CNR), 20133 Milan, Italy; valentina.lanzani@stiima.cnr.it (V.L.); alessandro.scano@stiima.cnr.it (A.S.)
* Correspondence: cristina.brambilla@stiima.cnr.it

Abstract: Kinematics, kinetics and biomechanics of human gait are widely investigated fields of research. The biomechanics of locomotion have been described as characterizing muscle activations and synergistic control, i.e., spatial and temporal patterns of coordinated muscle groups and joints. Both kinematic synergies and muscle synergies have been extracted from locomotion data, showing that in healthy people four–five synergies underlie human locomotion; such synergies are, in general, robust across subjects and might be altered by pathological gait, depending on the severity of the impairment. In this work, for the first time, we apply the mixed matrix factorization algorithm to the locomotion data of 15 healthy participants to extract hybrid kinematic–muscle synergies and show that they allow us to directly link task space variables (i.e., kinematics) to the neural structure of muscle synergies. We show that kinematic–muscle synergies can describe the biomechanics of motion to a better extent than muscle synergies or kinematic synergies alone. Moreover, this study shows that at a functional level, modular control of the lower limb during locomotion is based on an increased number of functional synergies with respect to standard muscle synergies and accounts for different biomechanical roles that each synergy may have within the movement. Kinematic–muscular synergies may have impact in future work for a deeper understanding of modular control and neuro-motor recovery in the medical and rehabilitation fields, as they associate neural and task space variables in the same factorization. Applications include the evaluation of post-stroke, Parkinson's disease and cerebral palsy patients, and for the design and development of robotic devices and exoskeletons during walking.

Keywords: mixed matrix factorization; kinematic–muscular synergies; EMG; gait; muscle synergies; motor control; kinematics

Citation: Lanzani, V.; Brambilla, C.; Scano, A. Kinematic–Muscular Synergies Describe Human Locomotion with a Set of Functional Synergies. *Biomimetics* **2024**, *9*, 619. https://doi.org/10.3390/biomimetics9100619

Academic Editor: Rafhael Milanezi de Andrade

Received: 23 September 2024
Revised: 9 October 2024
Accepted: 11 October 2024
Published: 13 October 2024

1. Introduction

Human locomotion is the result of a coordinated action of a large number of muscles. The number of muscles largely exceeds the number of degrees of freedom and, hence, the musculoskeletal system is highly redundant [1]. An accepted theory of motor control suggests that the central nervous system (CNS) produces movement through a combination of a limited number of spatial and/or temporal patterns, referred to as muscle synergies [2]. Muscle synergies can be extracted through the decomposition of electromyographic (EMG) signals in spatial [3], temporal [4], spatiotemporal [5] or space-by-time [6] motor modules. The spatial synergy model, which is the most used, describes the generation of motor commands as the combination of task-independent muscle weights (synergies) and task-dependent temporal activations.

So far, human gait has been analyzed in the framework of synergistic control in the muscle and in the kinematic domains separately. Studies agree in finding four to five motor synergies as fundamental blocks underlying locomotion, corresponding to biomechanical functions, such as weight acceptance, push off, swing, and leg deceleration [4,7–9]. In particular, the first two synergies and the last synergy (in chronological order with respect to a stride) are usually confirmed by most of the literature: weight acceptance at heel strike is mainly characterized by the activity of glutei muscles, tensor fascia latae, vastus medialis and rectus femoris, to guarantee the stabilization of the joints during weight transfer; the push off synergy is represented by the muscles needed to provide propulsion, i.e., gastrocnemii, soleus and peroneus longus; at the end of the swing phase, leg deceleration is controlled by the hamstring muscles [4,7–9]. In the first part of the swing phase, one or two synergies are usually identified, in which the trunk muscles are also involved. If only one synergy is considered, the main muscles involved are trunk muscles, rectus femoris and tibialis anterior [7]; while if two synergies are considered, the activity of the trunk muscles and the tibialis anterior is divided into two distinct synergies in which one controls the trunk position and the other controls the foot movement [4,8,9]. Several studies examined muscle synergies in healthy participants in different conditions, such as at different speeds, in the shift from walking to running, and on curvilinear trajectories, and these synergies were shown to be robust across individuals and walking conditions [10], such as changes in walking speed [8,11], body weight support [4] or shifting from walking to running [12]. In pathology, the number of extracted synergies is, in general, lower and depends on the severity of the disease [13]. Stroke patients have less synergies on the paretic side and synergy composition becomes less selective and less sparse, indicating increased co-activation patterns [14–16]. Indeed, modules extracted from healthy controls can merge into only two or three modules, reducing the complexity of motor control. Patients with cerebral palsy (CP) showed fewer synergies with respect to healthy controls, similar to stroke patients, indicating the use of a simplified motor strategy compared to healthy people [17].

While representing the coordinated neural input to muscles (muscle synergies) and the coordinated joint motor output (kinematic synergies), conventional approaches, such as NMF or PCA, may fail in finding a direct, quantifiable link between the neural drive and the motor output they generate. Indeed, these conventional approaches have some limitations, such as the constraint of non-negativity for the NMF, which does not allow for modeling unconstrained data (such as kinematics) that can be positive or negative, indicating joint flexion or extension [18]. PCA may capture this type of data, but the constraint of orthogonality between the principal components makes this method not ideal for capturing the link with task space, as no physiological observation supports the hypothesis of orthogonality between the extracted synergies. It follows that synergistic analyses have been mostly based on single domains. The investigations so far assessed muscle synergies and kinematic synergies separately. When some approaches tried to combine the two domains, they were usually based on qualitative or separate assessments. Indeed, biomechanical functions were related to synergies only on the base of the correlation of temporal activations during the gait cycle, but no direct combination between muscle synergies and kinematic outputs was performed [19]. However, the neural and motor output components were not integrated into a single model. The recently designed mixed matrix factorization (MMF) allows the factorization of any combination of positively constrained data (such as processed EMG envelopes) and unconstrained data (such as kinematics data, which can either be positive or negative) and finds common activations between the two domains in a unique set of kinematic–muscular synergies [18,20]. Kinematic–muscular synergies, that incorporate both the muscular activation and the functional joint output, provide a more comprehensive, biomechanically oriented characterization of motor control. MMF was already applied to study hand coordination, showing that more synergies are needed when both muscle and kinematic data are used to achieve the same reconstruction accuracy and that muscle activations can be related to different biomechanical functions [21]. Since in gait analysis, muscle synergies can usually be related to gait phases and, therefore, to biomechanical functions, gait is one natural field of application for kinematic–muscular synergies. Several limitations affect current analyses made with synergistic approaches so far. First, a limited number of muscles is analyzed for each leg (typically, eight muscles [22]).

Second, the available studies may provide a limited biomechanical interpretation: some synergies might show multiple peaks in temporal components that are associated with different task space functions, that are not captured and described clearly with standard muscle (or kinematic) synergies. Moreover, the role of some synergies may be equivocated as their functional role can be ambiguous (e.g., a synergy might be related to agonist action or to co-contraction to increase limb stiffness). Such effects could be particularly emphasized in pathology, where the mapping between muscle activations and motor output is not trivial [23].

Following this rationale, the primary aim of this study is to prove the feasibility and the added value of extracting kinematic–muscular synergies with respect to muscle synergies. Indeed, we wanted to demonstrate that kinematic–muscular synergies incorporate a more detailed description than muscle synergies alone when extracting the same number of muscle and kinematic–muscular synergies without altering the muscular part. A secondary objective is to show that a greater number of kinematic–muscular synergies are needed than muscle synergies to achieve the same reconstruction accuracy. Indeed, kinematic–muscular synergies embed a functional role into synergy weights and separate synergies that have the same structure but provide different biomechanical functions.

2. Materials and Methods

2.1. Participants

Data from fifteen subjects (7M, 8F; 23.8 (2.1) years; height: 1.69 (0.11) m; weight: 66.6 (10.8) kg) were considered for this study. Data are from a publicly available dataset recently published [24], which includes healthy subjects, with no neurological or musculoskeletal impairments.

2.2. OpenSim Model and Analysis Pipeline

Musculoskeletal simulations were performed in OpenSim v4.4 [25], using the available 3D Gait2392 model that simulates human gait [26,27]. The model includes 23 degrees of freedom of trunk and lower limbs and 76 muscles of the lower limbs and trunk. Twenty-four markers were placed in the model to match the motion analysis assessment protocol followed by Moreira et al. [24]. Marker trajectories and ground reaction forces (GRFs) were provided with a sample frequency of 200 Hz. The dataset includes seven different gait speeds; however, for this study, we considered only trials at 4 km/h, which is a comfortable natural walking speed. First of all, the OpenSim model was scaled to meet each participant's anthropometry. Then, joint kinematics was computed with the inverse kinematics tool, giving the 3D marker trajectories as input to the OpenSim model [28]. Muscle forces were computed with the static optimization procedure provided in OpenSim [27], starting from the resulting joint kinematics and the GRFs. Muscle forces are calculated minimizing the instantaneous total square muscle activations needed to achieve the experimentally acquired trajectory and the model includes the muscle force–length–velocity relationships. Four strides performed with the right limb were considered for the analysis of each subject. The events for segmenting each stride (begin and end sample of each stride) were defined as two consecutive contacts of the right heel with the ground and they were identified based on the GRFs. Articular angles and muscle activations were filtered with a 3rd-order Butterworth low-pass filter with a cut-off frequency of 6 Hz. The degrees of freedom considered for the analysis were the pelvis flexion, the hip flexion, the knee flexion, and the ankle flexion in the sagittal plane. The angular acceleration for each joint was computed as the second derivative of the articular angles. Sixteen muscles often used in muscle synergy analysis were considered: Soleus, Gastrocnemius medialis, Gastrocnemius lateralis, Tibialis anterior, Vastus medialis, Vastus lateralis, Rectus femoris, Biceps femoris long head, Biceps femoris short head, Semimembranosus, Semitendinosus, Tensor fascia latae, Gluteus medius, Gluteus maximum, Adductor longus, Psoas. Both joint accelerations and muscle activations were segmented into strides, and they were resampled at 101 samples for each stride. Data were rearranged in a matrix for synergy extraction. For extracting muscle synergies, the matrix data had 16 rows, one for each muscle, and 404 columns, corresponding to 4 strides of 101 samples (per subject). For extracting kinematic–muscular synergies, the matrix data had 20 rows, representing 4 joint accelerations and 16 muscles, and 404 columns, corresponding to 4 strides of 101 samples. To allow inter-subject comparisons, kinematic data from each degree of freedom and EMG data from each muscle were normalized by the maximum found in all strides for each subject. The scheme summarizing the pipeline of the study is shown in Figure 1.

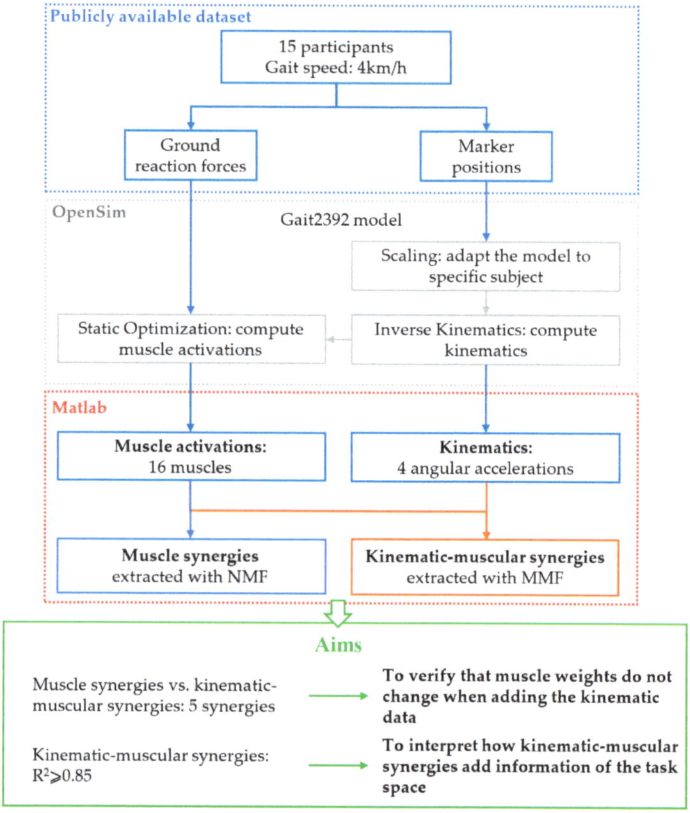

Figure 1. Scheme of the work. Markers' position and ground reaction forces from a publicly available dataset are used as input for musculoskeletal simulations in OpenSim. The outputs of the model are kinematics and muscle activations. In total, 16 muscle activations are used for extracting muscle synergies with NMF and the same muscle activations with 4 angular accelerations are used for extracting kinematic–muscular synergies with MMF. Then, five kinematic–muscular synergies are compared to five muscle synergies to demonstrate that the muscular weights do not change when adding kinematic data. Finally, a number of kinematic–muscular synergies achieving $R^2 \geq 0.85$ are extracted to show that they add information from the task space.

2.3. Synergy Extraction and Clustering

Methods for synergy extraction decompose the input signal $x(t)$ (generally the EMG) as the product of n time-invariant synergy vectors $\boldsymbol{w}_i$ shared across all stride repetition and corresponding time-varying activation coefficients $c_i(t)$ specific for each stride repetition, as follows:

$$x(t) = \sum_{i=1}^{n} \boldsymbol{w}_i c_i(t) \quad (1)$$

The algorithm iteratively decomposes x to minimize the error between the original signal and the reconstructed signal, obtained from the product between the synergies and the temporal coefficients. Spatial synergies and temporal coefficients were extracted from data of each subject with two algorithms. The NMF iterative algorithm based on multiplicative updates was used for muscle synergy extraction, giving the EMG signals as input x [29]. The MMF algorithm was used for kinematic–muscular synergy extraction, giving the EMG and kinematics signals as input x [18]. MMF extends the standard NMF, removing the constraint of non-negativity of signals to be factorized (and extracted), and is based on a gradient descent update rule. For the MMF algorithm, we chose the following set-up values: $\lambda = 200$ and $\mu = 0.05$. These parameters are selected because they offer a good trade-off between an accurate EMG reconstruction and fast algorithm execution [18]. The quality of reconstruction R^2 of the original signal was defined for both the extractions as $1 - SSE/SST$, where SSE is the sum of the squared residuals and SST is the sum of the squared differences with the mean input vector [30]. The algorithm was repeated 20 times and the solution achieving the highest R^2 was considered for the analysis. The procedure was performed increasing the number of extracted synergies in the factorization from 1 to 16 for muscle synergies and from 1 to 20 for kinematic–muscular synergies. A linear mixed-effects model was fitted in order to assess the differences between the R^2 obtained for muscle and kinematic–muscular synergies [31]. First, both R^2 were tested for normality at each number of synergies using the Kolmogorov–Smirnov test. Then, the R^2 was modelled as follows: $R^2 \sim 1 + n_{syn} \cdot model + (1 \mid subject)$ where model (*muscle/kinematic muscular*) and n_{syn} (number of extracted synergies) were fixed effects with interaction, n_{syn} was a categorical variable, and subjects were included as random effects on intercept. The level of significance (α) was set at 0.05.

Five muscle synergies were extracted for each subject since it is the higher number of synergies typically extracted from one subject in gait analysis [4]. In the first step of the analysis, the same number of kinematic–muscular synergies were extracted to allow direct comparison between muscle and kinematic–muscular synergies. For each participant, muscle and kinematic–muscular synergies were matched for similarity, computed as cosine angle between pairs of matched synergies. To perform this comparison, the muscle weights of the kinematic–muscular synergies were re-normalized to have unit norm and then they were matched for similarity with the paired muscle synergies.

Moreover, observing that five muscle synergies achieved $R^2 \geq 0.85$, in the second part of our analysis we extracted a number of kinematic–muscular synergies that achieved the same threshold to provide an assessment based on the same reconstruction accuracy.

Synergies from all participants were grouped with k-means clustering algorithm to reduce them to a small set of synergies shared by subjects that represent the repertoire of synergies available to healthy people and evaluate their variability [32]. The clustering procedure was repeated 100 times with new initial random cluster centroid estimates with the same number of clusters and the result with the lowest sum of Euclidean distances of each element in the cluster to the centroid was selected. This pipeline was repeated until the average inter-cluster similarity was greater than 0.70 in order to guarantee a limited number of clusters and a good intra-cluster similarity level [33,34]. As a measure of the robustness of the clustering, the intra-cluster similarity was computed with the cosine angle comparing all pairs of synergies in a cluster [35]. Five clusters were defined for muscle synergies and the same number of clusters were used to compare the kinematic–muscular synergies when extracting 5 synergies. Seven clusters were needed instead when extracting 6 kinematic–muscular synergies per subject.

3. Results

In Figure 2, the averaged muscle activations from the 15 subjects are shown (average values and deviations).

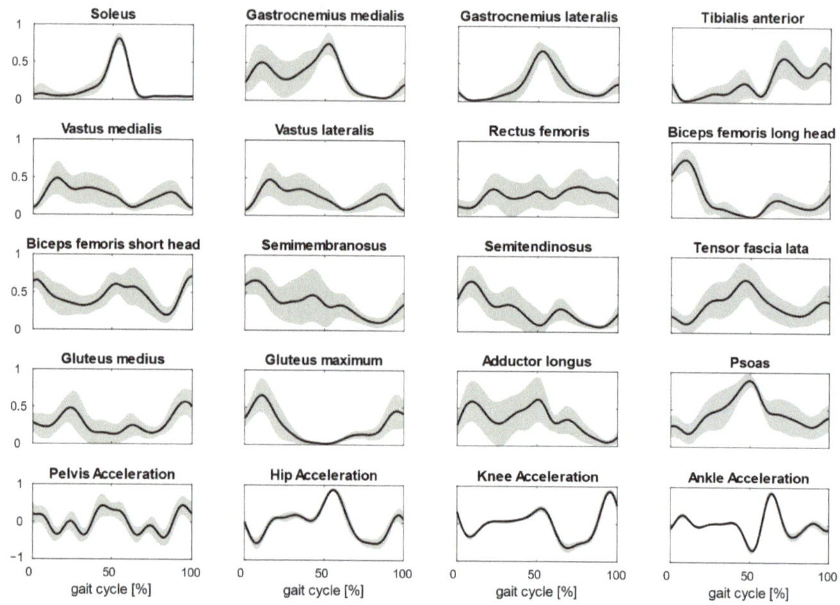

Figure 2. Plots show the averaged normalized activations of the 16 muscles considered during gait. The muscle activations are averaged on four steps and for all subjects. In the last row, joint accelerations used for MMF are shown too.

3.1. *Reconstruction R^2*

The mean reconstruction R^2 across participants is reported in Figure 3 for kinematic–muscular synergies and muscle synergies.

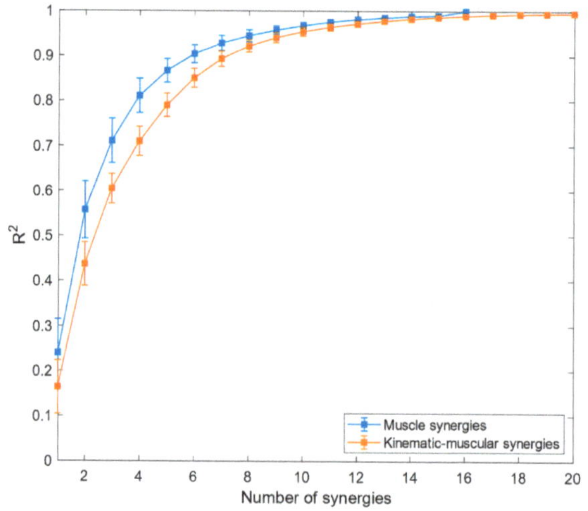

Figure 3. Reconstruction R^2 for muscle synergies (blue graph) and kinematic–muscular synergies (orange graph). Means and standard deviations across subjects are reported.

Referring to Figure 3, the reconstruction of R^2 for muscle synergies has higher values than the R^2 for the kinematic–muscular synergies. Moreover, the R^2 for the kinematic–muscular synergies requires more synergies to reach $R^2 = 1$ because the MMF includes 20 channels instead of 16. The linear-mixed effect analysis showed that model type (*muscle/kinematic–muscular*) has significant effects on R^2 from 1 ($p < 0.001$, $\beta = -0.065$) to 7 ($p = 0.03$, $\beta = -0.025$). In the first part of the experiment, the R^2 threshold to select the number of synergies in the muscle and in the kinematic–muscular configurations was set = 0.85; five muscle synergies were sufficient to reach this threshold, while for the kinematic–muscular synergies, six synergies were needed to reach the threshold.

3.2. Comparison between Kinematic–Muscular Synergies and Muscle Synergies

To highlight the first characteristics of kinematic–muscular synergies, in this study, we compared the muscle weights of muscle synergies and kinematic–muscular synergies. Their mean similarity is reported in Table 1. The similarity between the muscle weights of the models was consistently higher than 0.80 for all subjects and was significantly higher than the similarity between randomly paired synergies ($p < 0.001$). These results indicated that the muscular part is only minimally affected when extracting kinematic–muscular synergies. Thus, adding kinematic weights only minimally modified the muscle synergies extracted with NMF.

Table 1. Mean similarities between matched muscle and kinematic–muscular synergies are reported for all subjects. Standard deviations are reported in brackets.

Similarity between Matched Muscle and Kinematic–Muscular Synergies		
Subject	**Mean**	**Random Similarity**
S01	0.898 (0.141)	0.493 (0.276)
S02	0.881 (0.132)	0.507 (0.279)
S03	0.824 (0.154)	0.465 (0.259)
S04	0.970 (0.029)	0.532 (0.251)
S05	0.823 (0.245)	0.422 (0.299)
S06	0.878 (0.116)	0.488 (0.257)
S07	0.926 (0.094)	0.532 (0.245)
S08	0.939 (0.073)	0.487 (0.283)
S09	0.809 (0.251)	0.489 (0.263)
S10	0.868 (0.170)	0.470 (0.283)
S11	0.908 (0.069)	0.547 (0.255)
S12	0.845 (0.062)	0.510 (0.264)
S13	0.878 (0.152)	0.523 (0.264)
S14	0.873 (0.169)	0.516 (0.253)
S15	0.889 (0.217)	0.519 (0.255)
Total	**0.881 (0.045)**	**0.500 (0.032)**

Clustered muscle synergies and kinematic–muscular synergies were paired by similarity and are shown in Figure 4. All the kinematic–muscular synergies include weights that indicate which joints were accelerated due to muscle activity and could clarify whether each joint acceleration contributed to flexing or extending the joints. Referring to Figure 4, and denoting as M_i the ith muscle synergy and K_i the ith kinematic–muscular synergy, M1 and K1 were active at the beginning of the stance phase and show the activation of the gluteus and the hamstrings muscles, which extend the hip, and the activation of the vastus medialis and the lateralis that extend the knee as shown in kinematic–muscular synergies. M2 and K2 are active during the stance phase. M2 is characterized by the activation of the adductor, psoas, hamstring muscles and vastus medialis and lateralis. The kinematic–muscular synergy K2, instead, recruits many muscles with small activations, including the psoas, biceps femoris, rectus femoris. These muscles contribute to the flexing of the hip and to the posterior movement of the pelvis. M3 and K3 represent the push off of the gait cycle and are characterized by the psoas and the adductor, which flex the hip, and the gastrocnemii and the soleus that perform plantarflexion of the ankle and the flex of the knee. M4 and K4 are active during the swing phase and show tibialis anterior, psoas, biceps femoris, rectus femoris that extend the knee and the hip. The kinematic–muscular synergy K4 shows a higher magnitude of kinematic weights, reducing the muscle weights, probably as the limb is exploiting previous activation to perform the swing phase. Finally, M5 and K5 are active at the end of the gait cycle and are characterized by the activation of psoas, glutei and biceps femoris, flexing the knee and tilting the pelvis anteriorly.

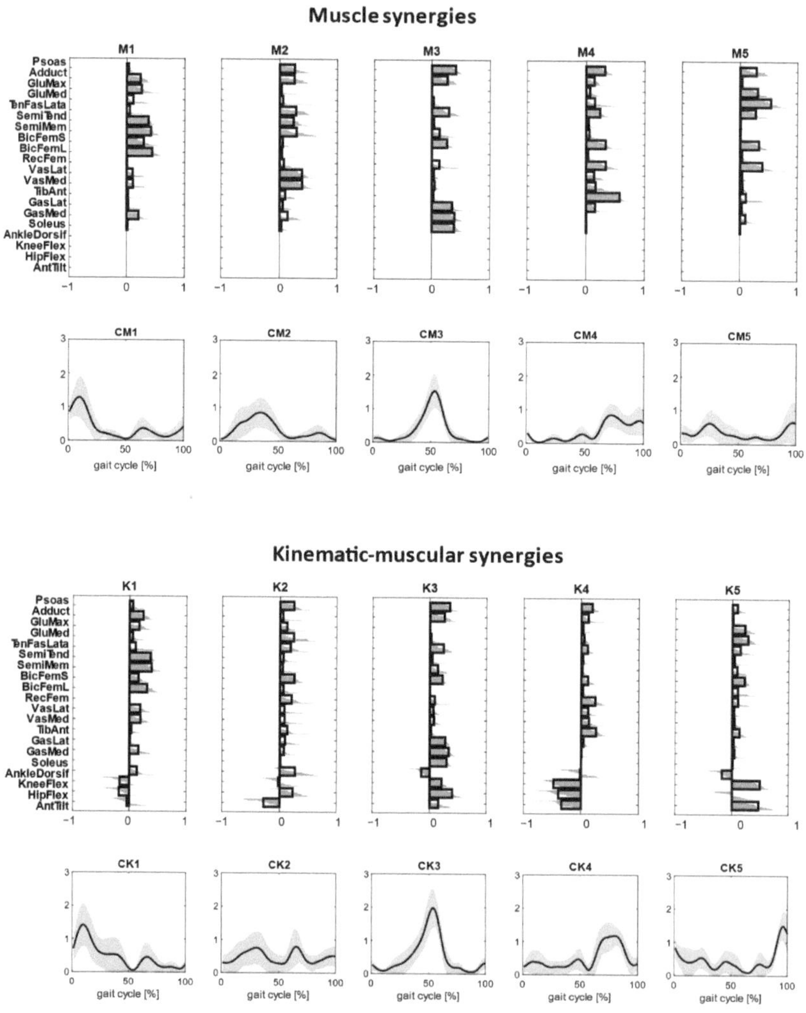

Figure 4. Clustered muscle synergies and corresponding temporal coefficients are reported in the top first panel. Clustered kinematic–muscular synergies and corresponding temporal coefficients are reported in the lower panel. Clusters are ordered based on synergy recruitment timings in the gait cycle.

3.3. Extraction of Kinematic–Muscular Synergies with $R^2 > 0.85$

To investigate the second characteristics of kinematic–muscular synergies, the study quantifies the assessments provided with kinematic–muscular synergies when their number is not equal to the number of muscle synergies but selected with a fixed R^2 threshold ($R^2 > 0.85$), as is usually carried out in experimental studies. We noted that the number of kinematic–muscular synergies required to reach the R^2 threshold was six. Thus, more kinematic–muscular synergies were needed than muscle synergies to obtain the same level of reconstruction achieved with muscle synergies. Figure 5 shows the result of the clustering procedure applied to the whole dataset of the extracted kinematic–muscular synergies when $R^2 > 0.85$. Seven clusters were found. They are presented in Figure 5 in chronological order following synergy recruitment in the gait cycle and denoted with a W_i label. The first synergy cluster W1 is associated with the beginning of the stance phase and the synergy cluster W7 is related to the end of the stride when the leg is repositioned on the ground before the new gait cycle.

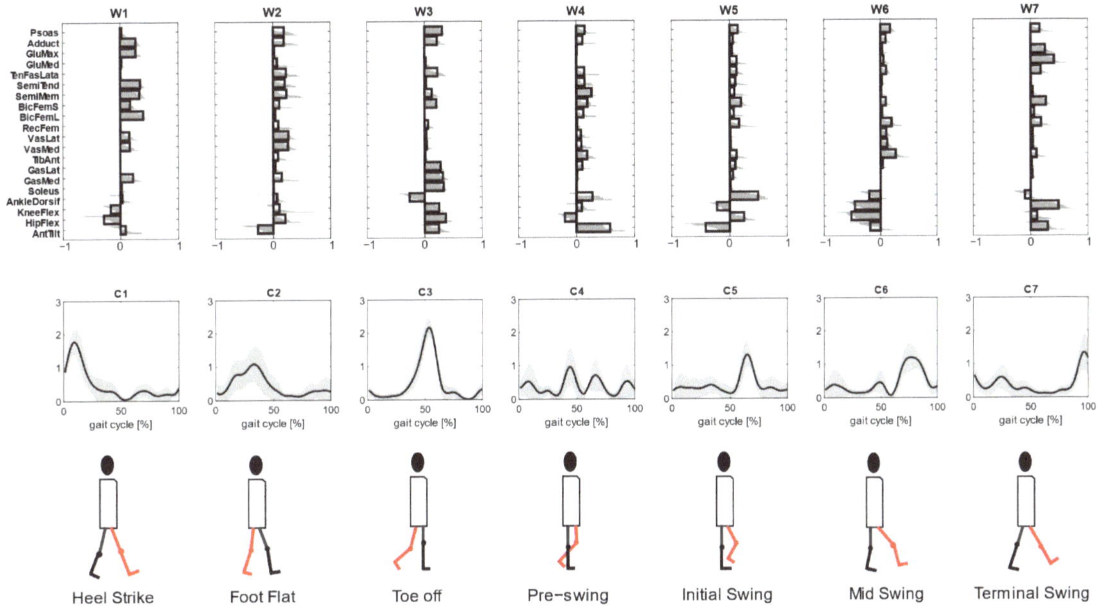

Figure 5. Kinematic–muscular synergies extracted with $R^2 \geq 0.85$ were grouped into 7 clusters so that the intra-cluster similarity is greater than 0.70 for all clusters (upper panel). The synergies activation coefficients are ordered following the gait cycle (lower panel). The third line represents the biomechanical function associated with the walking task.

Each cluster can be associated with biomechanical functionality within the walking task, depending on the muscles recruited, on the moving joints, and on the timings of activation. W1 shows a strong activation of the hamstring muscles accomplished with an activation of the adductor longus and the gluteus maximum that are responsible for hip extension and then shows the activation of the vastus medialis and the lateralis for knee extension. The stance phase continues with synergy W2 where an overall activation of many muscles with an initial posterior pelvic tilt can be observed. Then, W3 shows a strong activation of the soleus, gastrocnemius lateralis and medialis, which represent the part of the gait cycle in which the ankle is dorsiflexed, and the knee and hip are flexed during the late stance to provide propulsion to begin the swing phase and advance. Synergy W4 is activated just before the push off and at the beginning of the swing phase and shows the anterior pelvic tilt and the activations of the semitendinosus, semimembranosus, tensor fascia latae, biceps femoris and tibialis anterior. The swing phase is described in W5 where the pelvis changes its position, activating with a posterior tilt and the ankle is dorsiflexed. These kinematic activations are associated with both hamstring muscles, both biceps femoris, psoas and rectus femoris. The swing phase continues with full extension of all joints enabled by small activations of many muscles (W6). Finally, the gait cycle finishes with a knee flexion and anterior pelvic tilt that are needed to position the foot on the ground before a new gait cycle, as can be observed in synergy W7. In this synergy, the muscular part is characterized by a strong activation of the gluteus maximum and medialis, tensor fascia latae and short bicep femoris.

4. Discussion

4.1. Summary of the Findings

In this study, kinematic–muscular synergies were extracted with MMF during the locomotion of 15 participants for the first time. Kinematic–muscular synergies were compared to standard muscle synergies extracted with NMF. Gait data from a publicly available dataset were used to feed a musculoskeletal model in OpenSim, in which kinematics and muscle activations were computed. Muscle synergies were extracted from the activation of 16 muscles of the right lower limb with the standard NMF. Kinematic–muscular synergies were extracted from the same EMG activations and from the angular acceleration of four joints in the sagittal plane with the MMF algorithm. For a given number of extracted synergies, the reconstruction R^2 was higher for muscle synergies with respect to kinematic–muscular synergies and fewer synergies were needed to achieve $R^2 > 0.85$. Comparing the muscular part of the kinematic–muscular synergies to muscle synergies, the similarity was high for all the participants and significantly higher than the random similarity. This result showed that the muscular weights are minimally affected when adding the kinematic weights. Lastly, when a given R^2 is selected, more kinematic–muscular synergies are needed, and this effect is linked to a more accurate, functionally oriented description of synergistic control. Thus, including kinematics in the synergy analysis allowed us to highlight the link between muscle activation and their biomechanical function, which was described by individuating, for the first time, the repertoire of kinematic–muscular synergies available to healthy people.

4.2. Muscle Synergies vs. Kinematic–Muscular Synergies

All participants showed good similarity between muscle weights of the kinematic–muscle synergies and muscle synergies. This finding confirmed previous results [18] that showed that the addition of kinematic weights when using the MMF algorithm does not alter (or minimally alters, in the case of noisy data) the composition of standard muscle synergies originating from the neural structures. Indeed, muscle synergies and the muscle weights of kinematic–muscular synergies were mostly highly similar, and the main differences were found in M2 and K2 during the stance phase. In each cluster, the kinematic–muscular synergies add information on how joints move and associate it to standard muscle synergies. For example, gastrocnemii and soleus plantar flex the ankle, hamstring muscles flex the knee and extend the hip, while psoas, rectus femoris and vastus medialis and lateralis flex the hip. The differences between muscular and kinematic–muscular synergies may be related to the fixed chosen number of synergies. Indeed, the kinematic–muscular synergy model needs more synergies to reach the same reconstruction accuracy as the muscle synergy model and, therefore, five synergies are not enough to describe the biomechanics of gait in detail. In fact, when extracting six synergies, clusters W1, W2, W3, W6 and W7 are very similar to the muscle synergy clusters of Figure 4. The two added synergies, W4 and W5, describe the transition from the stance to the swing phase during the gait cycle. In particular, in W4, all muscles are involved in body stabilization, during ankle dorsiflexion and anterior pelvic tilt that characterize the exploitation of the propulsion produced in the late stance. W5, instead, represents the transition from back tilt to front tilt needed to prepare the body to heel strike the homolateral foot.

4.3. Kinematic–Muscular Synergies Add Functional Information to Muscle Synergies

With kinematic–muscular synergies, muscle activity can be associated with the kinematic accelerations that result from muscle contraction, enriching standard neural synergies with a functional role. Kinematic–muscular synergies incorporate task execution variables into muscle synergy extraction, providing a functional role to each synergy. In this way, functional synergies improve the interpretation of the results and their clinical use. Synergies from all the participants were clustered into seven mean synergies. Each synergy can be associated with a biomechanical functionality describing the phases of a stride depending on the muscles recruited and the timings of activation. Observing the kinematic part of kinematic–muscular synergies, one can unveil the functionality of each muscle module expanding the previous literature in which muscle or kinematic synergies were extracted. The first cluster of synergy M1, representing the early stance, shows a strong activation of the gluteus, tensor fascia latae, vastus medialis and lateralis, which are responsible for the stabilization of the hip joint during the heel strike and the load acceptance phase. In kinematic–muscular synergies (Figure 5), this gait phase was divided into two synergies: W1 with a strong activation of gluteus maximum and W2 with a strong activation of both tensor fascia latae and vastus lateralis muscles. In addition, in W1, there is also a high activation of semitendinosus and biceps femoris that contribute to the leg's stabilization during movement. In these synergies, the kinematic weights are reduced, and this finding reflects how, in this stride phase, the neuromuscular system works more to prepare to accept the load rather than to carry out the movement. The forward propulsion is well represented both considering muscle and kinematic–muscular synergies, in fact, in both cases, the synergies of M3 and W3 highlight a strong activity of the soleus and the gastrocnemius. Moreover, observing the kinematic–muscular synergy W3, other muscles useful to generate a propel propulsion are the adductor and the psoas that help to maintain balance [36].

The initial swing phase is represented in different ways in the literature because it depends on the muscles considered in each study. In fact, according to some studies, this phase principally involved the trunk muscles, such as the longissimus dorsi or the erector spinae, because, during the initial swing phase, the trunk position needs to be controlled in the frontal plane at the time of contralateral foot heel strike and ipsilateral foot lift [8]. In our study, we did not monitor trunk muscle activation, but we coherently observed in synergy W4 an activation of all the muscles involved in body stabilization, with a slightly accentuated activation of the tibialis anterior, semimembranosus, short biceps femoris and also a moderate activation of psoas. Therefore, the muscular part of kinematic–muscular synergy W4 has a lower magnitude, but it gives a clear view of joint motion showing ankle dorsiflexion and a strong anterior pelvic tilt that characterize the exploitation of the propulsion produced in the late stance [37]. In the kinematic–muscular synergies, another synergy is added, W5, that represents the swing phase when the pelvis makes the transition from back tilt to front tilt [38]. Indeed, we observed the activation of the tensor fascia latae, both the hamstring muscles and both bicep femoris that are responsible for this transition of the legs linked to the change of pelvic position, and the kinematic part shows a strong posterior pelvic tilt and ankle dorsiflexion and moderate hip extension and knee flexion that are needed to prepare the body to heel strike the homolateral foot. In kinematic–muscular synergy W6, this function is well represented, since the same muscle activations and the kinematic parts show an extension of all the joints of the leg and a posterior pelvic tilt that are typical joint movements to allow the deceleration of the foot.

Finally, in the muscle synergy associated with the late swing phase, the hamstrings are the key muscles, as they decelerate the leg [39]. The kinematic–muscular synergy associated with this phase (W7) also shows the activation of the bicep femoris and the rectus femoris, but it especially highlights a strong activation of the gluteus medialis and maximum and tensor fascia latae. Moreover, the kinematic part of this synergy is well accentuated and shows a strong knee flexion that is needed to control and attenuate the load due to the leg repositioning before a new gait cycle.

Kinematic–muscular synergies were shown to be consistent with the muscle modules that are usually identified in the literature, such as the activation of the glutei, tensor fascia latae, vastus medialis and rectus femoris in the stance phase; gastrocnemii and soleus activated during the push off; rectus femoris and tibialis anterior activated in the swing phase; and the hamstring muscles at the end of the swing phase [4,7–9]. This means that the kinematic–muscular synergy extraction adds information about how the movement is performed without altering information about the muscle activations. As an advantage, not only are the muscle activations clearly associated with a functional role, but also a fractionation effect is shown that helps to provide a more accurate description of synergistic control. For example, W5 and W6 are synergies that refer mainly to the swing phase and show a high predominance of kinematic coefficients. This is expected as the swing phase exploits inertial forces generated in the stance and late stance phases and thus contains more motion rather than neural drive. Coherently, synergies like W2 and W3 that represent the stance and late stance phase show a mixture of muscle and kinematic weights as muscle activation is needed to generate propulsion during the gait cycle. The extraction of kinematic–muscular synergies allows us to directly integrate the neural activations and the motor output in an efficient way with respect to other approaches based on the correlation of muscle and kinematic synergies, as performed by Esmaeili et al. [19]. In fact, we presented a joint analysis in which muscular and kinematic weights are extracted together in multidomain synergies. Moreover, synergies are extracted from joint accelerations and not from joint position, allowing us to directly couple the muscle activity and the joint movement on which they act. In Esmaeili et al. [19], instead, kinematic synergies were extracted from the joint position. For instance, the effect of the glutei muscles of the first synergy was visible in the second synergy, characterized by hip extension; the activity of the calf muscles during the push off generated the ankle dorsiflexion in the following synergy. Therefore, kinematic–muscular synergies provide a straightforward analysis to associate muscular activity and kinematic output, directly defining the functionality of each synergy.

4.4. Clinical Application of Kinematic–Muscular Synergistic Control

The use of synergies in motor control assessment has diffused in recent years, but the current synergy-based analysis methods do not yet exploit the full potential of the synergistic approaches. In fact, the standard synergistic models have some limitations, such as the inability to consider the task space variables. Incorporating task execution variables into muscle synergy extraction links muscle synergies to the motor function they produce. This process reflects the task space output due to muscle synergies and improves the interpretation of the results and their clinical use [40]. The coupling between the two may help in fostering synergistic protocols as suggested in recent works [41,42]. Moreover, muscle synergy analysis is still scarcely used for evaluating neuromotor rehabilitation in clinical scenarios. However, it was shown that modifications in muscle synergies can measure the progression of the rehabilitation process in an interpretable and quantitative manner [43]. Further improvements in the standard synergy analysis are needed before it can be fully transferred to clinical practice. This forward step, which includes the task space into muscle synergy analysis, may improve the interpretability of the results and may help the introduction of synergy analysis in the clinical practice and clinical decisional process, providing clinicians and therapists with a novel instrument to assess the efficacy of a therapy analyzing movement kinematics associated to the underlying neural control strategies.

The first step for such applications is to provide reference databases of healthy people to create a repertoire of kinematic–muscular synergies and this study is a pilot investigation in such a sense. Using kinematic–muscle synergies may improve the understanding of neuromotor coordination in several diseases. The practical consequences of adding kinematic weights to synergies impact the understanding of how the underlying neural motor strategies are reflected at task space and biomechanical levels [18]. Therefore, functional synergies may open new perspectives in the analysis of motor control in clinical practice providing a more in-depth evaluation of pathological motor patterns, especially in patients with impaired control of movement coordination, as the kinematic–muscular synergies may link directly muscular and kinematic patterns allowing to elucidate the relationship between the neural drive and motor outcomes [21]. Some of the practical added values of including kinematic weights to standard muscle synergies are proposed below.

First, the exploitation of functional synergies can add significant details on the altered muscle synergies of patients who are affected by pathologies that compromise movement coordination. Altered coordination can be associated with an altered motor output in different ways. Indeed, the first achievement of kinematic–muscular synergies is to show if altered coordination is the result of altered synergies or if synergies similar to physiological ones are available but are recruited in a biomechanically abnormal way. Second, muscle synergies with similar compositions might be extracted from healthy and pathological subjects; however, their biomechanical functions can be associated with different joint movements, which is fully resolved and clarified only with kinematic–muscular synergies. In such a sense, kinematic–muscular synergies might be a very useful tool to highlight the biomechanical function associated with each neural synergy. Third, in standard muscle-synergy assessment, there might be some muscle synergies that present multiple peaks of activation within a movement phase; kinematic–muscular synergies might instead associate such muscle coactivation peaks to specific kinematics, with the effect of producing a fractionation of muscle synergies that leads to a more specific interpretation of their biomechanical function. Fourth, using kinematic–muscular synergies may set a novel target for rehabilitation: restoring not only physiological, "neural" synergies but also promoting their efficacy at the task level.

For instance, it was shown that the locomotion of post-stroke patients with severe impairment is characterized by the merging of muscle synergies and patterns may change depending on the level of impairment [15]. In some cases, this effect is coupled with a reduction of the joint range of motion with respect to physiological walking. Kinematic–muscular synergies would naturally capture the neural–motor output relationship by associating reduced kinematic coefficients to the functional synergies of patients with respect to healthy people. Thus, abnormal muscle couplings would be interpreted with the support of a direct link to their effect on the task space. Hence, these synergies' responses manifesting at different levels of impairment in patients with stroke, may represent a precise and quantifiable marker of the physiological status of the patient in order to better understand the complex processes that follow accidents involving the cortical and spinal motor system and so guide the development of different rehabilitation approaches [16]. This result can be achieved by further classifying patients' synergistic control not only on the basis of abnormal coactivations but also by considering the biomechanical outcomes that are produced by such coactivations.

In Parkinson's disease (PD), motor disorders are the main symptoms that become the major source of disability with the progression of the disease and include tremors, bradykinesia and loss of balance. Several studies used muscle synergy analysis to characterize motor control in patients with Parkinson's disease and they revealed a lower number of extracted synergies compared to healthy subjects [44]. Therefore, a reduced number of recruited muscle synergies may be a sign of an alteration in muscle activation patterns that could explain changes in motor control in patients affected by neurological disorders [45]. These altered muscle activation patterns are often associated with a modified kinematic movement; by extracting kinematic–muscular synergies, investigators could better highlight the motor alterations due to motor disorders such as tremors and bradykinesia. Moreover, when analyzing postural control in patients with Parkinson's disease, it was noted that they are characterized by smaller indices of anticipatory synergy adjustments compared to healthy subjects because patients are affected by freezing of gait [46]. This symptom should be better investigated by observing the kinematic–muscular synergies that link the muscle activation patterns to the joint activated during motor movements so that researchers better understand how an alteration at the neural level is reflected in an alteration at the motor level.

Another pathological condition that affects movement is Spinal Cord Injury (SCI). Muscle synergies have been used in order to understand how the spinal circuitry reorganization after spinal cord injury reflects also in the modular organization [47]. SCI patients show significantly altered muscle synergy patterns [48,49] and directly linking the kinematic output to neural synergy alteration may shed light on functional recovery and neural plasticity. Moreover, it was found that despite consistent muscle synergies being extracted across SCI patients in upper limb movements, this outcome was coupled with a high level of variability in kinematic strategies [50]. Kinematic–muscular synergy analysis seems the natural choice to investigate this result and understand how similar muscle synergies may map into different joint movements.

Children with Cerebral Palsy (CP) show abnormal motor control resulting in altered coordination and stiff muscles. It was shown that locomotion is altered with different degrees of severity depending on the functional impairment, strength and presence of spasticity [17]. Usually, fewer muscle synergies are recruited by CP patients with respect to healthy people, with some abnormal synergies specific to the CP [22]. Moreover, wider temporal activation patterns are found in children with CP [51]. Synergy variability is high between CP children and synergy structure is more altered in patients with greater impairment [23]. However, providing a link between the muscle activation and the motor output is not always trivial [52] and kinematic–muscular synergies may support the understanding of pathological movement. The proposed method provides a direct link between the underlying neurophysiological mechanisms and their motor output, thus fostering the understanding of the progression of the pathology, the consequent choices regarding the selection of the best rehabilitation course, and the evaluation of the effects of surgical interventions.

The restoration of motor output is usually the primary objective of motor rehabilitation and approaches based on kinematic–muscular synergies can be used to implement individualized therapeutic strategies promoting a general restoration of motor control. It is demonstrated that muscle and kinematic synergies can be modified by targeted motor training developed using assistive approaches, such as functional electrical stimulation (FES) which are potentially useful to re-establish the normal ways of muscle activation that was impaired by neurological disease [53]. FES was developed to design a rehabilitation therapy based on muscle synergies because if the muscle stimulation reflects the healthy muscle activation the neuroplasticity may reorganize the neural circuitry of motor control to restore a normal pattern of muscle activation [54]. Kinematic–muscular synergies can be used to better customize FES therapy trying to make the therapies also task-specific so that the rehabilitation therapy becomes more focused on restoring specific motor control [55]. Furthermore, in rehabilitation, robotic devices and exoskeletons may be used to assist locomotion in patients with hemiparesis in order to elicit the restoration of motor control. Exoskeletons are robotic interactive wearables that actively assist motor-impaired individuals during walking and adjust abnormal gait by providing suitable assistive forces [56]. Muscle and kinematic synergies have been used not only for evaluating the effects of using exoskeletons on motor control but also for adjusting the level of assistance of the paretic limb based on the non-paretic limb synergies aiming at increasing the similarity between synergies [57,58]. Therefore, kinematic–muscular synergies provide a comprehensive evaluation of motor control, fusing both neural and kinematic information that can be used for assessing the restoration of motor control when using exoskeletons and improving the design of these devices.

5. Limitations and Future Work

Although the analytical approach and the potential applications proposed in this work are novel, the EMG data used were computed based on an OpenSim simulation starting from experimental kinematic data. Such simulated EMG data may be only partially representative of the real data and further work should be performed on the experimental data. Moreover, experimental input data were taken from healthy participants only. Thus, it would be interesting to investigate how this approach improves the understanding of the pathophysiology of several neurological disorders associated with motor control impairments, such as stroke, Parkinson's disease or cerebral palsy, and to study to what extent healthy people's control strategies are preserved in neurological patients and which novel insights can be obtained with kinematic–muscular synergies.

6. Conclusions

In this study, for the first time, functional kinematic–muscle synergies were extracted with the novel mixed matrix factorization from locomotion data. Data from 15 participants from a publicly available dataset were used to feed a musculoskeletal model in OpenSim used to compute lower limb muscle activations, muscle synergies and kinematic muscular synergies. We demonstrated that kinematic–muscle synergies can describe the biomechanics of motion to a better extent than muscle synergies alone and are increased in number to account for the different biomechanical roles that muscles have within a movement. Despite only healthy subjects being investigated in this study, the results suggest that applying this approach to patients could also be beneficial in improving the understanding of the pathophysiology of neurological disorders related to motor impairment. Therefore, this approach may have an impact on future work in improving the understanding of pathologies in rehabilitation.

Author Contributions: Conceptualization, A.S.; methodology, V.L., C.B. and A.S.; software, V.L., C.B. and A.S.; validation, V.L., C.B. and A.S.; formal analysis, V.L., C.B. and A.S.; investigation, V.L., C.B. and A.S.; data curation, V.L., C.B. and A.S.; writing—original draft preparation, V.L., C.B. and A.S.; writing—review and editing, V.L., C.B. and A.S.; visualization, V.L., C.B. and A.S.; supervision, A.S.; project administration, A.S.; funding acquisition, A.S. All authors have read and agreed to the published version of the manuscript.

Funding: This work was supported in part by the Italian Ministry of University and Research, under the complementary actions to the Plan of National Recovery and Resilience (PNRR) "Fit4MedRob-Fit for Medical Robotics" Grant (# PNC0000007).

Institutional Review Board Statement: Not applicable.

Data Availability Statement: Processed data are available from the corresponding author upon reasonable request.

Acknowledgments: The authors acknowledge the "Fit4MedRob-Fit for Medical Robotics" Grant (#PNC0000007) under the complementary actions to the Italian Plan of National Recovery and Resilience (PNRR).

Conflicts of Interest: Authors declare no conflicts of interest.

References

1. De Groote, F.; Jonkers, I.; Duysens, J. Task Constraints and Minimization of Muscle Effort Result in a Small Number of Muscle Synergies during Gait. *Front. Comput. Neurosci.* **2014**, *8*, 115. [CrossRef] [PubMed]
2. Bizzi, E.; Cheung, V.C.K.; d'Avella, A.; Saltiel, P.; Tresch, M. Combining Modules for Movement. *Brain Res. Rev.* **2008**, *57*, 125–133. [CrossRef] [PubMed]
3. Tresch, M.C.; Cheung, V.C.K.; d'Avella, A. Matrix Factorization Algorithms for the Identification of Muscle Synergies: Evaluation on Simulated and Experimental Data Sets. *J. Neurophysiol.* **2006**, *95*, 2199–2212. [CrossRef] [PubMed]
4. Ivanenko, Y.P.; Poppele, R.E.; Lacquaniti, F. Five Basic Muscle Activation Patterns Account for Muscle Activity during Human Locomotion. *J. Physiol.* **2004**, *556*, 267–282. [CrossRef] [PubMed]
5. d'Avella, A.; Saltiel, P.; Bizzi, E. Combinations of Muscle Synergies in the Construction of a Natural Motor Behavior. *Nat. Neurosci.* **2003**, *6*, 300–308. [CrossRef]

6. Delis, I.; Panzeri, S.; Pozzo, T.; Berret, B. A Unifying Model of Concurrent Spatial and Temporal Modularity in Muscle Activity. *J. Neurophysiol.* **2014**, *111*, 675–693. [CrossRef]
7. Chia Bejarano, N.; Pedrocchi, A.; Nardone, A.; Schieppati, M.; Baccinelli, W.; Monticone, M.; Ferrigno, G.; Ferrante, S. Tuning of Muscle Synergies During Walking Along Rectilinear and Curvilinear Trajectories in Humans. *Ann. Biomed. Eng.* **2017**, *45*, 1204–1218. [CrossRef]
8. Rimini, D.; Agostini, V.; Knaflitz, M. Intra-Subject Consistency during Locomotion: Similarity in Shared and Subject-Specific Muscle Synergies. *Front. Hum. Neurosci.* **2017**, *11*, 586. [CrossRef]
9. Ghislieri, M.; Agostini, V.; Knaflitz, M. Muscle Synergies Extracted Using Principal Activations: Improvement of Robustness and Interpretability. *IEEE Trans. Neural Syst. Rehabil. Eng.* **2020**, *28*, 453–460. [CrossRef]
10. Chvatal, S.A.; Ting, L.H. Common Muscle Synergies for Balance and Walking. *Front. Comput. Neurosci.* **2013**, *7*, 48. [CrossRef]
11. Marino, G.; Scano, A.; Beltrame, G.; Brambilla, C.; Marazzi, A.; Aparo, F.; Molinari Tosatti, L.; Gatti, R.; Portinaro, N. Influence of Backpack Carriage and Walking Speed on Muscle Synergies in Healthy Children. *Bioengineering* **2024**, *11*, 173. [CrossRef] [PubMed]
12. Cappellini, G.; Ivanenko, Y.P.; Poppele, R.E.; Lacquaniti, F. Motor Patterns in Human Walking and Running. *J. Neurophysiol.* **2006**, *95*, 3426–3437. [CrossRef]
13. Bowden, M.G.; Clark, D.J.; Kautz, S.A. Evaluation of Abnormal Synergy Patterns Poststroke: Relationship of the Fugl-Meyer Assessment to Hemiparetic Locomotion. *Neurorehabil. Neural Repair* **2010**, *24*, 328–337. [CrossRef]
14. Barroso, F.O.; Torricelli, D.; Molina-Rueda, F.; Alguacil-Diego, I.M.; Cano-de-la-Cuerda, R.; Santos, C.; Moreno, J.C.; Miangolarra-Page, J.C.; Pons, J.L. Combining Muscle Synergies and Biomechanical Analysis to Assess Gait in Stroke Patients. *J. Biomech.* **2017**, *63*, 98–103. [CrossRef]
15. Clark, D.J.; Ting, L.H.; Zajac, F.E.; Neptune, R.R.; Kautz, S.A. Merging of Healthy Motor Modules Predicts Reduced Locomotor Performance and Muscle Coordination Complexity Post-Stroke. *J. Neurophysiol.* **2010**, *103*, 844–857. [CrossRef]
16. Cheung, V.C.K.; Turolla, A.; Agostini, M.; Silvoni, S.; Bennis, C.; Kasi, P.; Paganoni, S.; Bonato, P.; Bizzi, E. Muscle Synergy Patterns as Physiological Markers of Motor Cortical Damage. *Proc. Natl. Acad. Sci. USA* **2012**, *109*, 14652–14656. [CrossRef] [PubMed]
17. Steele, K.M.; Rozumalski, A.; Schwartz, M.H. Muscle Synergies and Complexity of Neuromuscular Control during Gait in Cerebral Palsy. *Dev. Med. Child Neurol.* **2015**, *57*, 1176–1182. [CrossRef] [PubMed]
18. Scano, A.; Mira, R.M.; d'Avella, A. Mixed Matrix Factorization: A Novel Algorithm for the Extraction of Kinematic-Muscular Synergies. *J. Neurophysiol.* **2022**, *127*, 529–547. [CrossRef]
19. Esmaeili, S.; Karami, H.; Baniasad, M.; Shojaeefard, M.; Farahmand, F. The Association between Motor Modules and Movement Primitives of Gait: A Muscle and Kinematic Synergy Study. *J. Biomech.* **2022**, *134*, 110997. [CrossRef]
20. Russo, M.; Scano, A.; Brambilla, C.; d'Avella, A. SynergyAnalyzer: A Matlab Toolbox Implementing Mixed-Matrix Factorization to Identify Kinematic-Muscular Synergies. *Comput. Methods Programs Biomed.* **2024**, *251*, 108217. [CrossRef]
21. Scano, A.; Jarque-Bou, N.; Brambilla, C.; Atzori, M.; D'Avella, A.; Müller, H. Functional Synergies Applied to a Publicly Available Dataset of Hand Grasps Show Evidence of Kinematic-Muscular Synergistic Control. *IEEE Access* **2023**, *11*, 108544–108560. [CrossRef]
22. Tang, L.; Li, F.; Cao, S.; Zhang, X.; Wu, D.; Chen, X. Muscle Synergy Analysis in Children with Cerebral Palsy. *J. Neural Eng.* **2015**, *12*, 046017. [CrossRef] [PubMed]
23. Goudriaan, M.; Papageorgiou, E.; Shuman, B.R.; Steele, K.M.; Dominici, N.; Van Campenhout, A.; Ortibus, E.; Molenaers, G.; Desloovere, K. Muscle Synergy Structure and Gait Patterns in Children with Spastic Cerebral Palsy. *Dev. Med. Child Neurol.* **2022**, *64*, 462–468. [CrossRef] [PubMed]
24. Moreira, L.; Figueiredo, J.; Fonseca, P.; Vilas-Boas, J.P.; Santos, C.P. Lower Limb Kinematic, Kinetic, and EMG Data from Young Healthy Humans during Walking at Controlled Speeds. *Sci. Data* **2021**, *8*, 103. [CrossRef]
25. Seth, A.; Sherman, M.; Reinbolt, J.A.; Delp, S.L. OpenSim: A Musculoskeletal Modeling and Simulation Framework for *in Silico* Investigations and Exchange. *Procedia IUTAM* **2011**, *2*, 212–232. [CrossRef]
26. Delp, S.L.; Anderson, F.C.; Arnold, A.S.; Loan, P.; Habib, A.; John, C.T.; Guendelman, E.; Thelen, D.G. OpenSim: Open-Source Software to Create and Analyze Dynamic Simulations of Movement. *IEEE Trans. Biomed. Eng.* **2007**, *54*, 1940–1950. [CrossRef]
27. Anderson, F.C.; Pandy, M.G. Dynamic Optimization of Human Walking. *J. Biomech. Eng.* **2001**, *123*, 381–390. [CrossRef]
28. Brambilla, C.; Beltrame, G.; Marino, G.; Lanzani, V.; Gatti, R.; Portinaro, N.; Molinari Tosatti, L.; Scano, A. Biomechanical Analysis of Human Gait When Changing Velocity and Carried Loads: Simulation Study with OpenSim. *Biology* **2024**, *13*, 321. [CrossRef]
29. Lee, D.D.; Seung, H.S. Learning the Parts of Objects by Non-Negative Matrix Factorization. *Nature* **1999**, *401*, 788–791. [CrossRef]
30. d'Avella, A.; Portone, A.; Fernandez, L.; Lacquaniti, F. Control of Fast-Reaching Movements by Muscle Synergy Combinations. *J. Neurosci. Off. J. Soc. Neurosci.* **2006**, *26*, 7791–7810. [CrossRef]
31. McLean, R.A.; Sanders, W.L.; Stroup, W.W. A Unified Approach to Mixed Linear Models. *Am. Stat.* **1991**, *45*, 54–64. [CrossRef]
32. Hartigan, J.A. *Clustering Algorithms*; Wiley: Hoboken, NJ, USA, 1975; ISBN 978-0-471-35645-5.
33. Huang, B.; Chen, W.; Liang, J.; Cheng, L.; Xiong, C. Characterization and Categorization of Various Human Lower Limb Movements Based on Kinematic Synergies. *Front. Bioeng. Biotechnol.* **2022**, *9*, 793746. [CrossRef]
34. Torres-Oviedo, G.; Ting, L.H. Subject-Specific Muscle Synergies in Human Balance Control Are Consistent Across Different Biomechanical Contexts. *J. Neurophysiol.* **2010**, *103*, 3084. [CrossRef]

35. Banks, C.L.; Pai, M.M.; McGuirk, T.E.; Fregly, B.J.; Patten, C. Methodological Choices in Muscle Synergy Analysis Impact Differentiation of Physiological Characteristics Following Stroke. *Front. Comput. Neurosci.* **2017**, *11*, 78. [CrossRef]
36. Leighton, R.D. A Functional Model to Describe the Action of the Adductor Muscles at the Hip in the Transverse Plane. *Physiother. Theory Pract.* **2006**, *22*, 251–262. [CrossRef] [PubMed]
37. Lichtwark, G. The Role of the Tibialis Anterior Muscle and Tendon in Absorbing Energy during Walking. *J. Sci. Med. Sport* **2014**, *18*, e129. [CrossRef]
38. Lewis, C.L.; Laudicina, N.M.; Khuu, A.; Loverro, K.L. The Human Pelvis: Variation in Structure and Function during Gait. *Anat. Rec.* **2017**, *300*, 633–642. [CrossRef]
39. Arnold, A.S.; Schwartz, M.H.; Thelen, D.G.; Delp, S.L. Contributions of Muscles to Terminal-Swing Knee Motions Vary with Walking Speed. *J. Biomech.* **2007**, *40*, 3660–3671. [CrossRef]
40. Alessandro, C.; Delis, I.; Nori, F.; Panzeri, S.; Berret, B. Muscle Synergies in Neuroscience and Robotics: From Input-Space to Task-Space Perspectives. *Front. Comput. Neurosci.* **2013**, *7*, 43. [CrossRef]
41. Park, J.-H.; Shin, J.-H.; Lee, H.; Roh, J.; Park, H.-S. Relevance of Upper Limb Muscle Synergies to Dynamic Force Generation: Perspectives on Rehabilitation of Impaired Intermuscular Coordination in Stroke. *IEEE Trans. Neural Syst. Rehabil. Eng.* **2023**, *31*, 4851–4861. [CrossRef]
42. Scano, A.; Lanzani, V.; Brambilla, C.; d'Avella, A. Transferring Sensor-Based Assessments to Clinical Practice: The Case of Muscle Synergies. *Sensors* **2024**, *24*, 3934. [CrossRef] [PubMed]
43. Borzelli, D.; De Marchis, C.; Quercia, A.; De Pasquale, P.; Casile, A.; Quartarone, A.; Calabrò, R.S.; d'Avella, A. Muscle Synergy Analysis as a Tool for Assessing the Effectiveness of Gait Rehabilitation Therapies: A Methodological Review and Perspective. *Bioengineering* **2024**, *11*, 793. [CrossRef] [PubMed]
44. Rodriguez, K.L.; Roemmich, R.T.; Cam, B.; Fregly, B.J.; Hass, C.J. Persons with Parkinson's Disease Exhibit Decreased Neuromuscular Complexity during Gait. *Clin. Neurophysiol. Off. J. Int. Fed. Clin. Neurophysiol.* **2013**, *124*, 1390–1397. [CrossRef] [PubMed]
45. Safavynia, S.A.; Torres-Oviedo, G.; Ting, L.H. Muscle Synergies: Implications for Clinical Evaluation and Rehabilitation of Movement. *Top. Spinal Cord Inj. Rehabil.* **2011**, *17*, 16–24. [CrossRef]
46. Falaki, A.; Cuadra, C.; Lewis, M.M.; Prado-Rico, J.M.; Huang, X.; Latash, M.L. Multi-Muscle Synergies in Preparation for Gait Initiation in Parkinson's Disease. *Clin. Neurophysiol.* **2023**, *154*, 12–24. [CrossRef]
47. Sun, S.Y.; Giszter, S.F.; Harkema, S.J.; Angeli, C.A. Modular Organization of Locomotor Networks in People with Severe Spinal Cord Injury. *Front. Neurosci.* **2022**, *16*, 1041015. [CrossRef]
48. Milosevic, M.; Yokoyama, H.; Grangeon, M.; Masani, K.; Popovic, M.R.; Nakazawa, K.; Gagnon, D.H. Muscle Synergies Reveal Impaired Trunk Muscle Coordination Strategies in Individuals with Thoracic Spinal Cord Injury. *J. Electromyogr. Kinesiol.* **2017**, *36*, 40–48. [CrossRef]
49. Cheng, R.; Sui, Y.; Sayenko, D.; Burdick, J.W. Motor Control After Human SCI Through Activation of Muscle Synergies Under Spinal Cord Stimulation. *IEEE Trans. Neural Syst. Rehabil. Eng.* **2019**, *27*, 1331–1340. [CrossRef]
50. Bellitto, A.; De Luca, A.; Gamba, S.; Losio, L.; Massone, A.; Casadio, M.; Pierella, C. Clinical, Kinematic and Muscle Assessment of Bilateral Coordinated Upper-Limb Movements Following Cervical Spinal Cord Injury. *IEEE Trans. Neural Syst. Rehabil. Eng.* **2023**, *31*, 3607–3618. [CrossRef]
51. Cappellini, G.; Ivanenko, Y.P.; Martino, G.; MacLellan, M.J.; Sacco, A.; Morelli, D.; Lacquaniti, F. Immature Spinal Locomotor Output in Children with Cerebral Palsy. *Front. Physiol.* **2016**, *7*, 478. [CrossRef]
52. d'Avella, A.; Ivanenko, Y.; Lacquaniti, F. Muscle Synergies in Cerebral Palsy and Variability: Challenges and Opportunities. *Dev. Med. Child Neurol.* **2022**, *64*, 404–405. [CrossRef] [PubMed]
53. Roh, J.; Rymer, W.Z.; Perreault, E.J.; Yoo, S.B.; Beer, R.F. Alterations in Upper Limb Muscle Synergy Structure in Chronic Stroke Survivors. *J. Neurophysiol.* **2013**, *109*, 768–781. [CrossRef]
54. Niu, C.M.; Bao, Y.; Zhuang, C.; Li, S.; Wang, T.; Cui, L.; Xie, Q.; Lan, N. Synergy-Based FES for Post-Stroke Rehabilitation of Upper-Limb Motor Functions. *IEEE Trans. Neural Syst. Rehabil. Eng.* **2019**, *27*, 256–264. [CrossRef] [PubMed]
55. Dipietro, L.; Krebs, H.I.; Fasoli, S.E.; Volpe, B.T.; Stein, J.; Bever, C.; Hogan, N. Changing Motor Synergies in Chronic Stroke. *J. Neurophysiol.* **2007**, *98*, 757–768. [CrossRef] [PubMed]
56. Yeung, L.-F.; Ockenfeld, C.; Pang, M.-K.; Wai, H.-W.; Soo, O.-Y.; Li, S.-W.; Tong, K.-Y. Design of an Exoskeleton Ankle Robot for Robot-Assisted Gait Training of Stroke Patients. *IEEE Int. Conf. Rehabil. Robot. Proc.* **2017**, *2017*, 211–215. [CrossRef]
57. Rinaldi, L.; Yeung, L.-F.; Lam, P.C.-H.; Pang, M.Y.C.; Tong, R.K.-Y.; Cheung, V.C.K. Adapting to the Mechanical Properties and Active Force of an Exoskeleton by Altering Muscle Synergies in Chronic Stroke Survivors. *IEEE Trans. Neural Syst. Rehabil. Eng.* **2020**, *28*, 2203–2213. [CrossRef]
58. Hassan, M.; Kadone, H.; Ueno, T.; Hada, Y.; Sankai, Y.; Suzuki, K. Feasibility of Synergy-Based Exoskeleton Robot Control in Hemiplegia. *IEEE Trans. Neural Syst. Rehabil. Eng.* **2018**, *26*, 1233–1242. [CrossRef]

biomimetics

Article

Finger Prosthesis Driven by DEA Pairs as Agonist–Antagonist Artificial Muscles

Alexandre B. S. da Silva [1,2], Gabriel E. P. Mendes [1], Eduardo S. Bragato [1], Guilherme L. Novelli [1,2], Marina Monjardim [3] and Rafhael M. Andrade [1,2,*]

1 Department of Mechanical Engineering, Universidade Federal do Espírito Santo, Vitoria 29075-910, Brazil
2 Graduate Program of Mechanical Engineering, Universidade Federal do Espírito Santo, Vitoria 29075-910, Brazil
3 Graduate Program of Animal Biology, Universidade Federal do Espírito Santo, Vitoria 29075-910, Brazil
* Correspondence: rafhael.andrade@ufes.br

Abstract: Loss of an upper limb exerts a negative influence on an individual's ability to perform their activities of daily living (ADLs), reducing quality of life and self-esteem. A prosthesis capable of performing basic ADLs functions has the capability of restoring independence and autonomy to amputees. However, current technologies present in robotic prostheses are based on rigid actuators with several drawbacks, such as high weight and low compliance. Recent advances in robotics have allowed for the development of flexible actuators and artificial muscles to overcome the limitations of rigid actuators. Dielectric elastomer actuators (DEAs) consist of a thin elastomer membrane arranged between two compliant electrodes capable of changing dimensions when stimulated with an electrical potential difference. In this work, we present the design and testing of a finger prosthesis driven by two DEAs arranged as agonist–antagonist pairs as artificial muscles. The soft actuators are designed as fiber-constrained dielectric elastomers (FCDE), enabling displacement in just one direction as natural muscles. The finger prosthesis was designed and modeled to show bend movement using just one pair of DEAs and was made of PLA in an FDM 3D printer to be lightweight. The experimental results show great agreement with the proposed model and indicate that the proposed finger prosthesis is promising in overcoming the limitations of the current rigid based actuators.

Keywords: finger; prosthetics; dielectric elastomer; artificial muscle

Citation: da Silva, A.B.S.; Mendes, G.E.P.; Bragato, E.S.; Novelli, G.L.; Monjardim, M.; Andrade, R.M. Finger Prosthesis Driven by DEA Pairs as Agonist–Antagonist Artificial Muscles. *Biomimetics* **2024**, *9*, 110. https://doi.org/10.3390/biomimetics9020110

Academic Editor: Jinyou Shao

Received: 15 January 2024
Revised: 2 February 2024
Accepted: 8 February 2024
Published: 13 February 2024

1. Introduction

According to the Global Health Data Exchange of 2019 [1], for every 100,000 people worldwide, approximately 224 undergo the amputation of at least one upper limb, and 120 of them experience bilateral amputations of their limbs. The loss of an upper limb in the human body negatively impacts an individual's ability to perform their activities of daily living (ADLs), reducing their quality of life and self-esteem. A prosthetic capable of carrying out basic daily functions becomes crucial in restoring independence and autonomy to amputees [2].

Current technologies in prosthetics and orthotics are based on rigid actuators [3], with limitations in the number of degrees of freedom, high weight, low flexibility, and low compliance with the prosthesis user [4,5]. Recent advances in robotics have allowed for the development of flexible actuators and artificial muscles with the intention of creating alternatives to overcome the limitations of rigid actuators [6].

Flexible actuators constitute a category of materials that respond to stimuli such as an electric field, temperature variation, concentration, and pH, thus altering their shape or dimensions [7]. Within the class of flexible actuators, there are electroactive polymers (EAPs), which are a type of polymer that changes its shape through electrical stimulation. Among EAPs, there are dielectric elastomers (DEs), which consist of a thin elastomer membrane positioned between two compliant electrodes (Figure 1), changing their dimensions when stimulated with an electric potential difference [8].

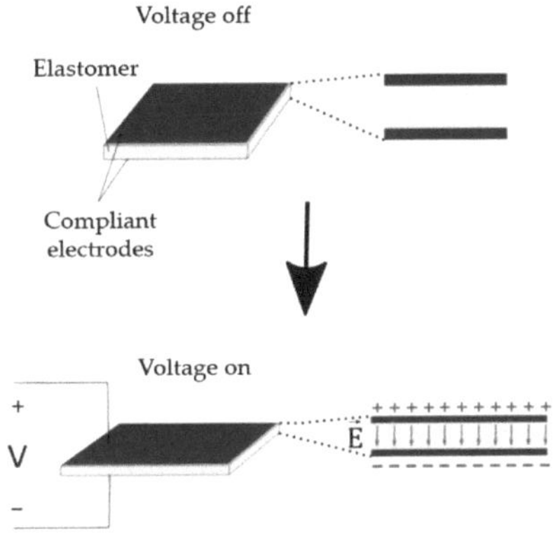

Figure 1. Rectangular dielectric elastomer.

When a potential difference is applied between two electrodes, Maxwell stress arises in the elastomer, which begins to act as a dielectric. As the thickness of the elastomer can be reduced and the electrodes can approach each other, DEs act as electromechanical transducers, converting electrical energy into mechanical energy or vice versa. Among the DE applications, the most common are actuators, which convert electrical energy into mechanical energy [9], but there are also generators that transform mechanical energy into electrical energy [10], and sensors.

Compared to other flexible actuator technologies, DEs standout for their high deformation capabilities, high work densities (work done per actuation cycle normalized by the volume of the actuator), high specific power (output power value normalized by the mass of the actuator), and high efficiency. However, there are still practical difficulties that hinder the applications of DEs. The need for high electrical voltages for actuation; manufacturing difficulties; issues related to durability and maintenance; and the non-linearity of the transducer, which has a viscoelastic behavior, are the main constraints with this material [11].

Most of the muscles responsible for finger and wrist movements originate in the forearm and are called extrinsic muscles. These muscles are larger and provide force to movement, divided into the extensor and flexor muscles of the fingers, arranged in agonist–antagonist pairs. Although external, these muscles have insertions in the hand region to perform finger movements. Intrinsic muscles originate in the hand and are responsible for secondary movements, allowing fine and precise control of each finger [12,13].

Artificial muscles are defined as materials or devices that reversibly change shape and dimensions through external stimuli [7]. Among the external stimuli are the electric field for electroactive polymers and piezoelectric actuators, the temperature for shape memory alloys, the pressure for pneumatic actuators, and the magnetic field for magneto-rheological actuators [14], among other stimuli. Compared to other flexible actuators, dielectric elastomers stand out for their high deformation capacity and high work density [11].

In comparison with human striated skeletal muscles (Table 1), dielectric elastomer actuators (DEAs) exhibit superior characteristics in terms of maximum deformation, maximum tension, maximum deformation rate, work density, specific power, and efficiency [8,11,15–18].

Table 1. Comparison between the performance of natural muscles and maximum values found in dielectric elastomers.

Metric	Skeletal Striated Muscle	Dielectric Elastomer
Maximum deformation (%)	40	142 (linear)
Maximum tension (MPa)	0.35	7.7
Maximum deformation rate (%/s)	50	450
Work density (kJ/m^3)	40	3500
Specific power (kW/kg)	0.28	3.6
Efficiency (%)	40	80

The literature on artificial hands and fingers commonly addresses the design of artificial fingers that combine mechanics with embedded electronics, comprising tactile sensors for normal force and shear force [19,20]. Despite the use of new materials, sensors, and manufacturing processes in recent hand and finger prosthesis designs, such as the use of pneumatic artificial muscles [21], 3D-printed hand prosthetics [22], artificial muscles made from nylon threads [23], and prosthetics using stretchable optical waveguides [24], the use of flexible actuators to simulate agonist and antagonist muscles has not yet been explored and could benefit hand prosthesis design by simulating the movement of the natural limb [25].

This work presents the modeling, design, fabrication, and experimental validation of a finger prosthesis actuated by two fiber-constrained dielectric elastomer actuators (FCDEA) arranged in agonist–antagonist pairs to simulate natural finger movement. We designed an underactuated mechanism composed of two coupled four-bar chain mechanisms to allow for just one pair of FCDEAs to drive the finger. Since FCDEAs are linear expanding actuators [9,11], once an FCDEA expands, the agonistic pair contracts, thus rotating the driving rod of the mechanism. This approach mimics the skeletal striated muscles of the forearm to move the finger and introduces some advantages compared to other robotic hand prostheses, such as easier operation, noiselessness, mechanical compliance, and low weight. In Section 2, we introduce the materials and methods used to model, design, and manufacture the finger prototype and artificial muscles. Section 3 presents the results obtained with the physical prototype of the prosthetic finger. Finally, Section 4 discusses the conclusions derived from the project's development and outlines the next steps in the project.

2. Materials and Methods

2.1. Finger FCDEA Setup

The proposed finger prosthesis is driven by two FCDEA displayed in antagonistic pairs, as shown in Figure 2a. When activated, the DE membrane contracts in thickness and expands in area. The fibers constraint the expansion in one direction, thereby allowing the actuator increase length in just one direction, mimicking a skeletal striated muscle (Figure 2b).

In the human body, the movement of the limbs is provided by skeletal striated muscle arranged in antagonistic pairs. Once one muscle contracts, the antagonistic pair expands, thereby allowing for the joint moving. On the other hand, joint stiffness is controlled by the contraction intensity of the antagonistic pairs, i.e., greater joint stiffness is achieved by greater contraction of the muscles. The only difference from the proposed actuation system is that the FCDEAS expand upon actuation instead of contract. However, since it is arranged in antagonistic pairs, the functioning is the same. In other words, joint torque and displacement increases as the activation intensity difference of the antagonistic FCDEAS increases, and joint stiffness increases as the activation intensity of both antagonistic FCDEAS reduces.

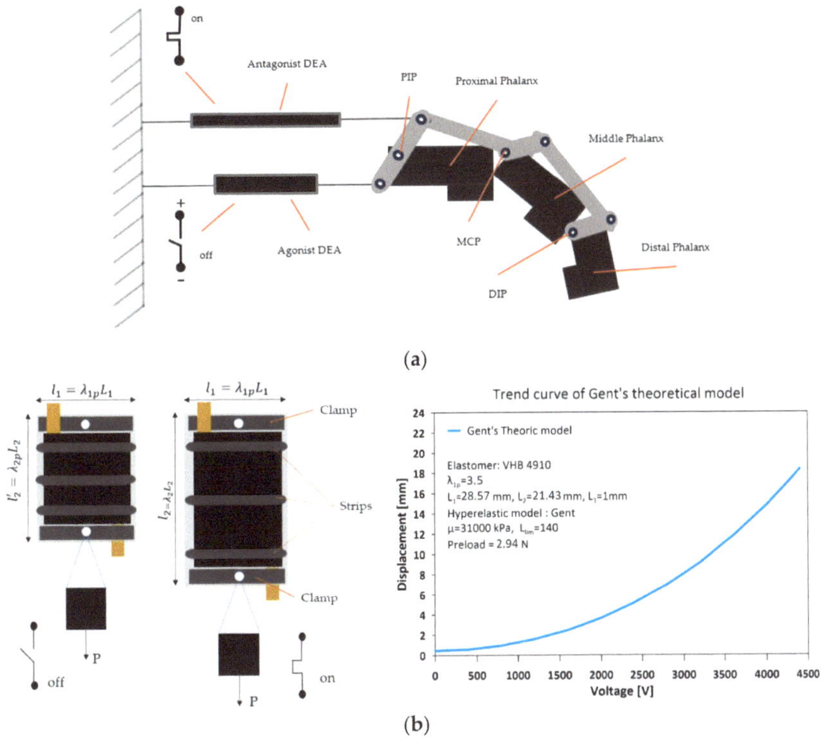

Figure 2. Finger prosthesis driven by two fiber-constrained dielectric elastomer actuators (FCDEA) displayed in antagonistic pairs. (**a**) Experimental setup for the FCDEAS composition and working principle of the finger prosthesis. (**b**) FCDEA activated and deactivated and displacement against voltage model.

The artificial muscle developed in this study was designed to replicate the contractile and extensible capacity of natural muscles, featuring a basic unit similar to a myofibril. Inspired by the efficient structure of sarcomeres, this type of artificial muscle aims to mimic the interaction between actin and myosin filaments to generate movement. In the FCDEA, the combination of two strips, along with the elastomer area, plays a functional role analogous to sarcomeres, while the strips mimic the Z-lines, allowing for an effective coordinated action of contraction and extension. Similarly to the muscular structure, increasing the number of FCDEA results in an increased tension that the artificial muscle can withstand. This innovative approach holds promising applications in fields such as robotics, biomechanical prosthetics, and assistive devices, where replicating muscular functionality is crucial for achieving precise and natural movements.

2.2. DEA Modeling

The strain energy density of the elastomer (WS) is obtained through material models known as hyperelastic models. For example, the Gent model considers the increase in stiffness at high deformations with a constant related to the stretch limit J_{lim}. Hoss and Marczak [26] cataloged more than twenty other hyperelastic models for various applications, while Wissler et al. [27] used some of these models to catalog the mechanical characteristics of an acrylic elastomer VHB 4010. Here, we highlight the Gent model [28], given by

$$W_s = -\frac{\mu J_{lim}}{2} \ln\left(1 - \frac{\lambda_1^2 + \lambda_2^2 + \lambda_3^2 - 3}{J_{lim}}\right) \tag{1}$$

where m is the shear modulus, J_{lim} is the stretch limit constant, and λ_i are the stretches in xyz directions.

To describe the state of the FCDEA, we should employ the hyperelastic Gent model and apply it to the original state equation of the standard elastomer. The constants used in these equations were determined for the VHB 4910, which was utilized in the experiments. Thus, the Gent model is redefined as follows:

$$\sigma_1 + \varepsilon\left(\lambda_{1p}\lambda_2\frac{V}{L_3}\right)^2 = \frac{\mu\left(\lambda_{1p}^2 - (\lambda_{1p}\lambda_2)^{-2}\right)}{1 - \left(\lambda_{1p}^2 + \lambda_2^2 + (\lambda_{1p}\lambda_2)^{-2} - 3\right)/J_{lim}} \tag{2}$$

$$\lambda_2\left(\frac{P}{L_1L_3}\right) + \varepsilon\left(\lambda_{1p}\lambda_2\frac{V}{L_3}\right)^2 = \frac{\mu\left(\lambda_2^2 - (\lambda_{1p}\lambda_2)^{-2}\right)}{1 - \left(\lambda_{1p}^2 + \lambda_2^2 + (\lambda_{1p}\lambda_2)^{-2} - 3\right)/J_{lim}} \tag{3}$$

where P is the applied load; σ_1 represents the actual mechanical stress in the L_1 direction; L_1 and L_3 are the initial dimensions of the VHB 4910; ε is the electrical permittivity; V is the electrical voltage obtained from an external source; and $l_1 = \lambda_{1p}L_1$, where λ_{1p} represents the pre-stretching in the L_1 direction.

The Gent model demonstrates an increase in stiffness as the stretch increases. Therefore, a higher load or electrical voltage is required to achieve the same variation $\delta\lambda_2$ as the stretch increases. To estimate the displacement of the DEA as a function of the applied electrical voltage using the hyperelastic Gent model (Figure 2b), the properties of the VHB 4910 elastomer were used based on previous works [29–31]. A relative electrical permittivity of $\varepsilon_r = 4.65$, a stretch limit constant of $J_{lim} = 140$, and a shear modulus $\mu = 31$ kPa were considered. The applied stretch load $P = 2.94$ N was adopted.

2.3. Finger Mechanism Modeling

The designed mechanism is an underactuated finger prothesis composed of two coupled four-bar chain mechanisms driven by two FCDEAs arranged in an antagonistic pair, as shown in Figure 3, where the expansion of the upper FCDEA provides clockwise rotation in θ, and the expansion of the lower FCDEA allows for anti-clockwise rotation in θ. The relationship between activation of the two FCDEAs is given by

$$\theta = \sin^{-1}\left(\frac{L_2}{2r}(\lambda_{2b} - \lambda_{2a})\right) \tag{4}$$

where q is the rotation of the first rod; L_2 is the nominal dimension of the FCDEA in the actuation direction; and l_{2a} and l_{2b} are the stretches of the upper and lower FCDEA, respectively, which can be calculated accordingly by Equations (2) and (3).

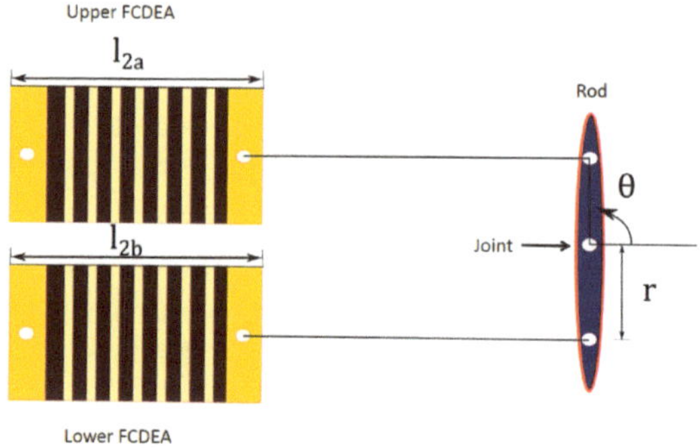

Figure 3. Underactuated mechanism to drive the finger prosthesis. The FCDAs are arranged in an antagonistic pair providing rod rotation.

The rotational movement of the first rod, as depicted in Figure 3, is responsible for driving the entire mechanism indicated in Figure 4. However, the complete finger movement has two components: the first one is responsible for moving the phalanges together (Figure 4a), and the second one is responsible for rotating the bar mechanisms around each phalange joint (Figure 4b).

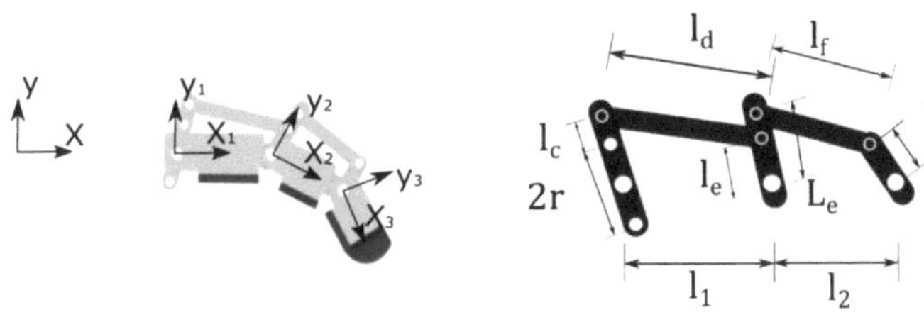

(**a**) (**b**)

Figure 4. Prosthetic finger mechanism. (**a**) Bar mechanism; (**b**) bar mechanism.

To describe this movement, we must first introduce the angles φ_i (Figure 5a), which indicate rod rotation relative to joint i, and the angles α_i, (Figure 5b), which show the rotation of the coordinate system (X_i, Y_i) relative to the inertial coordinate system (X, Y).

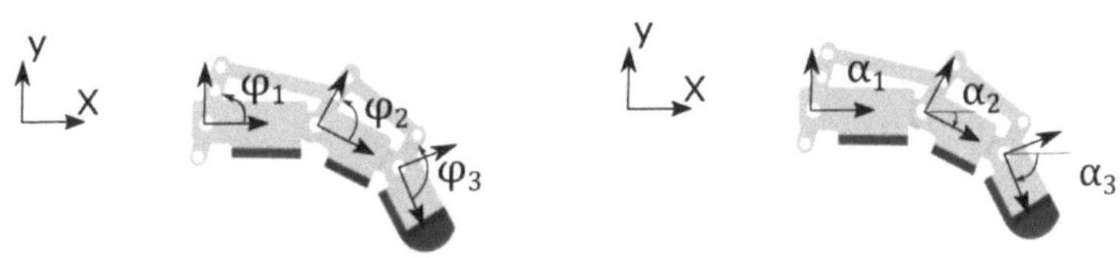

(**a**) (**b**)

Figure 5. (**a**) Angles φ_i and (**b**) angles α_i.

In this way, there are two possible extreme movements that can be performed by the mechanism. The first one is the rotation of the first rod altering only the angles φ_i (Figure 5a), and the second one is the rotation of the first rod, altering only the angles α_i (Figure 5b). To couple these two movements, a transmissibility coefficient τ_i is added to describe the amount of movement allocated to each motion. If this coefficient is defined as $\tau = 1$, the entire movement is carried out around the angles α, but if $\tau = 0$, the entire movement is carried out around the angles φ. Therefore,

$$\delta\alpha_i = \tau_i \delta\theta_i \tag{5}$$

$$\delta\varphi_i = (1 - \tau_i)\delta\theta_i \tag{6}$$

To perform the movement of a finger with 1 degree of freedom capable of flexing all phalanges simultaneously, it is necessary that $0 < \tau < 1$. However, the transmissibility coefficient does not necessarily need to be constant throughout the movement. As noted, assuming a finger with a transmissibility coefficient equal to 1, all the phalanges flex with the same angle α. However, upon contact with an object on one of these phalanges, the rotation in α is restricted, thereby reducing the transmissibility coefficient to 0 once it is entirely directed to move the rods relative to each phalanx, altering φ.

Considering a voltage of 3500 V applied, the expected displacement of the DEA, according to the hyperelastic Gent model and Figure 2b, is about 11 mm. This displacement is used to estimate the dimensions and positions of the bars presented in Figure 4b, as shown in the Table 2 below.

Table 2. Dimensions of the FCDEA.

Metric	Elastomer
Eletrical Voltage (V)	3500
λ_{1p}	3.50
L_1 (mm)	28.57
L_2 (mm)	21.43
l_2 (mm)	11.00

2.4. FCDEA Finger Prototype

The step-by-step fabrication process of the FCDEAs is presented in Figure 6. Fabricating the actuator involves preparing the necessary materials, such as cutting PET fibers with a thickness of 0.20 mm; PET clamps with a thickness of 0.75 mm; and the VHB 4910 (3M, Maplewood, MN, USA) elastomer, which should be cut into a square with at least 64 mm on each side. Next, the elastomer is pre-stretched, where $\lambda_{p1} = \lambda_{p2} = 3.5$. The acrylic frame is placed over the pre-stretched elastomer and pressed to improve adhesiveness. After that, the clamps, copper tapes, and PET fibers are attached onto the elastomer. Finally, carbon conductive grease (MG Chemicals, Burlington, ON, Canada) is applied to both surfaces of the VHB, and using a cutting tool, the DEA is detached from the acrylic frame. This results in a dielectric elastomer constrained by fibers. However, before its use, the elastomer needs to be suspended with a pre-load of 300 g for 30 min to enhance pre-stretching in the actuation direction. High voltage is provided by a DC-DC Converter 7000 V 3 W (E70, XP Power, Reading, Belgium).

Figure 6. The step-by-step fabrication process of the DEA.

The stretchable part of the elastomer membrane originally had dimensions of 64 mm × 64 mm. After stretching with $\lambda_{1p} = \lambda_{2p} = 3.5$, we obtained new dimensions of 224 mm × 224 mm. The acrylic frame, with internal dimensions of 120 mm × 160 mm, is positioned tightly adhered to the VHB so that, upon detaching the acrylic frame from the pre-stretcher, the pre-stretching remains constant. The clamps and fibers support lateral pre-stretching $\lambda_{1p} = 3.5$ throughout the actuation period of the FCDEA.

Two clamps, each 20 mm in height, and seven fibers, each 5 mm in height, are positioned, leaving approximately 10 mm of space (5 mm on each side) between the clamps and the acrylic frame. The dimensions of the free elastomer, without fibers or clamps, are 100 mm (160 mm − 2 × 5 mm − 2 × 20 mm − 7 × 5 mm = 100 mm) × 75 mm. These dimensions still have the same pre-stretching value $\lambda_{1p} = \lambda_{2p} = 3.5$. Therefore, upon detaching the FCDEA from the frame, it tends to contract immediately in the actuation direction. Hence, the new nominal dimensions of the elastomers are as follows:

$$L_1 = \frac{l_1}{\lambda_{1p}} \tag{7}$$

$$L_2 = \frac{l_2}{\lambda_{1p}} \tag{8}$$

Thus, $L_1 = 28.57$ mm and $L_2 = 21.43$ mm. For practical reasons, the length measurements of the elastomer during actuation were taken between the two clamps. Therefore, the sum of the heights of the fibers (7 × 5 mm = 35 mm) should be subtracted from L_2 to find the actual values of l_2 and λ_2.

The prosthetic finger was made of PLA by an FDM 3D printer (Sethi S3X, Sethi 3D, Campinas, SP, Brazil). Components were placed on an experimental bench to assess the bending and extension capacity of the prosthetic finger when actuated by the antagonistic pair of FCDEAs, as shown in Figure 7.

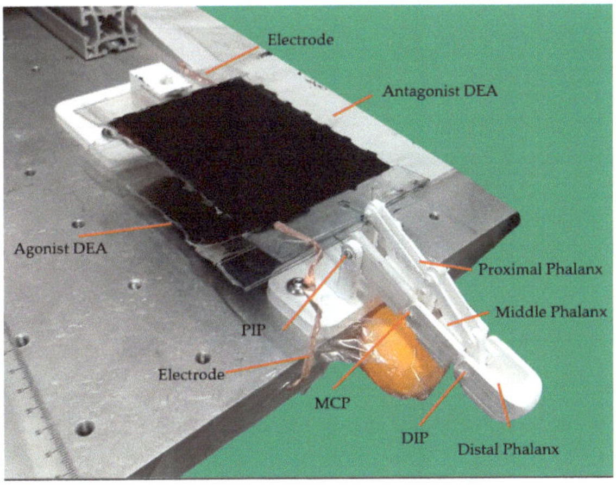

Figure 7. FCDEA finger prosthesis prototype assembled in a test bench.

3. Results and Discussion

3.1. Finger Mechanism Simulation

The movement of the underactuated prosthetic finger was estimated for different values of τ_i, as shown in Figure 8, considering a displacement of the upper elastomer of 11 mm. Figure 8a presents the movement for $\tau_i = 1$, while Figure 8b shows the movement for $\tau_i = 0$. Since the second four-bar mechanism reaches its rotation limit when the elastomer displacement is greater than 9.075 mm, $\tau_i = 0$ (Figure 8b), we did not provide accurate finger motion. For $\tau_i = 0.5$, as shown in Figure 8c, the movement was equivalent to a real finger, performing interphalangeal rotation. Figure 8d presents the result for values of $\tau_1 = 0$, $\tau_2 = 0.5$, $\tau_3 = 0.5$.

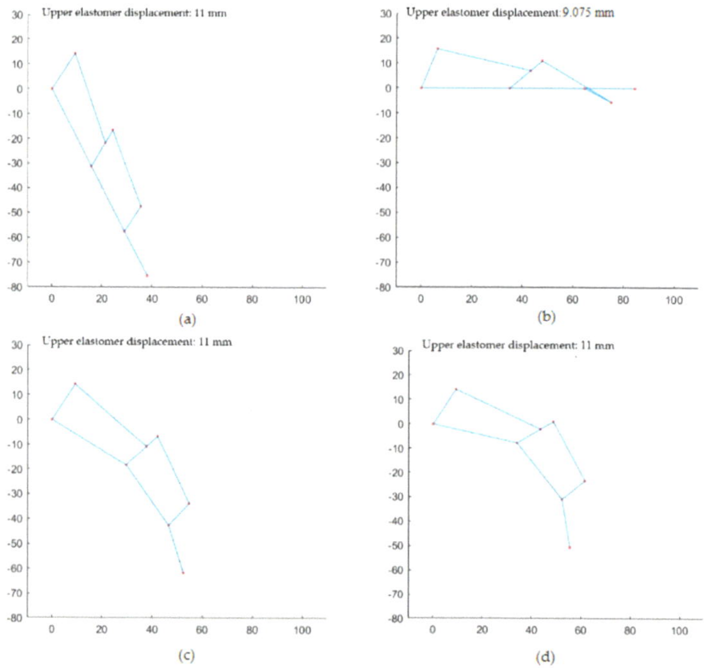

Figure 8. Prototype movement for (**a**) $\tau_i = 1$ (**b**) $\tau_i = 0$ (**c**) $\tau_i = 0.5$ (**d**) $\tau_1 = 0.2$, $\tau_2 = 0.5$, $\tau_3 = 0.5$.

3.2. Expandable Linear Actuator

Five FCDEAs were manufactured for the purpose of testing the elongation as a function of the applied voltage. Figure 9 represents the average curve obtained from the manufactured actuators and the curve from the theoretical Gent model. After the 30 min period with preload, the actuators had an average initial length $l_{2_0} = 96.4$ mm, and the maximum average displacement obtained was 27.4 mm for a supplied voltage of 4.8 kV, resulting in a maximum length of 123.8 mm.

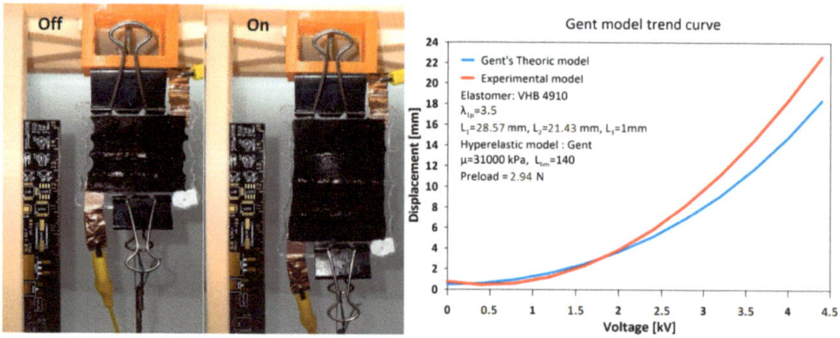

Figure 9. Electrical voltage curve at the source x average elongation of the actuators.

Thus, the maximum and minimum stretches obtained in the actuation direction were

$$\lambda_{2max} = \frac{123.8 - 35}{21.43} = 4.14 \tag{9}$$

$$\lambda_{2min} = \frac{96.4 - 35}{21.43} = 2.87 \quad (10)$$

The maximum stretch provided by electrical actuation was

$$\lambda_{actuator max} = \frac{123.8 - 96.4}{96.4} = 0.287 \quad (11)$$

Thus, the maximum deformation capacity of the FCDEA prototype was 28.4%, and the hyperelastic properties of the material was applied between elongations of 2.87 and 4.14. Since the manufactured FCDEAs had a greater displacement capacity than the Gent model, a voltage of 2700 V was considered to activate the FCDEA to obtain a displacement of 11 mm.

3.3. Physical Prototype

The physical prototype of the prosthetic finger was tested on an experimental bench, presented in Figure 10, where the movement of the upper FCDEAs was activated with a voltage of 2700 V, resulting in a displacement of 11 mm.

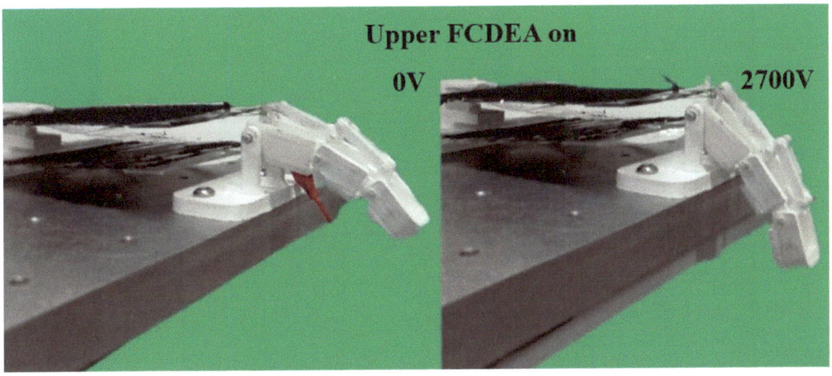

Figure 10. FCDEA finger prosthesis motion when the upper actuator was active.

Bending movement of the finger prototype was used to estimate values for τ_i, by minimizing the angular error between the joint angle α_i and the experimental joint angle $\alpha_{i_{exp}}$:

$$minimize\left(Error = \sum_{i=1}^{3} \alpha_i - \alpha_{i_{exp}}\right) \quad (12)$$

The movement performed by the finger actuated by the antagonist pair closely resembles the model presented in Figure 11 when the values of the transmissibility coefficients $\tau_1 = 0.7$, $\tau_2 = 0.6$, and $\tau_3 = 1$. The simulation with these coefficients is depicted in Figure 11a, and the comparison of the actual movement with the simulated movement is shown in Figure 11b.

Using the estimated transmissibility coefficients $\tau_1 = 0.7$, $\tau_2 = 0.6$, and $\tau_3 = 1$, it is possible to estimate the joint angles as a function of the input voltage, which may be used to control the position of the finger. Figure 12 presents the estimated the joint angle α_i, filled symbols, and the experimental joint angle $\alpha_{i_{exp}}$, empty ones. The proposed model was quite accurate in estimating the joint angles of the finger prosthesis.

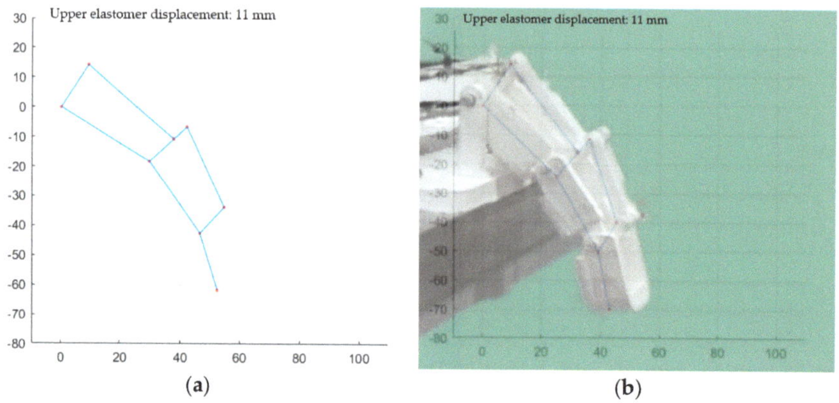

Figure 11. Model and prototype of the FCDEA finger prosthesis. (**a**) Model with transmissibility coefficients $\tau_1 = 0.7$, $\tau_2 = 0.6$, and $\tau_3 = 1$; (**b**) comparison between experiment and simulation movement.

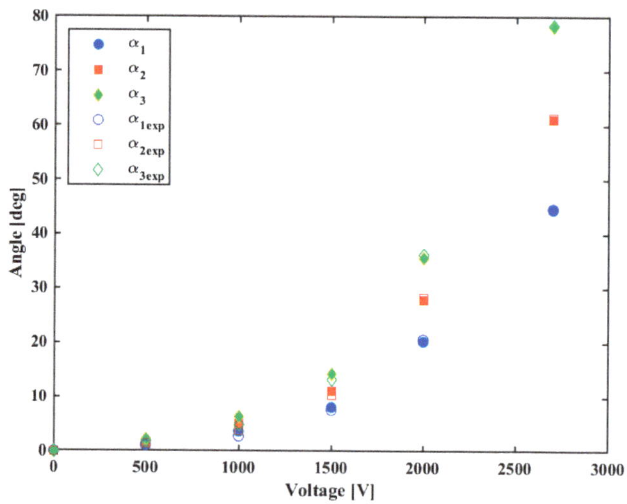

Figure 12. Estimated joint angles α_i and the experimental joint angles $\alpha_{i_{exp}}$.

3.4. Gripping an Object

Experiments were carried out to reproduce the flexion and extension motion of the finger around a small object, in this case, a table tennis ball. Figure 13a represents the complete flexion movement of the prototype around the object, including the initial and final positions, and Figure 13b illustrates the complete extension movement of the prototype. For flexion movements, the upper FCDEA was activated, while for extension movements, supply voltage was applied to the lower EDCF.

Figure 13 shows that the finger prothesis prototype driven by the FCDEA pairs as artificial agonist–antagonist muscles was able to mimic the behavior of a sound finger during flexion and extension movements around an object. Therefore, the same design concept can be used to build an FCDEA-driven full hand prosthesis to overcome some of the barriers faced by the current robotic hand prosthesis design, such as rigid actuators, noise, high weight, low flexibility, and low mechanical compliance [32,33]. Another possible design for the finger prosthesis could employ artificial tendons to move joints [13], where FCDEA can pull the tendons in an agonist–antagonist arrangement, as proposed here.

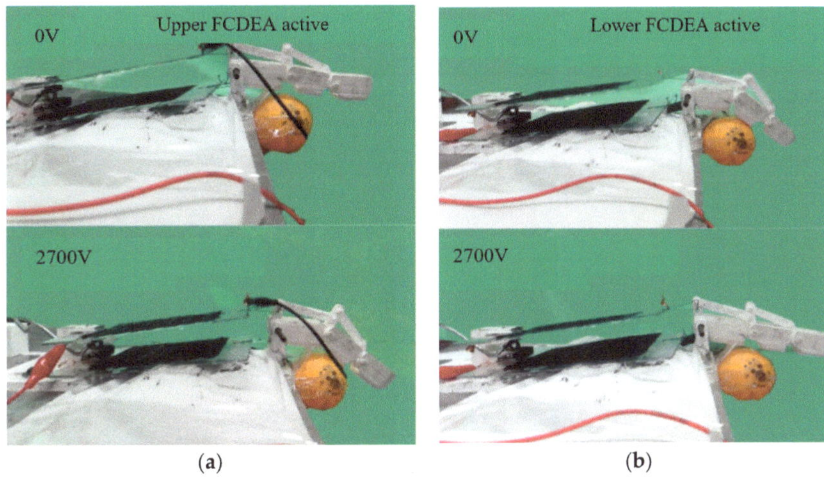

Figure 13. (**a**) Initial and final points of the flexion movement and (**b**) initial and final points of the extension movement.

However, this study presents some limitations that are important to highlight. Our experimental approach considered only one FCDEA activated at a time to model the finger motion. But an agonist–antagonist pair of skeletal striated muscle achieves precise position, force, and stiffness control through co-contraction [34]. Moreover, the ability of human hands to perform precise operations is the result of multiple muscle synergies [35]. However, here, only one pair of FCDEAs was used to move the whole finger mechanism through the PIP joint. Therefore, to have a finger prosthesis model that is more faithful to the natural finger, it is necessary to regulate the activation of both agonist and antagonist FCDEAs at the same time and provide other pairs of artificial muscles to move the MCP and DIP joints independently.

Finally, dielectric elastomers possess promising properties in the field of rehabilitation robotics, such as being low weight, having a high energy density, and having a significant deformation capacity. However, there are still some barriers to their applications. The durability and maintenance capabilities of actuators that use dielectric elastomers remain extremely limiting factors for real-world applications. Additionally, the need for high voltages to actuate the elastomers is also a concern for human use. Recent advances have been overcoming these barriers and improving the design features of these actuators, suggesting that dielectric elastomers will likely find practical applications in robotics in the near future [7].

4. Conclusions

This work presented the modeling, design, fabrication, and experimental testing of a prosthetic finger actuated by two fiber-constrained dielectric elastomer actuators (FCDEA) arranged in antagonistic pairs as artificial muscles. The finger's flexion movement is achieved when the upper FCDEA is activated, while the extension movement occurs when the lower actuator is engaged. The Gent hyperelastic model provided an appropriate estimated displacement, but some adjustments to the material parameters are still needed to precisely replicate the displacement of the manufactured FCDEA. A voltage of 2700 V was applied to achieve an 11 mm displacement of the elastomer and allow the full flexion and extension movement of the finger. The simulation results of the prosthetic finger showed results similar to those of the manufactured prototype. However, the actuator's parameterization still needs optimization, conducting tests with different elastomer dimensions, applied loads, and pre-stretching, among others, to maximize deformation and load capacity and reach the ideal size for the proposed application. Additionally, dielectric elastomer actuators exhibit complex dynamic response properties that should be explored through oscillatory tests under different loading conditions to enable their use in conjunction with force and position controllers, something we intend to explore in future works.

Author Contributions: Conceptualization, A.B.S.d.S., G.E.P.M., E.S.B., G.L.N., M.M. and R.M.A.; formal analysis, A.B.S.d.S., G.E.P.M., E.S.B., G.L.N. and R.M.A.; funding acquisition, R.M.A.; investigation, A.B.S.d.S., G.E.P.M., E.S.B., G.L.N. and R.M.A.; methodology, G.E.P.M., E.S.B., G.L.N. and R.M.A.; project administration, R.M.A.; resources, R.M.A.; supervision, R.M.A.; validation, A.B.S.d.S., G.E.P.M., E.S.B., G.L.N. and R.M.A.; writing—original draft, A.B.S.d.S., M.M. and R.M.A.; writing—review and editing, R.M.A. All authors have read and agreed to the published version of the manuscript.

Funding: This project was partially funded by the Fundação de Amparo à Pesquisa e Inovação do Espírito Santo (FAPES) TO 151/2021, project 2021-8GJZ6, and TO: 460/2021, project 2021-L7SZ4, and TO: 946/2023, project 2023-F0GWQ, FINEP (Financiadora de Estudos e Projetos com recursos do FNDCT ref 2784/20), Brazilian National Council for Scientific and Technological Development (CNPq) project 402317/2023-8 and project 442445/2023-7, and by the Coordenação de Aperfeiçoamento de Pessoal de Nível Superior—Brasil (CAPES)—Finance Code 001.

Institutional Review Board Statement: The study was conducted in accordance with the Declaration of Helsinki, and approved by the In-stitutional Review Board of Universidade Federal do Espírito Santo (protocol code 41368820.3.0000.5542 with approval on 22 March 2021.

Informed Consent Statement: Not applicable.

Data Availability Statement: Data is unavailable due to privacy or ethical restrictions.

Conflicts of Interest: The authors declare no conflict of interest. The funders had no role in the design of the study; in the collection, analyses, or interpretation of data; in the writing of the manuscript; or in the decision to publish the results.

References

1. Institute for Health Metrics and Evaluation. Global Burden of Disease Study 2019 (GBD 2019) Data Resources | GHDx 2019. Available online: https://ghdx.healthdata.org/gbd-2019 (accessed on 23 July 2023).
2. Light, C.M.; Chappell, P.H. Development of a Lightweight and Adaptable Multiple-Axis Hand Prosthesis. *Med. Eng. Phys.* **2000**, *22*, 679–684. [CrossRef]
3. Cao, W.; Chen, C.; Wang, D.; Wu, X.; Chen, L.; Xu, T.; Liu, J. A Lower Limb Exoskeleton with Rigid and Soft Structure for Loaded Walking Assistance. *IEEE Robot. Autom. Lett.* **2022**, *7*, 454–461. [CrossRef]
4. Näf, M.B.; Junius, K.; Rossini, M.; Rodriguez-Guerrero, C.; Vanderborght, B.; Lefeber, D. Misalignment Compensation for Full Human-Exoskeleton Kinematic Compatibility: State of the Art and Evaluation. *Appl. Mech. Rev.* **2019**, *70*, 050802. [CrossRef]
5. Andrade, R.M.; Sapienza, S.; Mohebbi, A.; Fabara, E.E.; Bonato, P. Overground Walking with a Transparent Exoskeleton Shows Changes in Spatiotemporal Gait Parameters. *IEEE J. Transl. Eng. Health Med.* **2024**, *12*, 182–193. [CrossRef]
6. Youn, J.-H.; Jeong, S.M.; Hwang, G.; Kim, H.; Hyeon, K.; Park, J.; Kyung, K.-U. Dielectric Elastomer Actuator for Soft Robotics Applications and Challenges. *Appl. Sci.* **2020**, *10*, 640. [CrossRef]
7. Zhang, J.; Sheng, J.; O'Neill, C.T.; Walsh, C.J.; Wood, R.J.; Ryu, J.-H.; Desai, J.P.; Yip, M.C. Robotic Artificial Muscles: Current Progress and Future Perspectives. *IEEE Trans. Robot.* **2019**, *35*, 761–781. [CrossRef]
8. Brochu, P.; Pei, Q. Advances in Dielectric Elastomers for Actuators and Artificial Muscles. *Macromol. Rapid Commun.* **2010**, *31*, 10–36. [CrossRef] [PubMed]
9. Novelli, G.L.; Andrade, R.M. Towards an Active Ankle-Foot Prosthesis Powered by Dielectric Elastomer Actuators in Antagonistic Pairs. In Proceedings of the 2021 International Symposium on Medical Robotics (ISMR), Atlanta, GA, USA, 17–19 November 2021; pp. 1–6.
10. Lai, H.; Reid, K. Investigation of Dielectric Elastomer Human Energy Harvesting to Reduce Knee Joint Torque Deviation Due to Bracing. *J. Rehabil. Assist. Technol. Eng.* **2019**, *6*. [CrossRef] [PubMed]
11. Novelli, G.L.; Vargas, G.G.; Andrade, R.M. Dielectric Elastomer Actuators as Artificial Muscles for Wearable Robots. *J. Intell. Mater. Syst. Struct.* **2023**, *34*, 1007–1025. [CrossRef]
12. Maw, J.; Wong, K.Y.; Gillespie, P. Hand Anatomy. *Br. J. Hosp. Med.* **2016**, *77*, C34–C40. [CrossRef] [PubMed]
13. Silva, R.C.; Lourenço, B.G.; Ulhoa, P.H.F.; Dias, E.A.F.; da Cunha, F.L.; Tonetto, C.P.; Villani, L.G.; Vimieiro, C.B.S.; Lepski, G.A.; Monjardim, M.; et al. Biomimetic Design of a Tendon-Driven Myoelectric Soft Hand Exoskeleton for Upper-Limb Rehabilitation. *Biomimetics* **2023**, *8*, 317. [CrossRef]
14. de Andrade, R.M.; Fabriz Ulhoa, P.H.; Fragoso Dias, E.A.; Filho, A.B.; Vimieiro, C.B.S. Design and Testing a Highly Backdrivable and Kinematic Compatible Magneto-Rheological Knee Exoskeleton. *J. Intell. Mater. Syst. Struct.* **2023**, *34*, 653–663. [CrossRef]

15. Hunter, I.W.; Lafontaine, S. A Comparison of Muscle with Artificial Actuators. In Proceedings of the Technical Digest IEEE Solid-State Sensor and Actuator Workshop, Hilton Head, SC, USA, 22–25 June 1992; pp. 178–185.
16. Mirvakili, S.M.; Hunter, I.W. Artificial Muscles: Mechanisms, Applications, and Challenges. *Adv. Mater.* **2018**, *30*, 1704407. [CrossRef]
17. Liang, W.; Liu, H.; Wang, K.; Qian, Z.; Ren, L.; Ren, L. Comparative Study of Robotic Artificial Actuators and Biological Muscle. *Adv. Mech. Eng.* **2020**, *12*, 1687814020933409. [CrossRef]
18. Lu, T.; Shi, Z.; Shi, Q.; Wang, T.J. Bioinspired Bicipital Muscle with Fiber-Constrained Dielectric Elastomer Actuator. *Extreme Mech. Lett.* **2016**, *6*, 75–81. [CrossRef]
19. Weiner, P.; Neef, C.; Shibata, Y.; Nakamura, Y.; Asfour, T. An Embedded, Multi-Modal Sensor System for Scalable Robotic and Prosthetic Hand Fingers. *Sensors* **2019**, *20*, 101. [CrossRef]
20. De Arco, L.; Pontes, M.J.; Segatto, M.E.V.; Monteiro, M.E.; Cifuentes, C.A.; Díaz, C.A.R. Soft-Sensor System for Grasp Type Recognition in Underactuated Hand Prostheses. *Sensors* **2023**, *23*, 3364. [CrossRef] [PubMed]
21. Nemoto, Y.; Ogawa, K.; Yoshikawa, M. F3Hand II: A Flexible Five-Fingered Prosthetic Hand Using Curved Pneumatic Artificial Muscles. In Proceedings of the 2020 IEEE/SICE International Symposium on System Integration (SII), Honolulu, HI, USA, 12–15 January 2020; pp. 99–104.
22. Liu, S.; Van, M.; Chen, Z.; Angeles, J.; Chen, C. A Novel Prosthetic Finger Design with High Load-Carrying Capacity. *Mech. Mach. Theory* **2021**, *156*, 104121. [CrossRef]
23. Saharan, L.; Wu, L.; Tadesse, Y. Modeling and Simulation of Robotic Finger Powered by Nylon Artificial Muscles. *J. Mech. Robot.* **2019**, *12*, 1–24. [CrossRef]
24. Zhao, H.; O'Brien, K.; Li, S.; Shepherd, R.F. Optoelectronically Innervated Soft Prosthetic Hand via Stretchable Optical Waveguides. *Sci. Robot.* **2016**, *1*, eaai7529. [CrossRef] [PubMed]
25. Allen, D.P.; Little, R.; Laube, J.; Warren, J.; Voit, W.; Gregg, R.D. Towards an Ankle-Foot Orthosis Powered by a Dielectric Elastomer Actuator. *Mechatronics* **2021**, *76*, 102551. [CrossRef]
26. Hoss, L.; Marczak, R.J. A New Constitutive Model for Rubber-Like Materials. *Mecánica Comput.* **2010**, *29*, 2759–2773.
27. Wissler, M.; Mazza, E. Mechanical Behavior of an Acrylic Elastomer Used in Dielectric Elastomer Actuators. *Sens. Actuators Phys.* **2007**, *134*, 494–504. [CrossRef]
28. Gent, A.N. A New Constitutive Relation for Rubber. *Rubber Chem. Technol.* **1996**, *69*, 59–61. [CrossRef]
29. Vo, V.T.K.; Ang, M.H.; Koh, S.J.A. Maximal Performance of an Antagonistically Coupled Dielectric Elastomer Actuator System. *Soft Robot.* **2021**, *8*, 200–212. [CrossRef] [PubMed]
30. Zhu, J.; Kollosche, M.; Lu, T.; Kofod, G.; Suo, Z. Two Types of Transitions to Wrinkles in Dielectric Elastomers. *Soft Matter* **2012**, *8*, 8840–8846. [CrossRef]
31. Kaltseis, R.; Keplinger, C.; Koh, S.J.A.; Baumgartner, R.; Feng Goh, Y.; Hoe Ng, W.; Kogler, A.; Tröls, A.; Chiang Foo, C.; Suo, Z.; et al. Natural Rubber for Sustainable High-Power Electrical Energy Generation. *RSC Adv.* **2014**, *4*, 27905–27913. [CrossRef]
32. Cordella, F.; Ciancio, A.L.; Sacchetti, R.; Davalli, A.; Cutti, A.G.; Guglielmelli, E.; Zollo, L. Literature Review on Needs of Upper Limb Prosthesis Users. *Front. Neurosci.* **2016**, *10*, 209. [CrossRef]
33. Kumar, D.K.; Jelfs, B.; Sui, X.; Arjunan, S.P. Prosthetic Hand Control: A Multidisciplinary Review to Identify Strengths, Shortcomings, and the Future. *Biomed. Signal Process. Control* **2019**, *53*, 101588. [CrossRef]
34. Osu, R.; Franklin, D.W.; Kato, H.; Gomi, H.; Domen, K.; Yoshioka, T.; Kawato, M. Short- and Long-Term Changes in Joint Co-Contraction Associated with Motor Learning as Revealed from Surface EMG. *J. Neurophysiol.* **2002**, *88*, 991–1004. [CrossRef]
35. Ajiboye, A.B.; Weir, R.F. Muscle Synergies as a Predictive Framework for the EMG Patterns of New Hand Postures. *J. Neural Eng.* **2009**, *6*, 036004. [CrossRef] [PubMed]

biomimetics

Article

A Semi-Autonomous Hierarchical Control Framework for Prosthetic Hands Inspired by Dual Streams of Human

Xuanyi Zhou [1,2,3], Jianhua Zhang [4], Bangchu Yang [1,2], Xiaolong Ma [1,2], Hao Fu [1,2], Shibo Cai [1,2] and Guanjun Bao [1,2,*,†]

1 College of Mechanical Engineering, Zhejiang University of Technology, Hangzhou 310023, China; zhouxuanyi@zjut.edu.cn (X.Z.); mxlslf@126.com (X.M.); fhneversettle@163.com (H.F.); ccc@zjut.edu.cn (S.C.)
2 Key Laboratory of Special Purpose Equipment and Advanced Processing Technology, Ministry of Education and Zhejiang Province, Zhejiang University of Technology, Hangzhou 310023, China
3 State Key Laboratory of Chemical Engineering, College of Chemical and Biological Engineering, Zhejiang University, 38 Zheda Road, Hangzhou 310027, China
4 School of Mechanical Engineering, Beijing University of Science and Technology, Beijing 100083, China; jhzhang@ustb.edu.cn
* Correspondence: gjbao@zjut.edu.cn
† Member of IEEE.

Abstract: The routine use of prosthetic hands significantly enhances amputees' daily lives, yet it often introduces cognitive load and reduces reaction speed. To address this issue, we introduce a wearable semi-autonomous hierarchical control framework tailored for amputees. Drawing inspiration from the visual processing stream in humans, a fully autonomous bionic controller is integrated into the prosthetic hand control system to offload cognitive burden, complemented by a Human-in-the-Loop (HIL) control method. In the ventral-stream phase, the controller integrates multi-modal information from the user's hand–eye coordination and biological instincts to analyze the user's movement intention and manipulate primitive switches in the variable domain of view. Transitioning to the dorsal-stream phase, precise force control is attained through the HIL control strategy, combining feedback from the prosthetic hand's sensors and the user's electromyographic (EMG) signals. The effectiveness of the proposed interface is demonstrated by the experimental results. Our approach presents a more effective method of interaction between a robotic control system and the human.

Keywords: prosthetic hand; control strategy; grasp; manipulation; human inspired

Citation: Zhou, X.; Zhang, J.; Yang, B.; Ma, X.; Fu, H.; Cai, S.; Bao, G. A Semi-Autonomous Hierarchical Control Framework for Prosthetic Hands Inspired by Dual Streams of Human. *Biomimetics* **2024**, *9*, 62. https://doi.org/10.3390/biomimetics9010062

Academic Editors: Rafhael Milanezi de Andrade and Xiaojie Wang

Received: 11 December 2023
Revised: 3 January 2024
Accepted: 6 January 2024
Published: 22 January 2024

1. Introduction

It is anticipated that future humanoid robots will perform various complex tasks through communication with human users [1]. Despite significant advancements in prosthetics technology in recent years, only 50 to 60 percent of amputees are willing to wear prosthetics [2], with rejection rates as high as 40 percent [3]. Brain dynamics experiments may provide insight into why this "prosthetic rejection" occurs [4]. Since prosthetic hands are less comfortable than human hands, using them imposes great cognitive burden on amputees, leading to brain fatigue and psychological frustration that ultimately results in prosthetic hand rejection [5].

Recent studies shed new light on the concept of "cognitive load" when using prosthetic hands. Prolonged use of prosthetic hands while handling tools can increase the strength of (electroencephalographic) EEG alpha waves in the brain [6]. This process can lead to an increase in cognitive load, resulting in fatigue and reduced responsiveness to other objects [7], which is the key reason for the high rejection rate of prostheses. The significant increase in EEG alpha-wave power indicates that users consciously exert greater control over their prosthetic hands [8]. This phenomenon is not observed in ordinary individuals when using their hands naturally and skillfully. User experience research reports confirm the existence of "cognitive load" [9]. Equipping prosthetic hands with sensor feedback to reduce the visual dependence of amputees is a promising solution to alleviate cognitive load [10].

An inspiration for reducing the "cognitive load" is the ventral stream and dorsal stream, which are two visual systems hypotheses of the human brain. Recently, evidence was provided for a functional segregation of dorsal and ventral streams, supporting the hypothesis [11]. Anatomical studies have proved the interaction between ventral and dorsal streams, especially for skilled grasping [12]. As the demanding precision of the grasp increases, these physiological interconnections gradually become more active [13,14].

The ventral stream, known as the "what pathway," is primarily responsible for object recognition and perception. The dorsal stream, referred to as the "where pathway," is primarily involved in the processing of spatial awareness and movement guidance [15]. The primary function of the dual stream is shown in Table 1. In the context of controlling prosthetic hands, the ventral stream can be utilized to extract information about the user's hand–eye coordination and their viewing field. By analyzing the information, the controller can determine the intention behind the user's movements and enable the manipulation of primitive switches in the variable domain of view. The dorsal stream plays a crucial role in integrating visual information with motor control. In the context of controlling prosthetic hands, the dorsal stream enables precise force control. Moreover, the dorsal stream needs to obtain detailed information about the identity of the object stored in the ventral-stream region when object attributes require complex fine-tuning of the grasp. Correspondingly, the ventral stream may obtain the latest grab-relevant information from the dorsal-stream region to improve the internal representation of the object. Based on the hypothesis, incorporating both the ventral- and dorsal-stream principles into the control framework of prosthetic hands makes it possible to create a more sophisticated and intuitive control system. The ventral stream contributes to the perception and recognition of objects, while the dorsal stream facilitates movement guidance and precise force control. This framework should improve the performance of prosthetic hands and user experience.

Table 1. The primary function of the dual stream.

	The Ventral Stream	The Dorsal Stream
Function	Identification	Visually guided movement
Sensitive features	High sensitivity to spatial	High sensitivity to time
Memory features	Long-term memory	Short-term memory
Reaction speed	Slow	Quick
Comprehension	Very fast	Very slow
Reference frame	Object-Centric	Human-Centric
Visual input	Fovea or parafovea	Entire retina

2. Related Work

Prosthetic hands primarily rely on EMG signals for control [16,17]. Although this method effectively utilizes residual muscles in the amputated limb, the absence of corresponding tactile feedback necessitates users to rely on visual compensation, resulting in an increased cognitive load. This phenomenon has emerged as a significant factor in prosthetic rejection among amputees. To address this issue, researchers have developed a gaze-training method known as Gazing Training. This method assists amputees in adapting to prosthetic hand usage by reducing the need for conscious control and alleviating cognitive load [18]. However, it is important to note that while Gazing Training demonstrates partial success in mitigating cognitive load during the rehabilitation process, it does not completely eliminate the underlying challenge [19].

Therefore, some researchers aspire to combine intelligent autonomous control methods with human users' electromyographic (EMG) signals by integrating control techniques from the field of robotics [20,21]. This integration aims to establish a semi-autonomous controller that combines autonomous control capabilities with human EMG signals. The objective of this semi-autonomous control approach is to partially or even entirely transfer the cognitive load to the controller, enabling autonomous assistance for amputee patients in accomplishing daily tasks. With advancements in both the understanding of the human brain and computer technology in recent years, this is possibility gradually transforming into reality [22].

In 2015, Markovic proposed a framework for a semi-autonomous controller, which consists of an autonomous control unit and an EMG control unit [23]. The autonomous control unit employs computer vision sensors to capture depth and red green blue (RGB) information, as well as proprioceptive feedback from the prosthetic hand, for data processing and information fusion. On the other hand, the EMG control unit utilizes electromyographic signals to reflect user intentions and facilitate manual control of the prosthetic hand. The combination of the EMG control unit and the autonomous control unit forms a semi-autonomous controller that integrates autonomous control capabilities with human EMG signals, allowing for switching between control modes [24].

Bu proposed a visually guided approach to assist patients in achieving semi-autonomous manipulation of prosthetic hands, with the primary goal of alleviating users' cognitive load [25]. Chunyuan introduced global visual information based on EMG signals [26]. Machine vision was employed to extract object shape, size, type, and appearance information, which was then integrated with pre-hand shape for joint motion planning using visual and dual antagonistic-channel EMG signals. Wang proposed an RNN network by incorporating features of the user's gaze point [27]. The semi-autonomous control of the prosthetic hand is achieved by the automatic recognition of the used tools and motion primitives. The integration of computer vision into the semi-autonomous control of prosthetic hands imposed significant computation demands on the controller [28]. To address the issue of computational load in prosthetic hand operations, Fukuda proposed a distributed control system to enhance the real-time performance of the semi-autonomous controller [29]. Moreover, the inertial measurement unit (IMU) was employed to obtain the prosthetic hand's state, while the visual system was utilized to perceive object states [30]. Vorobev proposed a semi-autonomous control method for prosthetic hands. Motion commands were transmitted to the prosthetic hand's main controller using sensors triggered by foot movement inside the shoe [31].

In summary, addressing the issue of cognitive load, the Gazing Training approach is proposed as a rehabilitation training method for amputees to adapt to prosthetic hands [32]. The Gazing Training partially mitigates the cognitive load but does not completely eliminate the problem. In recent years, some researchers have proposed a semi-autonomous control strategy from the perspective of robotics engineering. In these approaches, the controller acquires information about object shape, size, etc., and translates it into corresponding motion primitives, while the user's EMG signals generate grasping commands for the prosthetic hand. This semi-autonomous hierarchical control strategy aims to transfer a portion of the user's cognitive load to the controller, thereby reducing the user's alpha-wave power. Building upon the foundation of traditional semi-autonomous controllers, this paper primarily presents two optimizations to the semi-autonomous hierarchical control strategy for prosthetic hands. The innovations of the paper are summarized as follows:

(1) A controller is constructed based on the pathway of the human ventral–dorsal nerves. Object semantic segmentation and convolutional neural network (CNN) recognition are categorized as the ventral stream, while the motion tracking of the limb is introduced as the control of the dorsal stream. Moreover, the dorsal stream and ventral stream are integrated to ensure accurate motion primitives.
(2) In order to reduce the cognitive burden, a semi-autonomous controller is proposed. Feedback of the prosthetic hand is integrated to enhance the perceived experience. EMG signals of the user are obtained to realize the human in the loop control.

3. Methodology

In the guidance stage of the human visual neural stream, the perception-motion guidance is carried out by the ventral stream to recognize and locate objects. The ventral-stream information is matched to the related object memory and long-term action primitives. Inspired by the ventral stream, the prosthetic hand control system is applied to locate the object. Moreover, the initial user intentions are obtained through the posture of the grasping task. The visual information of objects can be further obtained through the CMOS image sensor of the head-mounted device. Since the CMOS image sensor in the headset follows the vision of the human user, the headset is able to locate the objects and prosthetic hands.

In the human dorsal-stream guidance stage, the human user guides the prosthetic hand to grasp and manipulate in a vision-aided manner. At the same time, the CMOS image sensor of the headset will obtain the real-time distance between the prosthetic hand and the object. When the prosthetic hand approaches the object, the controller will drive the prosthetic hand to perform grasp motions with the EMG signal of the human body simultaneously. At this stage, humans mainly use the information of the dorsal neural stream for guidance. Based on the characteristics of dorsal-stream information, the position and force cloud of the grasping process are fed back to the human user. Finally, the EMG signals of the human are combined to realize the human in the loop control. The framework of two nerve streams of visual information is shown in Figure 1.

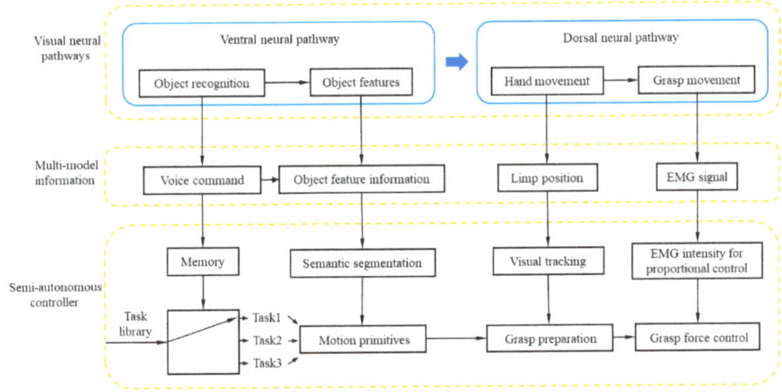

Figure 1. The framework of two nerve streams of visual information.

3.1. Task-Centric Planning

The task planning is divided into two parts: motion primitive and sequence planning. The controller is proposed in the task-centric task, which is inspired by the two visual streams, realizing the sets and independent planning control of prosthetic hand motion primitives. Firstly, multiple motion primitives are stored in the semi-autonomous controller. Different motion primitives are applied in the semi-autonomous controller for the grasping features of different objects. The object's category and the prosthetic hand's configuration are input into the controller as decision factors. The controller drives the prosthetic hand to complete the grasping task. In the decision-making stage, it is necessary to analyze the object character and obtain the location information in the scene. The decision is calculated by the convolutional neural network and semantic analysis. In order to realize real-time control, the SSD-Mobilenet-V2 convolutional neural network is used to realize semantic segmentation and object recognition for visual images. SSD-Mobilenet-V2 is used for semantic segmentation while Mobilenet is used for object recognition. The Single-Shot Multi-box Detector (SSD) neural network generates constraint squares of fixed size through the forward-propagation CNN network, compared to the intersection over union (IOU) of different Anchor boxes generated on the feature graph, obtaining boundary boxes close to 0.5.

Images captured from CMOS image sensors are segmented and associated with semantics. Specifically, after obtaining the boundary boxes through the SSD method mentioned above, a mobilenet-V2 convolution network is introduced to identify objects in the boundary boxes and generate index information. To improve the operational efficiency of the neural network for embedded devices, the mobilenet-V2 convolutional network is adopted to simplify the calculation process through a deeply separable convolutional operation method. The main advantage of this method is that Linear Bottleneck replaces ReLU to activate a function, achieving channel reductions depth wise and dimension reductions point wise, respectively. At the same time, to reduce the effect of feature reduction caused by linear bottleneck dimension reduction, the addition and nonlinearity of features are realized by inverted residuals. Similarly, an inverted residual neural network has the property of "short-circuit". The procedure has been simplified as the lightweight software requirement of mobile devices.

In different scenes, objects will correspond to a different set of motion primitives. Taking three grasping and manipulating tasks as examples, the varieties of manipulated primitives involved in this paper are shown in Table 2.

Table 2. The sets of motion primitives.

	Toggle Switch	Screw Cap	Grasp Cap
The rest gesture	gesture 1	gesture 1	gesture 1
Pre-shape gesture	gesture 12	Prepare and pre-envelope	Prepare and pre-envelope
Manipulate gesture	gesture 12	gesture 2	gesture 2

For different grasp and manipulate tasks, motion primitives are required for the different manipulate time sequence planning. Gesture 1 is no contact and no motion of the hand. Gesture 2 is motion of the hand without and within the hand. Gesture 12 is the within-hand movement [33]. To ensure the suitable sequence of motion primitive switching, sequential planning uses the end state of the previous primitive as the beginning state for the next primitive. Considering a common motion in daily life, "object grasping", the sequence of manipulation can be planned as follows:

(1) The beginning of the task. The initial motion primitive is in the free state, which is gesture 1.
(2) Object recognition stage. When the target recognition is completed, the prosthetic hand forms the pre-grasp posture according to the feature information of the object. If the target image is lost, it is estimated that the user gives up grasping, and the prosthetic hand restores to the free primitive state.
(3) Pre-grasp stage. The head-mounted CMOS image sensor obtains the spatial position relationship between the prosthetic hand and the object in real time. When the space distance between the hand and object is less than the threshold value, it is judged that the hand and object are in contact and enter the grasping primitive stage. When the space distance between the hand and object is greater than the threshold value, the hand and object are considered to be separated and return to the pre-hand type stage. The user can change his view field to restore the original free primitive at this stage.
(4) Grasp stage. Control is performed on the prosthetic hand after fusing EMG signal and object vision information feedback.
(5) After grasping, the prosthetic hand ends the grasping task and returns to the initial stage. After completing the grasping task, the prosthetic hand is controlled to separate from the object and restored to the initial free primitive state, changing the view field of the CMOS sensor. In case of unsuccessful separation of the hand and object or an emergency, voice command can perform an emergency reset. We restrict the prosthetic hand from switching directly to the free primitive state when performing grasp primitives for the user's safety. The aforementioned manipulate sequence planning for the controller is shown in Figure 2.

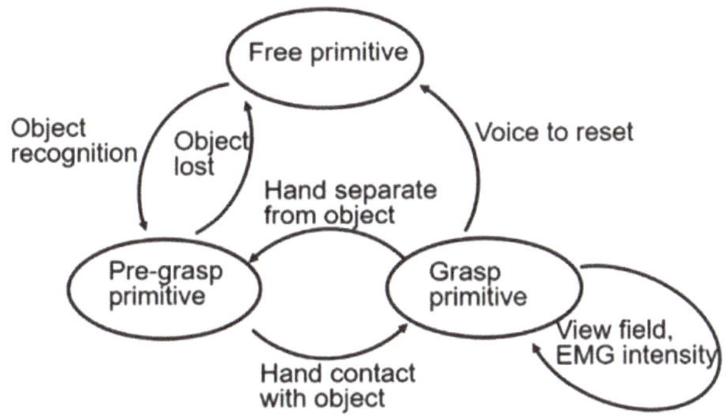

Figure 2. Manipulate sequence planning for the controller.

3.2. Precise Force Control Strategy

The precise force control of the prosthetic hand during contact is proposed for object grasping. In the grasping process, human users are able to participate in the control loop, forming the human-in-loop control mode. The pressure cloud image is shown in the headset. With the pressure feedback, the human users can adjust the grasping force to realize closed-loop control between the human and the prosthetic hand.

Human signals are generated by the flexor digitorum profundus. Signal preprocessing is carried out, including signal amplification, peak-to-peak detection, envelope processing, mean filtering, Fourier transform, and bandpass filtering. The value of the EMG signal is obtained in real time.

The STM32 microcontroller is applied as the central semi-autonomous controller. An Arduino microcontroller was used to obtain EMG values of flexor digitorum Profundo in real time. These two microcontrollers communicate through a serial port and transmit data in hexadecimal format. The central semi-autonomous controller will request for the intensity of the EMG signals obtained by the Arduino controller in real time. Considering the diversity of tasks, environments, and personal behaviors in grasping, the semi-autonomous controller will continuously detect the action potential of the user's digital-flexor deep muscle. A typical original EMG signal and bandpass filtering EMG intensity are shown in Figure 3. The proportional control output u will be obtained when the EMG intensity P exceeds the threshold, where k_1 is the proportional coefficient.

$$u = k_1 P \tag{1}$$

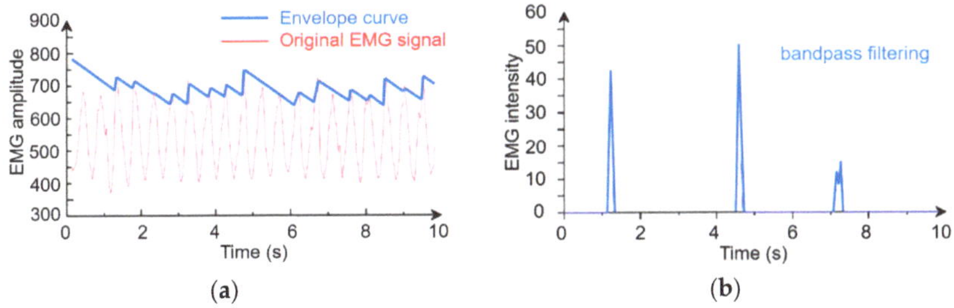

Figure 3. The myoelectric signal. (**a**) Original EMG signal. (**b**) Bandpass filtering.

The control output value u of the semi-autonomous controller is transmitted to the prosthetic hand system through Bluetooth. The prosthetic hand system converts the proportional control output u into pulse-width modulation (PWM) signal. It controls five servo motors to realize the movement of the finger joint of the prosthetic hand. Pressure sensors are embedded in each finger to read the contact state between the prosthetic hand and the object. The haptic pressure distribution cloud is transmitted to a semi-autonomous controller via Bluetooth built into the prosthetic hand system. To tackle the challenge of providing grip force feedback in prosthetic hands, this study leverages computer graphics techniques utilizing the open-source graphics library OpenGL. Within the semi-autonomous controller, a pressure cloud map of the prosthetic hand is generated. This map is subsequently projected onto the user's retina through a head-mounted device, facilitating closed-loop control. The fundamental approach involves the transmission of pressure sensor information from the prosthetic hand system to the semi-autonomous controller via the embedded Bluetooth of the microcontroller unit. The semi-autonomous controller, in turn, employs OpenGL to render the pressure cloud map. The combined controller information and prosthetic hand data are then projected onto the user's retina using the micro display of the head-mounted device.

Users will dynamically adjust the incremental control output of the prosthetic hand system based on real-time feedback from the controller and visual perception. According to Equation (1), once the proportional control output u is generated, the semi-autonomous controller will continuously monitor the user's electromyographic signals. If the user perceives that a stable grasp has not yet been achieved, the proportional control output u will persistently accumulate control output increments.

$$u_i = u_{i-1} + k_2 P \tag{2}$$

where k_2 is the incremental coefficient, which plays a role in amplifying and shrinking the signal during the incremental proportional control. The procedures of the proportional control mode are shown in Figure 4.

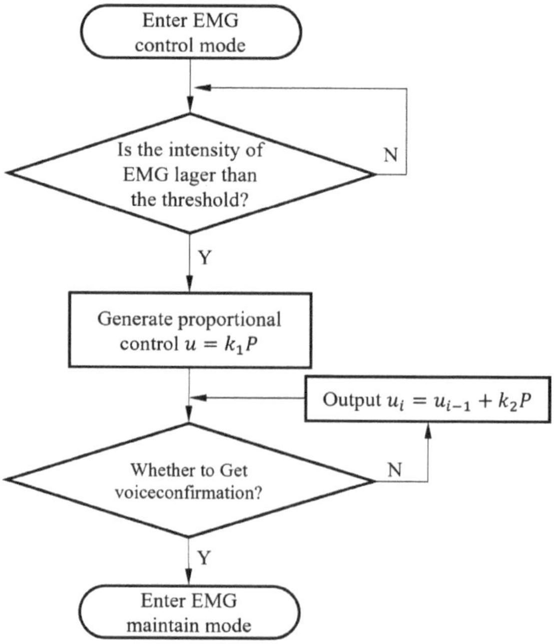

Figure 4. The procedures of the proportional control mode.

To solve the grasping force acquisition of prosthetic hands, we draw a pressure cloud map of prosthetic hands in the semi-autonomous controller, using computer graphics technology (OpenGL). This process costs less in terms of CPU resources. The pressure cloud figure is displayed in a headset and projected onto the user's retina. In this way, the human brain is connected to the control loop of the prosthetic hand. Rendering the pressure cloud image in the semi-autonomous controller mainly uses the GLWight library under the Qt framework; the generated rendering image is transmitted to the head-mounted device through an HDMI cable. The micro-display device in the headset uses a Vufine+ wearable display for visual projection.

OpenGL's geometric shader function interface is a four-channel array representing R (red region), G (green region), B (blue region), and α (transparency component). The following formula will convert the contact pressure information transmitted into the OpenGL into a cloud map. The pressure cloud figure will be displayed on the head mount display. When the contact pressure value c_i is less than 0.5-times the contact pressure threshold L, the data relationship of contact pressure information to the cloud map is shown in Formula (3):

$$\begin{cases} R_i = 0 \\ G_i = \frac{255c_i}{0.5L} \\ B_i = 255 - \frac{255c_i}{0.5L} \end{cases} \tag{3}$$

When the contact pressure value c_i is larger than 0.5-times of the contact pressure threshold L, the data relationship of contact pressure information to the cloud map is shown in Formula (4):

$$\begin{cases} R_i = \frac{255c_i}{0.5L} \\ G_i = 255 - \frac{255c_i}{0.5L} \\ B_i = 0 \end{cases} \tag{4}$$

where c_i is the contact force value of the ith finger; L is the contact force threshold; R_i, G_i, B_i represent the color components of the ith finger in red, green, and blue cloud images, respectively.

Tailored to specific grasping or operational tasks, the semi-autonomous controller dynamically monitors the real-time electromyographic signal strength of the user's deep flexor muscles. It triggers the electromyographic control phase when this intensity surpasses a predefined threshold. Considering that deep flexor muscle signals originate from the deeper muscle groups of the human body and surface electrode signal acquisition is susceptible to interference, various techniques, including differential amplification, peak detection, envelope processing, mean filtering, Fourier transformation, and bandpass filtering, are employed to extract real-time electromyographic signal intensity values. These intensity values are communicated in real time via serial communication to the STM32 embedded in the semi-autonomous controller for proportional control.

When the prosthetic hand contacts the target object, the contact force value and force distribution position of the tactile sensor embedded in the prosthetic hand are changed. This tactile information will be processed by the processor embedded in the prosthetic hand and then transmitted to the semi-autonomous controller, which uses OpenGL to draw the pressure cloud. A micro-display on the headset projects the pressure cloud image onto the user's retina, enabling the user to know the motion state and contact force distribution of the prosthetic hand system in real time for further incremental control until the user confirms the stable grasp, as shown in Figure 5.

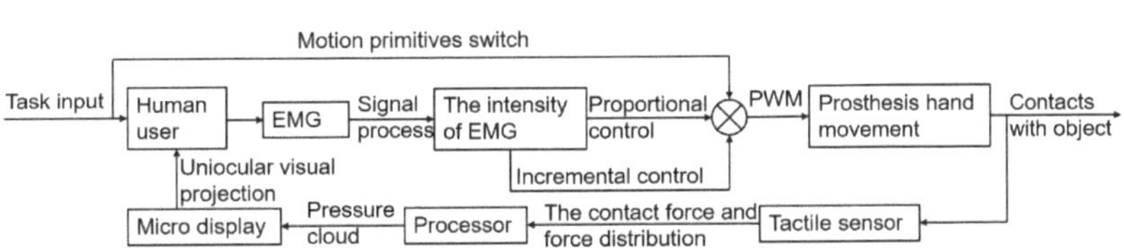

Figure 5. Block diagram for grasping force control of prosthetic hand with the human in the control loop.

4. Experiment

This section sets up a grasping platform for prosthetic hands to verify the feasibility of using the semi-autonomous controller and its robust performance. Users can realize grasping and manipulating tasks of a prosthetic hand using the semi-autonomous controller in the way of "hierarchical visual stream driven".

As shown in Figure 6 (from the recorder's perspective), the subject of this experiment is a 25-year-old Chinese adult male with normal vision. Elements of the semi-autonomous control framework are listed below:

- Jetson Nano (Nvidia, Santa Clara, CA, USA), which has 128-core NVIDIA Maxwell™ architecture GPU and Quad-core ARM® Cortex®-A57 MPCore processor and the semi-autonomous controller are integrated in the Jetson Nano.
- WX151HD CMOS image sensor (S-YUE, Shenzhen, China), which has 150-degree wide angle.
- ZJUT prosthetic hand (developed by Zhejiang University of Technology, Hangzhou, China) is equipped with 5 actuators.
- DYHW110 micro-scale pressure sensor (Dayshensor, Bengbu, China) is integrated in the prosthetic hand to obtain the touch force. It has a range of 5 kg, and the combined error is 0.3% of the full scale (F.S.).
- Vufine+ wearable display (Vufine, Sunnyvale, CA, USA) is a high-definition, wearable display that seamlessly integrates with the proposed control framework.

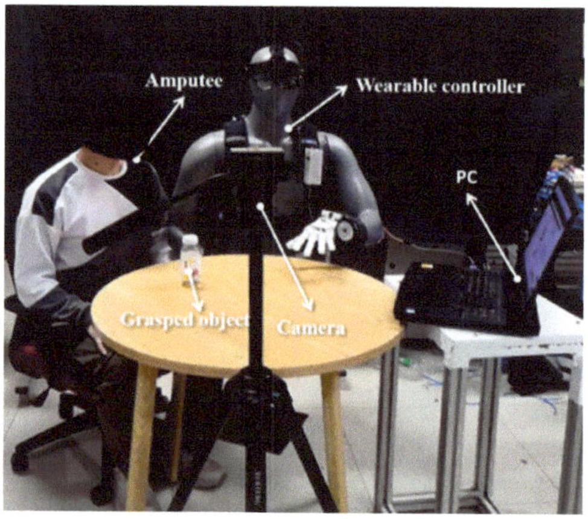

Figure 6. Experimental setup.

On the right side of the experimental setup is the semi-autonomous controller independently developed by the laboratory for this study. This controller follows the user's multi-modal information and facilitates the grasping function of the prosthetic hand. The detailed structure of the semi-autonomous controller was thoroughly discussed in Section 3. Additionally, the prosthetic hand system, also independently developed by the laboratory, is affixed to the damaged hand of the mannequin wearing the semi-autonomous controller. Other components within the experimental environment include a Graphic User Interface (GUI) on a personal computer. This GUI serves the convenience of subjects, experiment operators, safety officers, and recorders, enabling them to monitor the state of the semi-autonomous controller and make timely adjustments. The object being manipulated in the experiment is a plastic beverage bottle, mimicking the user's routine grasping actions for daily beverage needs. The experiment recorder controls and adjusts the process based on the recorded experimental footage, as illustrated in Figure 7.

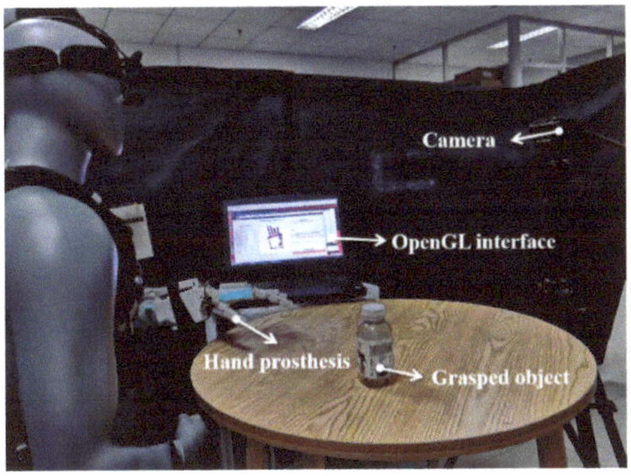

Figure 7. The view field from the subject's perspective.

4.1. Prosthetic Hand Grasping Experiment

In their daily lives, amputees often grasped objects with their prosthetic hands. The purpose of the experiment is that the semi-autonomous controller can assist the user in grabbing the "bottle" naturally under the multi-mode interaction between the semi-autonomous controller and the human user. The grasping process is as follows, which can be seen in Figure 8:

(1) The semi-autonomous controller was worn by a dummy model. A prosthetic hand, electromyographic electrode, and head-mounted device on the human subject are used to obtain the human EMG signal and project the signal to the human eye. (Figure 8.1).
(2) The human subject attached the prosthetic hand to his left hand and exhaled, "grab the bottle". The subject was looking at the bottle with his arm close to it. The head-mounted device will perform visual semantic segmentation and convolutional neural network recognition for the "bottle" in this process. According to the information returned by the CMOS image sensor and the speech task library, the prosthetic hand can switch the combination of motion primitives. (Figure 8.2).
(3) When the human subject's arm comes close to the "bottle", it generates an EMG signal. The semi-autonomous controller implements proportional and incremental control under a specific motion primitive, depending on the EMG strength, until the user is instructed to "determine" the prosthetic hand motion. During this period, users can observe the changes in the pressure cloud of the prosthetic hand in real time. (Figure 8.3–6).
(4) The experimental subjects grasp the "bottle" and place it in another position on the table. After placement, the prosthetic hand is released and reset by voice. (Figure 8.7–8).

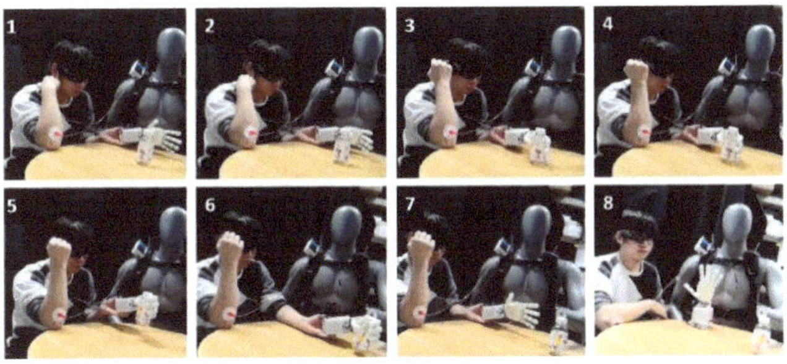

Figure 8. Arm and hand collaborative control experiment.

4.2. Prosthetic Hand and Human Hand Coordinative Manipulation Experiments

This experiment aims to verify that the semi-autonomous controller can assist humans in completing the manipulation of "screw the bottle cap" naturally under the multi-mode interaction between the semi-autonomous controller and the human user. The cooperative manipulation experiment of the prosthetic hand and human hand can be seen in Figure 9:

(1) The semi-autonomous controller was worn by the dummy. A prosthetic hand, EMG electrode, and head-mounted device were attached to the human subjects. (Figure 9.1).
(2) The experimental subject fixed the prosthetic hand on his left hand and exclaimed the command "screw the bottle cap" by voice. While the subject is looking at the "bottle", the user is grabbing the "bottle" with his right hand, and the prosthetic hand with his left arm is approaching the "bottle cap". The head-mounted device will perform visual semantic segmentation and convolutional neural network recognition for the "bottle" in this process. According to the information returned by the CMOS image sensor and the speech task library, the prosthetic hand can switch the combination of motion primitives. (Figure 9.2–3).
(3) When the distance between the prosthetic hand and the "bottle cap" is less than the threshold value, the prosthetic hand will grab the "bottle cap", and the user drives the prosthetic hand to rotate the "bottle cap" to the set angle through his left arm. (Figure 9.4–7).
(4) When the "bottle cap" is not unscrewed and the distance between the prosthetic hand and the "bottle cap" exceeds a certain distance, the prosthetic hand will return to the pre-hand type.
(5) Repeat Step 3 and Step 4 until the cap is unscrewed.
(6) Place the unscrewed bottle cap on the desktop and command the voice to reset the prosthetic hand. (Figure 9.8–10).

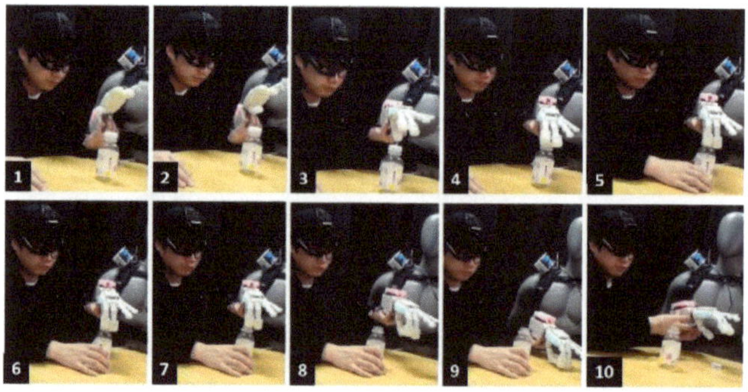

Figure 9. The cooperative manipulation experiment of a prosthetic hand and human hand.

4.3. Results

According to the results of two experiments, the proposed two visual-stream-driven manipulation strategy is in line with the natural manipulation rules of human beings and can effectively assist patients in completing familiar grasping and manipulation tasks in daily life. The experimental results are consistent with the expectations.

The experimental results shown in Table 3 combine the user experience and controller characteristics. The grasping task of a prosthetic hand and the human manipulation task described in this paper are similar in the structure form of the control method. In terms of the task layer, voice interaction is combined with the visual neural network. Due to differences in user intention and environment, the processing content will change the output result. At the planning level, the action primitive of the grasping task mainly switches between grasping hand type and pre-grasping hand type; the state transition condition is primarily used to trigger the user's man-in-loop control. The motion primitives of manipulation tasks switch between multiple manipulators, and the state transition condition is mainly used to trigger the arrangement of action primitives. In the motion control layer, the size of manipulated objects is usually tiny, so the switch of motion primitive is mainly realized through arm movement guidance. There are differences in the size, mass, and type of the objects in the grasping task, which requires higher grasping stability; thus, the force control strategy of the human-in-loop is introduced.

Table 3. The experimental results.

	Task Type	Task Characteristics
Coordinative movement of hand and arm	Grasp	Force control
Two hands coordination	manipulation	Movements switch

5. Conclusions and Future Works

5.1. Conclusions

Different from the traditional semi-autonomous controller design, this work is inspired by the two visual streams of the humanoid control strategy. The summary of this framework can be seen in Figure 10. The main contributions are as follows:

(1) In terms of the control layer of the motion primitive planning, the traditional object semantic segmentation and CNN recognition are classified as ventral flow, residual arm motion tracking is introduced as dorsal flow, inspired by the human brain. We optimized the information collection method of the motion primitive planning control layer and the state transfer strategy among the movement primitives according to the multimodal information in different stages of ventral flow and dorsal flow.

(2) In terms of the force control layer, the issue of the user's "cognitive burden" is reduced in the existing semi-autonomous control strategy, so this paper takes the user as a high-dimensional controller and the EMG strength of flexor digitorum profundus as the control quantity, giving feedback to the prosthetic hand body state, realizing precise force control with human-in-loop.

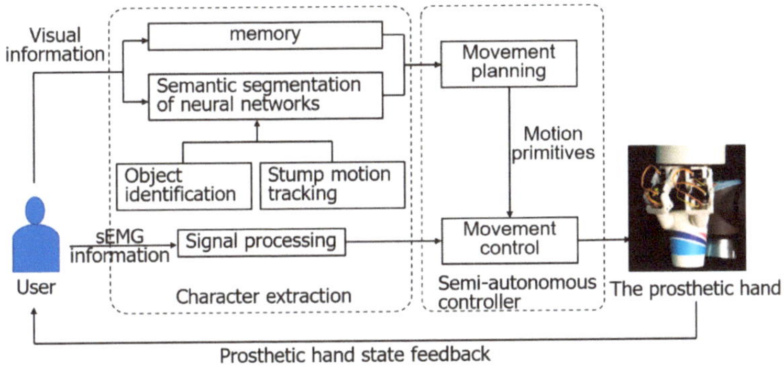

Figure 10. Framework of the semi-automatic controller for the prostheses hand.

5.2. Future Works

The proposed control framework, inspired by the human ventral–dorsal stream visual process, currently emphasizes functional grasping actions within the spectrum of human hand operations. However, this biological process represents just one aspect of the myriad ways humans execute gripping tasks, emphasizing functional manipulations. Ongoing neuroscientific efforts delve into understanding how different brain regions guide grasping operations through visual cues. Building upon the current ventral–dorsal stream, future work will explore bio-inspired control strategies, integrating the latest neuroscientific findings related to the user's dual visual neural pathways. The aim is to deepen the algorithmic sophistication and broaden the spectrum of multimodal information, fostering a more profound integration between humans and machines.

In future investigations, a pivotal area for exploration revolves around the usability and social acceptance of prosthetic hands in various environments, especially social situations. Current technology, while advancing rapidly, may pose challenges in social integration due to its external and conspicuous nature. To enhance user experience and societal acceptance, future work could focus on developing discreet, aesthetically pleasing designs that seamlessly integrate prosthetic hands into social settings. This involves not only technical advancements but also a nuanced understanding of user preferences, comfort levels, and societal perceptions. Exploring materials, form factors, and user-centric design principles could contribute significantly to reducing the stigma associated with prosthetic devices, fostering a more inclusive and socially integrated environment for users. This research direction aligns with the broader goal of not only improving the functional aspects of prosthetic hands but also enhancing the overall quality of life and social experiences for individuals using this technology.

Author Contributions: Conceptualization, J.Z. and G.B.; methodology, X.Z.; software, B.Y.; validation, B.Y. and X.Z.; formal analysis, X.M.; data curation, H.F.; writing—original draft preparation, X.Z.; writing—review and editing, S.C. All authors have read and agreed to the published version of the manuscript.

Funding: This research was funded by the Key Research and Development Program of Zhejiang (Grant No. 2021C04015), the Key Research Program of Zhejiang (Grand No. LZ23E050005), Natural Science Foundation of Zhejiang Province (Grand No. Q23E050071).

Institutional Review Board Statement: Not applicable.

Data Availability Statement: Data are contained within the article.

Conflicts of Interest: The authors declare no conflicts of interest. The funders had no role in the design of the study; in the collection, analyses, or interpretation of data; in the writing of the manuscript; or in the decision to publish the results.

References

1. Mouri, T.; Kawasaki, H.; Ito, S. Unknown object grasping strategy imitating human grasping reflex for anthropomorphic robot hand. *J. Adv. Mech. Des. Syst. Manuf.* **2007**, *1*, 1–11. [CrossRef]
2. Stephens-Fripp, B.; Alici, G.; Mutlu, R. A review of non-invasive sensory feedback methods for transradial prosthetic hands. *IEEE Access* **2018**, *6*, 6878–6899. [CrossRef]
3. Biddiss, E.A.; Chau, T.T. Upper limb prosthesis use and abandonment: A survey of the last 25 years. *Prosthet. Orthot. Int.* **2007**, *31*, 236–257. [CrossRef]
4. Ortiz, O.; Kuruganti, U.; Blustein, D. A Platform to Assess Brain Dynamics Reflective of Cognitive Load during Prosthesis Use. MEC20 Symposium. 2020. Available online: https://www.researchgate.net/publication/373294777_A_PLATFORM_TO_ASSESS_BRAIN_DYNAMICS_REFLECTIVE_OF_COGNITIVE_LOAD_DURING_PROSTHESIS_USE (accessed on 10 December 2023).
5. Parr, J.V.V.; Vine, S.J.; Wilson, M.R.; Harrison, N.R.; Wood, G. Visual attention, EEG alpha power and T7-Fz connectivity are implicated in prosthetic hand control and can be optimized through gaze training. *J. Neuroeng. Rehabil.* **2019**, *16*, 52. [CrossRef] [PubMed]
6. Deeny, S.; Chicoine, C.; Hargrove, L.; Parrish, T.; Jayaraman, A. A simple ERP method for quantitative analysis of cognitive workload in myoelectric prosthesis control and human-machine interaction. *PLoS ONE* **2014**, *9*, e112091. [CrossRef] [PubMed]
7. Parr, J.V.; Wright, D.J.; Uiga, L.; Marshall, B.; Mohamed, M.O.; Wood, G. A scoping review of the application of motor learning principles to optimize myoelectric prosthetic hand control. *Prosthet. Orthot. Int.* **2022**, *46*, 274–281. [CrossRef] [PubMed]
8. Ruo, A.; Villani, V.; Sabattini, L. Use of EEG signals for mental workload assessment in human-robot collaboration. In Proceedings of the International Workshop on Human-Friendly Robotics, Delft, The Netherlands, 22–23 September 2022; pp. 233–247.
9. Cordella, F.; Ciancio, A.L.; Sacchetti, R.; Davalli, A.; Cutti, A.G.; Guglielmelli, E.; Zollo, L. Literature review on needs of upper limb prosthesis users. *Front. Neurosci.* **2016**, *10*, 209. [CrossRef]
10. Park, J.; Zahabi, M. Cognitive workload assessment of prosthetic devices: A review of literature and meta-analysis. *IEEE Trans. Hum. Mach. Syst.* **2022**, *52*, 181–195. [CrossRef]
11. Foster, R.M.; Kleinholdermann, U.; Leifheit, S.; Franz, V.H. Does bimanual grasping of the Muller-Lyer illusion provide evidence for a functional segregation of dorsal and ventral streams? *Neuropsychologia* **2012**, *50*, 3392–3402. [CrossRef]

12. Cloutman, L.L. Interaction between dorsal and ventral processing streams: Where, when and how? *Brain Lang.* **2013**, *127*, 251–263. [CrossRef]
13. Van Polanen, V.; Davare, M. Interactions between dorsal and ventral streams for controlling skilled grasp. *Neuropsychologia* **2015**, *79*, 186–191. [CrossRef] [PubMed]
14. Brandi, M.-L.; Wohlschläger, A.; Sorg, C.; Hermsdörfer, J. The neural correlates of planning and executing actual tool use. *J. Neurosci.* **2014**, *34*, 13183–13194. [CrossRef] [PubMed]
15. Culham, J.C.; Danckert, S.L.; DeSouza, J.F.X.; Gati, J.S.; Menon, R.S.; Goodale, M.A. Visually guided grasping produces fMRI activation in dorsal but not ventral stream brain areas. *Exp. Brain Res.* **2003**, *153*, 180–189. [CrossRef] [PubMed]
16. Meattini, R.; Benatti, S.; Scarcia, U.; De Gregorio, D.; Benini, L.; Melchiorri, C. An sEMG-based human–robot interface for robotic hands using machine learning and synergies. *IEEE Trans. Compon. Packag. Manuf. Technol.* **2018**, *8*, 1149–1158. [CrossRef]
17. Lange, G.; Low, C.Y.; Johar, K.; Hanapiah, F.A.; Kamaruzaman, F. Classification of electroencephalogram data from hand grasp and release movements for BCI controlled prosthesis. *Procedia Technol.* **2016**, *26*, 374–381. [CrossRef]
18. Thomas, N.; Ung, G.; Ayaz, H.; Brown, J.D. Neurophysiological evaluation of haptic feedback for myoelectric prostheses. *IEEE Trans. Hum. Mach. Syst.* **2021**, *51*, 253–264. [CrossRef]
19. Cognolato, M.; Gijsberts, A.; Gregori, V.; Saetta, G.; Giacomino, K.; Hager, A.-G.M.; Gigli, A.; Faccio, D.; Tiengo, C.; Bassetto, F. Gaze, visual, myoelectric, and inertial data of grasps for intelligent prosthetics. *Sci. Data* **2020**, *7*, 43. [CrossRef]
20. Laffranchi, M.; Boccardo, N.; Traverso, S.; Lombardi, L.; Canepa, M.; Lince, A.; Semprini, M.; Saglia, J.A.; Naceri, A.; Sacchetti, R. The Hannes hand prosthesis replicates the key biological properties of the human hand. *Sci. Robot.* **2020**, *5*, eabb0467. [CrossRef]
21. Kilby, J.; Prasad, K.; Mawston, G. Multi-channel surface electromyography electrodes: A review. *IEEE Sens. J.* **2016**, *16*, 5510–5519. [CrossRef]
22. Mendez, V.; Iberite, F.; Shokur, S.; Micera, S. Current solutions and future trends for robotic prosthetic hands. *Annu. Rev. Control Robot. Auton. Syst.* **2021**, *4*, 595–627. [CrossRef]
23. Markovic, M.; Dosen, S.; Popovic, D.; Graimann, B.; Farina, D. Sensor fusion and computer vision for context-aware control of a multi degree-of-freedom prosthesis. *J. Neural Eng.* **2015**, *12*, 066022. [CrossRef] [PubMed]
24. Starke, J.; Weiner, P.; Crell, M.; Asfour, T. Semi-autonomous control of prosthetic hands based on multimodal sensing, human grasp demonstration and user intention. *Robot. Auton. Syst.* **2022**, *154*, 104123. [CrossRef]
25. Fukuda, O.; Takahashi, Y.; Bu, N.; Okumura, H.; Arai, K. Development of an IoT-based prosthetic control system. *J. Robot. Mechatron.* **2017**, *29*, 1049–1056. [CrossRef]
26. Shi, C.; Yang, D.; Zhao, J.; Liu, H. Computer vision-based grasp pattern recognition with application to myoelectric control of dexterous hand prosthesis. *IEEE Trans. Neural Syst. Rehabil. Eng.* **2020**, *28*, 2090–2099. [CrossRef]
27. Wang, X.; Haji Fathaliyan, A.; Santos, V.J. Toward shared autonomy control schemes for human-robot systems: Action primitive recognition using eye gaze features. *Front. Neurorobotics* **2020**, *14*, 567571. [CrossRef] [PubMed]
28. Guo, W.; Xu, W.; Zhao, Y.; Shi, X.; Sheng, X.; Zhu, X. Towards Human-in-the-Loop Shared Control for Upper-Limb Prostheses: A Systematic Analysis of State-of-the-Art Technologies. *IEEE Trans. Med. Robot. Bionics* **2023**, *5*, 563–579. [CrossRef]
29. He, Y.; Shima, R.; Fukuda, O.; Bu, N.; Yamaguchi, N.; Okumura, H. Development of distributed control system for vision-based myoelectric prosthetic hand. *IEEE Access* **2019**, *7*, 54542–54549. [CrossRef]
30. He, Y.; Fukuda, O.; Bu, N.; Yamaguchi, N.; Okumura, H. Prosthetic Hand Control System Based on Object Matching and Tracking. The Proceedings of JSME Annual Conference on Robotics and Mechatronics (Robomec), 2019; p. 2P1-M09. Available online: https://www.researchgate.net/publication/338151345_Prosthetic_hand_control_system_based_on_object_matching_and_tracking (accessed on 10 December 2023).
31. Vorobev, E.; Mikheev, A.; Konstantinov, A. A Method of Semiautomatic Control for an Arm Prosthesis. *J. Mach. Manuf. Reliab.* **2018**, *47*, 290–295. [CrossRef]
32. Ghosh, S.; Dhall, A.; Hayat, M.; Knibbe, J.; Ji, Q. Automatic gaze analysis: A survey of deep learning based approaches. *IEEE Trans. Pattern Anal. Mach. Intell.* **2023**, *46*, 61–84. [CrossRef]
33. Bullock, I.M.; Ma, R.R.; Dollar, A.M. A hand-centric classification of human and robot dexterous manipulation. *IEEE Trans. Haptics* **2012**, *6*, 129–144. [CrossRef]

biomimetics

MDPI

Article

Research on Multimodal Control Method for Prosthetic Hands Based on Visuo-Tactile and Arm Motion Measurement

Jianwei Cui * and Bingyan Yan

Institute of Instrument Science and Engineering, Southeast University, Nanjing 210096, China; 220213676@seu.edu.cn
* Correspondence: cjw@seu.edu.cn

Abstract: The realization of hand function reengineering using a manipulator is a research hotspot in the field of robotics. In this paper, we propose a multimodal perception and control method for a robotic hand to assist the disabled. The movement of the human hand can be divided into two parts: the coordination of the posture of the fingers, and the coordination of the timing of grasping and releasing objects. Therefore, we first used a pinhole camera to construct a visual device suitable for finger mounting, and preclassified the shape of the object based on YOLOv8; then, a filtering process using multi-frame synthesized point cloud data from miniature 2D Lidar, and DBSCAN algorithm clustering objects and the DTW algorithm, was proposed to further identify the cross-sectional shape and size of the grasped part of the object and realize control of the robot's grasping gesture; finally, a multimodal perception and control method for prosthetic hands was proposed. To control the grasping attitude, a fusion algorithm based on information of upper limb motion state, hand position, and lesser toe haptics was proposed to realize control of the robotic grasping process with a human in the ring. The device designed in this paper does not contact the human skin, does not produce discomfort, and the completion rate of the grasping process experiment reached 91.63%, which indicates that the proposed control method has feasibility and applicability.

Keywords: intention recognition; human–machine interaction; 2D Lidar; environmental perception; prosthetic hand

check for updates

Citation: Cui, J.; Yan, B. Research on Multimodal Control Method for Prosthetic Hands Based on Visuo-Tactile and Arm Motion Measurement. *Biomimetics* **2024**, *9*, 775. https://doi.org/10.3390/biomimetics9120775

Academic Editor: Rafhael Milanezi de Andrade

Received: 11 June 2024
Revised: 6 December 2024
Accepted: 12 December 2024
Published: 19 December 2024

1. Introduction

The prosthetic hand is an important tool to restore hand function for people with upper limb disabilities. Accurate recognition of the intention of upper limb actions is essential for controlling coordinated actions of a prosthetic hand. The coordinated actions of the human hand can be summarized in two parts: postural coordination and process coordination. The information that describes upper limb action patterns and intentions includes bioelectric, bionic, and kinematic signals [1].

Bioelectric signals are mainly composed of electroencephalography (EEG) and electromyography (EMG) signals. EEG signals can directly reflect intention information in the human brain about 100 to 150 milliseconds ahead of limb actions [2], which is suitable for the intention recognition and control of robotic arms and prosthetic hands. Baoguo Xu et al. studied EEG signals of humans during left, right, up, and down actions and successfully controlled the action of a robot and grasped the target object [3]. Kiyoshi Nagai et al. implemented EEG information to trigger robot actions [4]. However, EEG signals are usually very weak [5,6] and are affected by environmental noise. Hence, intention recognition using EEG signals has poor accuracy and stability. In addition, only simple action control can be accomplished.

Note that changes in electrical signals, EMG signals, can be measured on the surface of the skin when human muscles move a limb. These electrical signals can rapidly reflect limb action intentions and can be applied in the field of prosthetic hands. By using EMG signals, researchers have investigated strategies to control a pneumatic robotic arm to complete grasping actions [7], as well as methods to control a prosthetic hand to complete actions such as closing, opening, wrist flexion, and double wrist flexion using single-channel surface EMG signals [8]. Reference [9] proposed a new human–computer interaction for tetraplegic patients, which was evaluated using EMG in five patients with spinal cord injuries. The preliminary results provided a new solution for advanced spinal cord injury patients. In addition, EMG signals can also be used to control a prosthetic hand to perform certain fine actions, such as axial hole assembly [10]. Based on EMG and eye action signals, Keisuke Shima et al. controlled a prosthetic hand to help a disabled person eat [11]. Segil et al. fitted four EMG electrodes to the missing forearm region to control a prosthetic hand, achieving an average of 55% physiological function in patients with limb loss [12].

Previous studies on prosthetic hands have indicated that EMG signals are superior to EEG signals. However, there are many problems with EMG signals. Although reference [12] provided examples of electrodes attached to the amputation site of an arm, the muscles of the arm of a disabled individual often show varying degrees of atrophy. Muscle atrophy leads to weaker and noisier EMG signals, making it difficult to acquire and process them. Many previous works did not consider how to coordinate the actuation of multiple muscles, which not only makes the acquisition of EMG signals difficult but also makes them time-consuming and difficult to use. In addition, EMG electrodes can cause skin irritation and serious malfunctions due to interference from factors such as humidity.

It is worth noting that body actions are the result of the human brain's intentions, i.e., they reflect human intentions. Body motion information includes velocity, acceleration, posture, position, etc. This body motion information can be measured using inertial sensors. With the development of MEMS, inertial sensor technology has matured [13] and has become a commonly used sensor to collect human motion parameters, gradually becoming applied in the field of prosthetic control. For example, Ben et al. introduced an inertial measurement unit (IMU)-based method to recognize the action intention of human lower limbs [14]. Samuel et al. illustrated the drawbacks of EMG signals and proposed an algorithm for human gesture recognition integrating an IMU and EEG signals [15]. Noccaro et al. recorded arm actions for human-like upper limb actions using a robotic arm based on an IMU [16]. Merad et al. designed a coordinated elbow motion prediction method based on IMU upper limb kinematics, which provided ideas for flexible control of prosthetic limbs [17]. The prosthetic hand research team at Southeast University investigated the use of inertial sensors to control shoelace tying with prosthetic hands, with good results [18].

In recent years, multimodal signals such as visual, haptic, and speech signals have been widely used for prosthetic limb control to achieve precise manipulation of prosthetic limbs. Martin Harold et al. designed an upper limb prosthesis control method integrating visual, inertial, and EMG signals [19]. Ghazaei et al. proposed a control method based on the fusion of visual and EMG information, which successfully realized the task of grasping 500 different types of objects [20]. It can be seen that visual information is important for the control of a prosthetic hand, but a prosthetic hand is only suitable for monocular vision, and it is difficult to accurately obtain the dimensions of objects due to a lack of depth information, coupled with the facts that there are many different types of objects and the dimensions of the same type of object can also be very different, which makes monocular-vision-based control methods cumbersome and difficult to adapt to complex environments and tasks. In order to solve this problem, the use of Lidar as a visual perception sensor, which can provide rich depth information, can effectively compensate for the shortcomings of monocular vision and is able to efficiently acquire features such as object dimensions [21]. There have also been studies on controlling a prosthetic hand with voice signals, which had better results in quiet situations, but this voice control method can often be embarrassing for users.

Therefore, in this paper, we propose a multimodal control algorithm for prosthetic hands based on opto-haptic and arm kinematic measurements, which is used to solve the cross-sectional shape and size recognition and dynamic gesture adjustment of target objects in complex scenes. The main contributions can be categorized into the following three aspects:

(1) In this paper, a new type of prosthetic hand control system integrating visual and tactile sensing and arm motion measurement is constructed, and a cross-sectional object shape and size recognition method based on the fusion of a monocular camera and two-dimensional Lidar is proposed in a pioneering manner. At the same time, a dynamic grasping control algorithm with multimodal information fusion is designed, which combines the upper limb motion state and lesser toe tactile information to realize intelligent recognition and dynamic gesture adjustment for objects in the middle of the target area in a complex environment, providing a brand-new technological path for the intelligent control of prosthetic hands.

(2) In this study, an algorithm for recognizing the cross-sectional shape and size of objects based on the fusion of a monocular camera and 2D Lidar data is proposed for the first time. The algorithm recognizes and classifies the shape of an object through the YOLOv8 model and extracts geometric features, and then combines 2D Lidar with the DBSCAN algorithm to cluster the objects, with the DTW algorithm used to further recognize the cross-sectional shape and size of the grasped part of the object, realizing grasping attitude control of a prosthetic hand. The method in this paper overcomes the limitations of a single sensor in complex environments, significantly improves the accuracy and robustness of object recognition, and reduces costs at the same time.

(3) In this study, a grasping action control algorithm based on the fusion of an IMU and haptic sensor data is proposed. The upper limb motion state is identified using a sliding window attitude angle variance calculation algorithm, a palm motion model is established, and the palm position is classified using the MLP (multi-layer perceptron) algorithm; the foot haptic data are processed using Kalman filtering, and the pressing action of the lesser toes is identified by combining with the peak extraction algorithm. By integrating the upper limb kinematic state, hand position, and foot haptic information, precise optimization of the grasping action is achieved. Compared with traditional EMG, EEG, or voice control methods, this algorithm enhances the flexibility, covertness, and comfort of operation through natural human interaction.

The rest of this paper is organized as follows: in Section 2, we present the overall design of the prosthetic hand control system; Section 3 introduces an algorithm for recognizing the shape and size of an object cross-section based on the fusion of a monocular camera and 2D Lidar data; Section 4 details a grasping action control algorithm based on the fusion of an IMU and haptic sensor data; Section 5 is the experimental part, which confirmed the feasibility of the methodology proposed in this paper through experiments, and Section 6 provides the conclusions and future perspectives.

2. System Design

The overall structure of the prosthetic hand control system is shown in Figure 1 and consists of the main control module (Raspberry Pi), Lidar, pinhole camera, IMU, flexible pressure sensing insole, and prosthetic hand. Among them, the pinhole camera has a size of $\varphi 3.9 \times 20$ mm and the laser of the Lidar has a laser size of $6 \times 6 \times 8$ mm, which can be integrated and mounted on the finger.

The grasping action of the human hand is a complex process that requires the synergistic cooperation of multiple organs and is the result of the fusion processing of human intention and environmental perception. Based on an analysis of the complete grasping action process, this paper designed a complete control system for the prosthetic hand action process, as shown in Figure 2. In this system, IMUs are worn on the large and small arms to detect the movement of the upper limbs in real time and determine the position of the end of the hand, the Lidar and camera are mounted on the prosthetic hand for recognizing the size and type of the object, and flexible pressure sensing insoles are placed on the soles of the feet to detect the user's active manipulation signals. The design concepts of this system include the following: first, an IMU is used to obtain the velocity information of the upper limb to judge its motion state, and a D-H (Denavit–Hartenberg) model of the upper limb is established to solve the position of the hand relative to the body, so as to judge the grasping timing; second, unlike a traditional prosthetic hand system that recognizes the specific class of an object, this paper uses camera vision to recognize the shape of the object; third, this paper innovates with the use of 2D Lidar to recognize the object size, designing point cloud preprocessing and size calculation algorithms; fourth, the foot pressing action is used as a human–computer interaction signal; fifth, a grasping strategy for the prosthetic hand based on the shape and size of the object is designed, and the above information is fused to control the action process of the prosthetic hand.

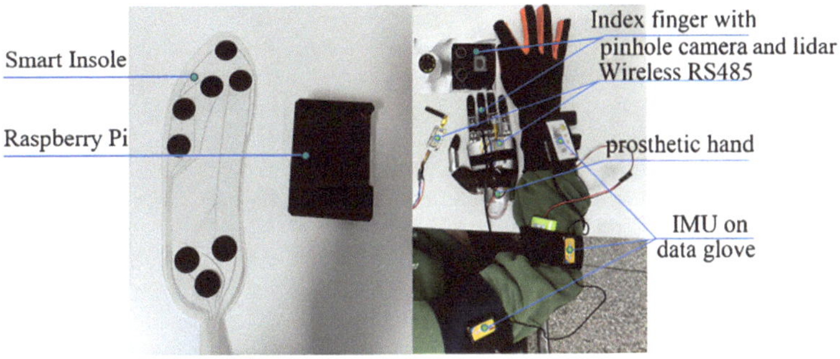

Figure 1. Hardware composition of the prosthetic hand control system.

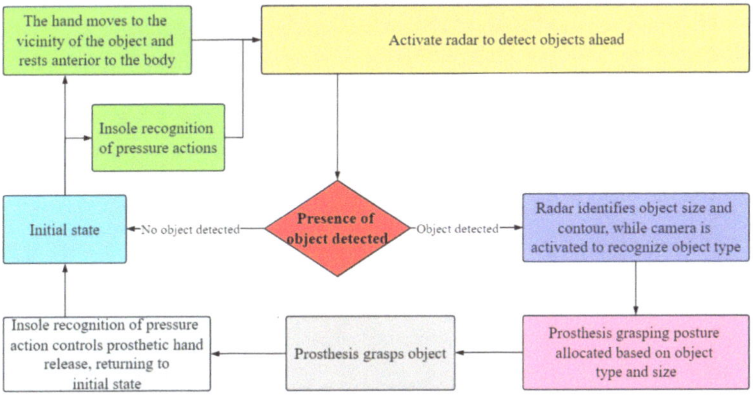

Figure 2. Prosthetic hand control system flow.

The control process of the prosthetic hand designed in this system is as follows: The initial state of the system is that the upper limbs are static and naturally drooping, when it is necessary to grasp an object, the arm drives the robotic hand to move towards the object. When the IMU detects that the upper limbs have stopped moving and the hand is located in front of the body, this means that the robotic hand can reach the vicinity of the object. At this time, the Lidar is used to detect whether there is an object in front of the body, if no object exists, this means that no movement of the prosthetic hand is needed, if there is an object, the camera acquires an environment image, and together with the Lidar, it recognizes the type and size of the object, and then the system controls the prosthetic hand to adjust the posture for grasping. When the object needs to be released, the flexible pressure sensing insole recognizes the user's foot pressing action, allowing the prosthetic hand to release the object and return to the initial state.

3. Monocular Camera and 2D Lidar-Based Object Cross-Section Shape and Size Identification Method

3.1. Overview of the Methodology

Monocular cameras are small, lightweight, and low-cost, but a lack of depth information makes it difficult to accurately obtain the size of an object, coupled with the fact that there are many types of objects and the size of the same type of object can vary greatly, which makes monocular-vision-based control methods cumbersome and difficult to adapt to complex environments and tasks.

Lidar is commonly used for slam navigation in mobile robots. Lidar illuminates the target in the sector plane and can obtain a distance value corresponding to the emission angle, and point cloud data are formed through multiple measurements. The point cloud data of Lidar are suitable for long-distance measurement and are often used to describe the traveling path of mobile robots. In this paper, the point cloud data of Lidar were used to measure the cross-sectional shape and size of the grasped object, and the main problems that needed to be solved were the small size of the measurement target, the short measurement time, the small amount of measurement data obtained, and the large error. In addition, the random jitter of a human hand increased the measurement error and difficulty.

In order to solve the above problems, a monocular camera was first used to recognize and classify the shape of the object using the YOLOv8 model to extract geometric features, and then the DBSCAN algorithm was utilized in combination with 2D Lidar to cluster the objects, with the DTW algorithm used to further identify the cross-sectional shape and size of the object's grasped part, in order to realize control of the grasping posture of the prosthetic hand.

3.2. Monocular Camera Object Shape Recognition Method

Most camera vision algorithms for prosthetic hands choose to recognize specific classes of objects. However, there are many types of objects and it is difficult to cover them comprehensively. Including more objects will lead to a wide range of classes, affecting the recognition accuracy and efficiency. Based on the above analysis and the fact that common relatively regular objects have relatively simple shapes, and different objects of the same shape have a great similarity in their corresponding grasping gestures, this section divides common objects into three categories, i.e., columns, balls, and slices, from the point of view of the design of the grasping gesture. YOLO is a widely used and validated target detection and recognition model. In this section, the shape of the object is recognized using YOLOv8 [22]. The images used in this method were sourced from Coco, supermarket, and fruit scenes, etc., and were labeled to form a new dataset. Figure 3 shows some of the objects in each category in the dataset. The dataset generated in this section consists of a total of about 21,000 images, including 54 types of columns, 113 types of balls, and 16 types of slices, and the dataset is divided into training, validation, and testing sets in a ratio of 8:1:1.

The training in this paper was caried out on the Ubuntu 20.4 + PyTorch platform, the hardware configuration of the machine is shown in Table 1, and the initial model chosen for training was YOLOv8. The loss curve of the training model is shown in Figure 4, *box_loss* represents the loss in recognizing the box, *cls_loss* represents the loss in classifying, the horizontal axis represents the number of iterations, and the vertical axis represents the value of the loss. It can be seen that the loss decreased faster in the first 20 rounds and slowly decreased later. mAP is an important metric for evaluating tasks such as target detection and segmentation in computer vision, and it is well suited for evaluating the accuracy of a model, which is the average of the AP (average value of precision). The AP responds to the area enclosed by the precision–recall curve and the horizontal and vertical axes, and the mAP is computed as shown in Equation (1),

$$mAP = \frac{\sum_{i=1}^{n} \frac{\sum_{k=1}^{R_i} P(k)\Delta Re(k)}{R_i}}{n} \tag{1}$$

where n is the number of categories in the detection task, R_i is the total number of true positive cases for category i, $P(k)$ is the precision of category i in the first k detections, and $\Delta Re(k)$ is the difference in recall when category i changes from $i-1$ to i. Figure 5 demonstrates the mAP variation curve during the training process of this training set, and it can be seen that the mAP value reached 0.91, which represents good performance for the model.

(a) Column **(b)** Ball **(c)** Slice

Figure 3. Dataset classification.

Table 1. Training hardware environment.

CPU	Graphics Card	Memory	Video Storage
11th Gen Intel(R) Core(TM) i7-11700k	NVIDIA GeForce RTX 3080	32 GB	10 GB

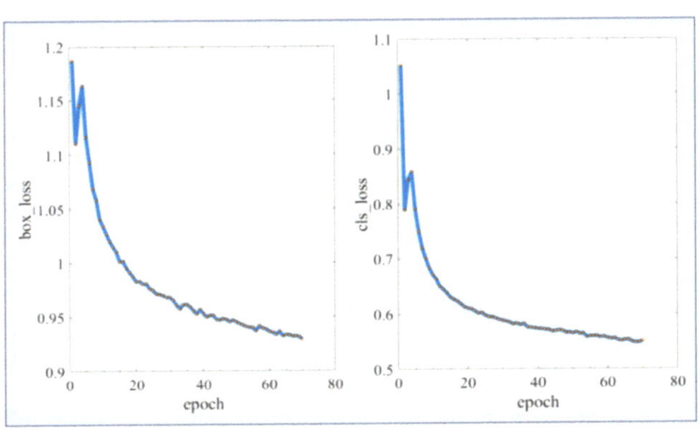

Figure 4. Training loss curve.

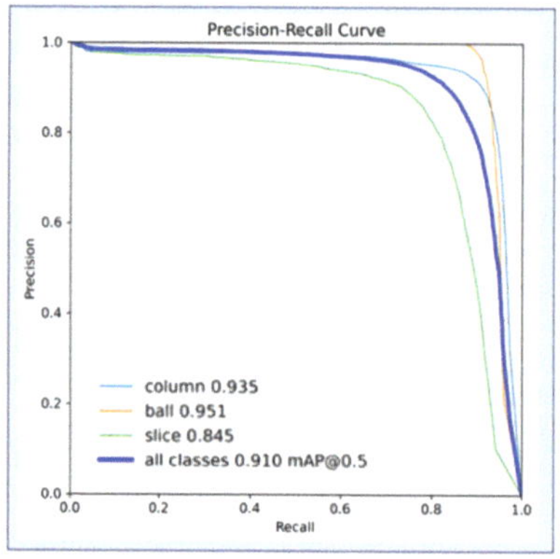

Figure 5. mAP curve.

3.3. Pre-Processing of Lidar Point Cloud

Firstly, the point cloud data needed to be filtered and put through a noise reduction process. Because the research object of this paper was a prosthetic hand, we could only use a small size of Lidar, the data density was low but the frame rate was high, and the point cloud data were in sparse matrix form. The slam algorithm point cloud filtering method commonly used in mobile robots is not ideal for a more sparse point cloud. In this paper, we proposed a multi-frame noise reduction algorithm for point cloud filtering by referring to the multi-frame noise reduction algorithm in image processing. This algorithm is utilized by continuously irradiating the same cross-section of an object in a split second to obtain multiple frames of a two-dimensional point cloud, and then the coordinates of the same position are averaged to obtain a set of two-dimensional point cloud data after noise reduction. The calculation formula is shown in Equation (2), where N is the number of frames, which can be obtained as a set of points $P_i(x_i, y_i), i = 1, 2, 3, \ldots$.

$$\begin{cases} x_i = \frac{1}{N} \sum_{j=1}^{N} x_{ij} \\ y_i = \frac{1}{N} \sum_{j=1}^{N} y_{ij} \end{cases} \tag{2}$$

Figure 6a shows a cylindrical object and the Lidar scans of the object. A multi-frame 2D point cloud of a certain cross-section was obtained, as shown in Figure 6b. The multi-frame point cloud in Figure 6b was processed using the above multi-frame noise reduction algorithm to obtain the result in Figure 6c. The processed points were connected to obtain the solid line in Figure 6c. The local area of the cylindrical object is zoomed in as shown in Figure 6d. The multi-frame point cloud is distributed close to both sides of the solid line, which indicates that the multi-frame noise reduction algorithm played a filtering role in the above error.

The Lidar used in the system emits a laser beam from the same point, similarly to a point light source. In addition, different objects illuminated by the Lidar are separated on the 2D point cloud map, and the same object forms a set of neighboring points. Therefore, this section used the DBSCAN (density-based spatial clustering of applications with noise) clustering algorithm to extract the objects and a method to also filter the points that were separated from the object caused by random errors, i.e., isolated point exclusion method. DBSCAN is a density-based clustering algorithm that clusters points in dense regions, while marking points in sparse areas as noise. The core idea of DBSCAN is to expand clusters from core points, which are points that have a sufficient number of neighboring points within a certain distance. The two main parameters of the algorithm are eps and $min\_samples$, where eps represents the maximum distance between two points to define the neighborhood, while *min_samples* represents the minimum number of neighboring points required to form a cluster.

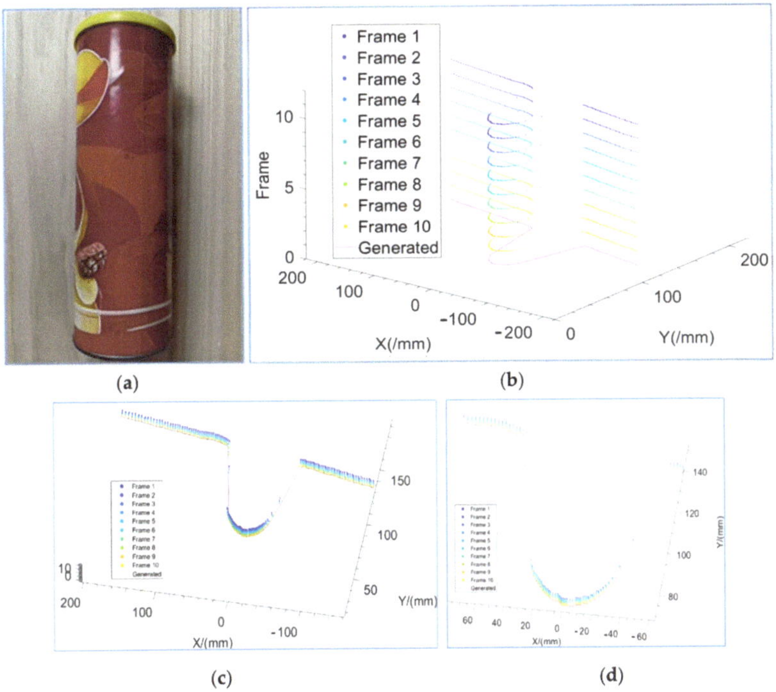

Figure 6. Noise reduction algorithm for object point cloud. (**a**) Environment. (**b**) Continuous multi-frame point cloud. (**c**) Continuous multi-frame point cloud overlay. (**d**) Continuous multi-frame point cloud overlay localized to the object.

Figure 7 demonstrates the process of filtering a fruit pile using the DBSCAN clustering algorithm. When the environment in Figure 7a was scanned using Lidar, a two-dimensional point cloud map of the cross-section could be obtained. Clustering this point cloud using the DBSCAN algorithm led to the result of Figure 7b, which contains multiple noise points, e.g., at the place circled by the circle. After the removal of the isolated noise points, the remaining point cloud consisted of two parts, i.e., the real object point cloud and the clustered noise point cloud, to filter out the clustered noise as much as possible. Each cluster of the point cloud was processed by using the curve fitting method, considering that the method in this chapter aims to identify relatively regular objects and noise is usually irregular. Moreover, these characteristics were reflected in the curve fitting polynomials, which were within the error tolerance. True objects have lower powers and noise has higher powers. After the above processing, Figure 7c yields the real objects, numbered 1 and 2 for the tabletop and numbered 3–5 for the spherical fruit.

As seen in Figure 7c, a complex environment containing multiple objects can be processed to obtain multiple real objects. However, the hand only wants to grasp one of them, and because the hand will be close to the object to be grasped, the real object to be grasped is the one that is close to the origin, such as object No. 4 in Figure 7c.

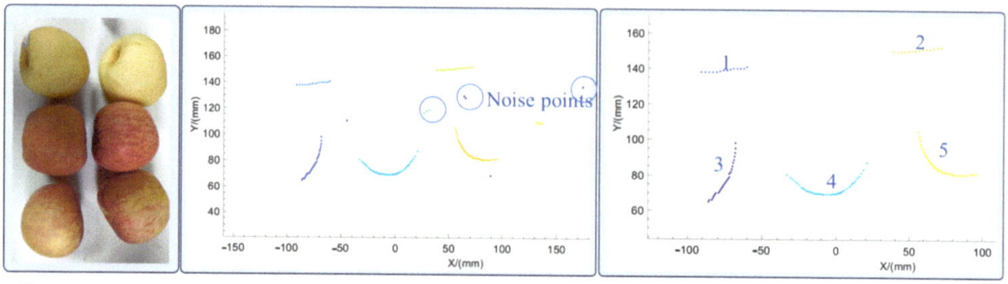

(**a**) Environment (**b**) Point cloud after DBSCAN clustering (**c**) Filtered point cloud

Figure 7. Multiple object DBSCAN filter.

3.4. Recognizing the Shape and Size of the Object's Cross-Section

After processing in the above subsections, the method in this research locates the target object. In this section, we identify the shape and size of the object cross-section. The identification of the cross-section shape is used to obtain the size of the object, because the sizing algorithm is different for different shapes of the cross-section. According to the common object shapes, this method divides the cross-section shapes into three categories, i.e., circular, straight-line segment, and L-shaped. The dimensions of this method were used to design the grasping posture of the prosthetic hand. Therefore, the dimensions are the effective grasping dimensions, i.e., for the circular cross-section, the dimension is the diameter, for the straight-line segment cross-section, the dimension is the straight-line segment length, and for the L-shape cross-section, the dimension is the length of the first and last point connecting lines.

The classification problem of cross-sectional shapes can be transformed into a similarity problem with respect to semicircles, lines, and L-shaped lines. The semicircles, lines, and L-shaped lines are generated based on the cross-section to be identified, which is referred to as the template in this paper, and the specific generation method is shown in Figure 8a. The black dotted line represents the 2D point cloud of the object's cross-section scanned by the Lidar system. The blue dotted line represents the generated semicircular point cloud, the red dotted line represents the L-shaped linear point cloud, and the green dotted line represents the linear point cloud. The 2D point cloud can be decomposed into an x-direction sequence $\{x_1, x_2, \ldots, x_n\}$ and y-direction sequence $\{y_1, y_2, y_3, \ldots, y_n\}$, and the template point cloud is generated as shown in Equation (3).

$$\begin{cases} y_i^s = \frac{y_n - y_1}{x_n - x_1}(x_i - x_1) + y_1 \\ y_i^c = \frac{y_1 + y_n}{2} - \sqrt{\frac{(x_n - x_1)^2 + (y_n - y_1)^2}{4} - \left(x_i - \frac{x_1 + x_n}{2}\right)^2} \\ y_i^l = \frac{y_2 - y_1}{x_2 - x_1}(x_i - x_1) + y_1 \ , \ x_i \leq \frac{y_n - y_1 + \frac{y_2 - y_1}{x_2 - x_1}x_1 - \frac{y_n - y_{n-1}}{x_n - x_{n-1}}x_n}{\frac{y_2 - y_1}{x_2 - x_1} - \frac{y_n - y_{n-1}}{x_n - x_{n-1}}} \\ y_i^l = \frac{y_n - y_{n-1}}{x_n - x_{n-1}}(x_i - x_n) + y_n, x_i > \frac{y_n - y_1 + \frac{y_2 - y_1}{x_2 - x_1}x_1 - \frac{y_n - y_{n-1}}{x_n - x_{n-1}}x_n}{\frac{y_2 - y_1}{x_2 - x_1} - \frac{y_n - y_{n-1}}{x_n - x_{n-1}}} \end{cases} \tag{3}$$

where y_i^s, y_i^c, y_i^l are straight-line, circular, and L-shaped point cloud templates, respectively. In this section, a dynamic time warping algorithm (DTW) is used to classify the cross-section shapes. The DTW algorithm is used to compare the similarity of two sequences, and it is able to compute the distance between the two sequences. The process of using the DTW algorithm to compute the cross-section point cloud sequences and template point cloud sequences is shown in Figure 8b, with the top part of the object's cross-section point cloud and the left side of the template point cloud of the generated straight-line segments. In the middle part of the process of distance calculation for the two sequences, as shown in Equation (4), D_{ij} is the cumulative distance from $pattern_i$ to $thing_j$, as shown in the red circle in Figure 8b, D_{ij} is obtained using the sum of the three worthwhile minimums in the upper left corner and the distance from the current position, and the cumulative distance between any two points of the template and the object point cloud can be obtained after traversal. Let n be the number of points in the object and template point clouds, and $D_{\{nn\}}$ be the shortest distance between two points from the current template and the object's cross-section. In Figure 8b, the white broken line represents the path that achieves the shortest distance. By calculating the distances between the points in the object's cross-section point cloud and the sequences of the three template point clouds, the template corresponding to the shortest distance determines the classification of the object's cross-section.

$$\begin{cases} D_{ij} = \min\left(D_{i-1,j}, D_{i-1,j-1}, D_{i,j-1}\right) + \left|thing_j - pattern_i\right|, i,j = 0,1,\dots \\ D_{-1,j} = \infty,\ D_{i,-1} = \infty \end{cases} \tag{4}$$

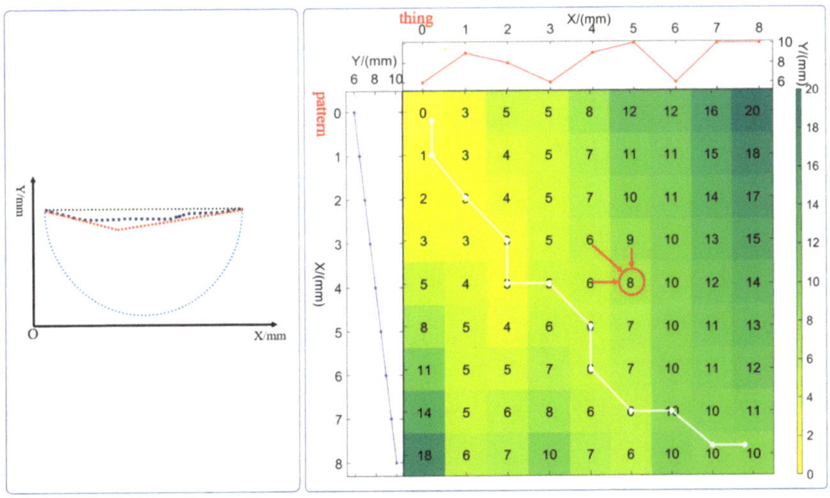

(**a**) Modelling calculations (**b**) Diagram of the DTW calculation process

Figure 8. Algorithm for point cloud DTW similarity calculation.

Using the above algorithm, the shape of the cross-section of the object to be recognized by the Lidar can be obtained in this section, and the size of the cross-section can be easily calculated based on the shape. If the object is a straight segment type or L-shape, its cross-section dimensions are presented in Equation (5), as follows:

$$size = \sqrt{(x_n - x_1)^2 + (y_n - y_1)^2} \tag{5}$$

where (x_1, y_1) is the start point of the cross-section point cloud of the object, (x_n, y_n) is the end point of the cross-section point cloud of the object, and size is the requested size. If the object is a circular, the steps for sizing its cross-section are as follows: first, the Equation of the circle is fitted using the least squares circular fitting method, and then its diameter is the requested size. The least squares fitting point cloud circular fitting method is as follows: first, determine the parameters of the circle to be fitted, i.e., $(x-a)^2 + (y-b)^2 = r^2$, where (a, b) is the center of the circle and r is the radius, and the error sum is calculated as shown in Equation (6), where N is the number of points,(x_i, y_i) are the coordinates of the cloud of points, and $Q(a, b, r)$ is the error sum of the cloud of points, so that the set of error sums which are minimum (a,b,r) is the parameter of the fitted circle. And the parameters can be obtained by taking the partial derivatives of (a,b,r) and assigning a value of 0.

$$\begin{cases} d_i^2 = (x_i - a)^2 + (y_i - b)^2 \\ \delta_i = d_i^2 - r^2 \\ Q(a,b,r) = \sum\limits_{i=1}^{N} \delta_i^2 = \sum\limits_{i=1}^{N} (x_i^2 + y_i^2 - 2ax_i - 2by_i + a^2 + b^2 - r^2)^2 \\ \frac{\partial Q}{\partial a} = 0, \frac{\partial Q}{\partial b} = 0, \frac{\partial Q}{\partial r} = 0 \end{cases} \quad (6)$$

4. Multi-Sensor-Based Control Method for Prosthetic Hand

4.1. IMU Recognizes Upper Limb Motion State and Hand Position

According to the analysis in Section 2, a prosthetic hand system needs to recognize the motion state of the upper limb, including the motion and static state, and the motion state of the upper limb is determined by the velocity, which can be represented by the difference in the attitude angle. The built-in accelerometer of the IMU can capture data of the roll angle, the yaw angle, and the pitch angle. Hence, wearing the IMU on the upper limb can determine the motion state of the upper limb. In this section, the difference in attitude angle is calculated by the following process:

(1) A sliding window is used to calculate the moving variance of the attitude angles acquired by the IMU, with a sampling rate of 50 Hz and a window length $l = 10$, i.e., 10 sampling points.

(2) Calculate the variance of the three-axis attitude angles of the current window as shown in Equation (7), where $\varnothing_{k-i}, \theta_{k-i}, \varphi_{k-i}$ are the values of the turn angle, yaw angle, and pitch angle of the $l - i + 1$st point of the current window, respectively, and $\overline{\varnothing}, \overline{\theta}, \overline{\varphi}$ is the mean value of the attitude angles in the current window.

$$Var_k = \begin{cases} 0 & , k \leq l \\ \frac{1}{l}\sum\limits_{j=1}^{l} \left[\left(\varnothing_{k+i} - \overline{\varnothing}\right)^2 + \left(\theta_{k+i} - \overline{\theta}\right)^2 + \left(\varphi_{k+i} - \overline{\varphi}\right)^2 \right], & k > l \end{cases} \quad (7)$$

Figure 9a shows the acceleration change during the process of drinking water. The change in angular velocity indicates that there is movement in the upper limb. In the process of picking up the cup, i.e., moving the cup to the mouth, putting down the cup, and putting down the hand, the acceleration and posture angle change obviously. The acceleration stays near 0 rad/s and the posture angle stays stable for the rest of the time. The movement variance change curve of putting on glasses is shown in Figure 9b, and the curve shows obvious wave peaks when the arm is in a motion state. Therefore, this can be considered as a stationary state by setting the threshold th, when $Var_k \leq th$, otherwise it is considered as a motion state. During the rest of the time, the variance value is close to 0. According to the analysis of the angular velocity curves of the above actions, it can be seen that at the moment of grasping or opening the palm, the upper limb is in a stationary state, and the angular velocity of the upper limb is close to 0 rad/s (e.g., segments C_1D_1, E_1F_1 and G_1H_1). Hence, the upper limb will only change the grasping and releasing state when it is at rest, which is the first rule of the upper limb motion state transition, i.e., the dynamic static rule.

According to the analysis in Section 2, the user acquires the end position of the hand when using the prosthetic hand to complete an action and divides the end position of the hand into the initial position, in front of the torso, and the position near the torso, as shown in Figure 10. To further analyze the end position of the hand, a D-H model of the upper limb was established for positive kinematic analysis in this subsection, as shown in Figure 11, with points O, A, B, C, and D representing the chest, shoulder, elbow, wrist, and palm positions of the human body, respectively. $O_O - X_OY_OZ_O$,$O_A - X_AY_AZ_A$,$O_B - X_BY_BZ_B$,$O_C - X_CY_CZ_C$,$O_C - X_DY_DZ_D$ denote the base coordinate system of the corresponding position, respectively. And $L_{OA}, L_{AB}, L_{BC}, L_{CD}$ denote the dimensions of the arm, i.e., half shoulder width, upper arm length, forearm length, and palm length. The D-H equation of the hand is shown in Equation (8), and the coordinates of the hand in the chest coordinate system are obtained by multiplying the four transformation matrices with the coordinates of the palm coordinate system. After the above positive kinematic analysis, the coordinates of the end of the hand could be obtained. To classify the position of the end of the hand, a multilayer perceptual machine model was introduced in this section. The x, y, z-axis coordinate values of the end of the hand, P_D^x, P_D^y, P_D^z, and the angles between the three-axis attitude angles, $\Delta\varnothing$, $\Delta\theta$, and $\Delta\varphi$, for the large and small arms were selected as the eigenvalues. The MLP had six input neurons and three output neurons. The input to the MLP consisted of P_D^x, P_D^y, P_D^z and $\Delta\varnothing$, $\Delta\theta$,$\Delta\varphi$, while the output was the position of the hand, representing the front-of-torso, near-torso, and initial positions. The training data were collected from input data obtained from different volunteers. The training dataset had nearly 5000 sets of data. The parameters of this model were chosen to have two hidden layers with 10 and 6 neurons, respectively, and the learning rate was 0.1. The accuracy of the classification of the 100 sets of data using the trained model was 98%.

$$\begin{bmatrix} P_O^D \\ 1 \end{bmatrix} = T_A^O T_B^A T_C^B T_D^C \begin{bmatrix} P_D^D \\ 1 \end{bmatrix} \quad (8)$$

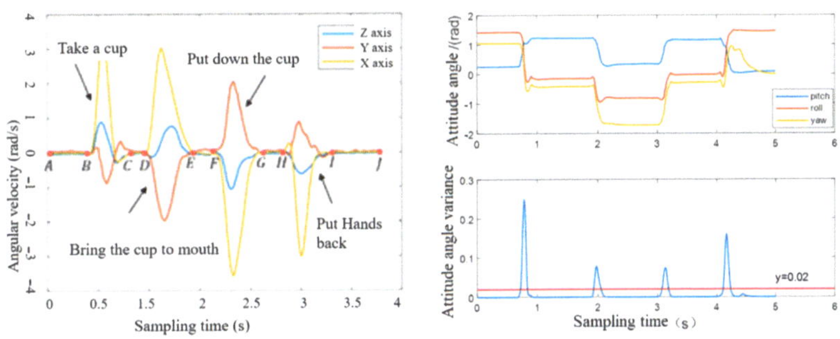

(**a**) (**b**)

Figure 9. (**a**) Acceleration changes while drinking water. (**b**) Attitude angle changes while wearing glasses.

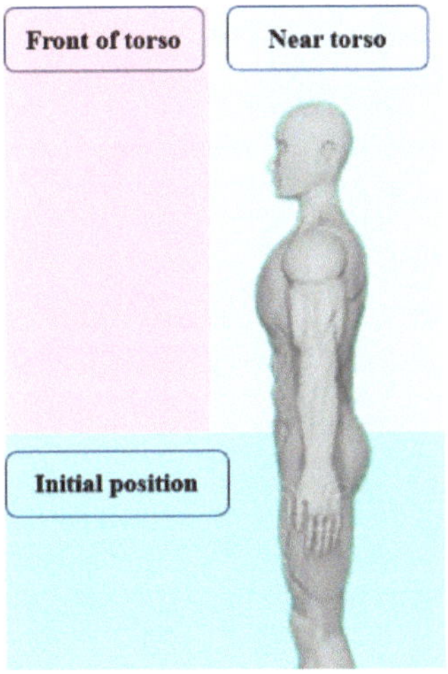

Figure 10. Division of the end position of the hand.

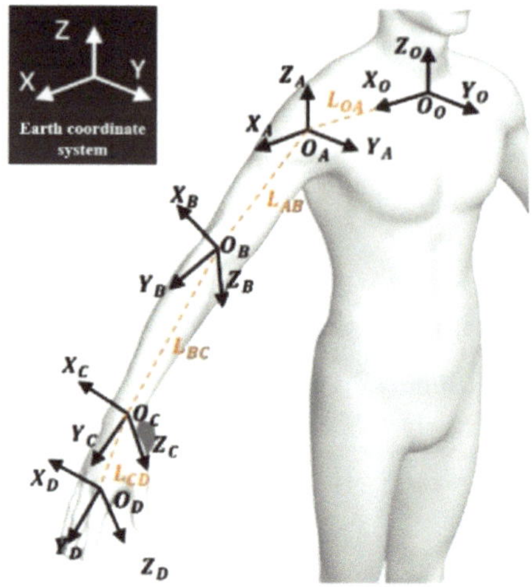

Figure 11. D-H model of the upper limb.

4.2. Foot Pressing Action Recognition

Research has shown that there are two states of the human foot during daily activities: the static state (e.g., sitting, standing, etc.) and the locomotion state (e.g., walking, running, etc.). In the above states, foot strength is mainly concentrated in the forefoot and heel, but increasing the tactile force in these two areas is very difficult and may also affect balance, making them unsuitable for characterization signals. In contrast, the lesser toe and the little toe can easily press a flexible pressure sensing insole, and their pressing action is more flexible and easy to control, and also does not affect the balance state. Since the lesser toe is more flexible than the little toe, and as the touch force generated by its pressing an insole is more obvious, the pressing action of the lesser toe was selected as the feature signal in this paper.

The most flexible big toe pressing action was selected as the characteristic signal, as shown by the blue dotted line in Figure 12, and the lesser toe forms an obvious wave peak after pressing the insole, while the rest of the time the pressure value is lower. Hence, by detecting the wave peak, it can be determined whether there is a pressing action or not. Due to the sensor's error and external interference, this section used average filtering and Kalman filtering to process the data, and the filtered results are shown as the orange solid line in Figure 12, and it can be seen from points such as A, D, etc., that the signal's smoothness characteristics were significantly improved; while from points such as B, C, etc., it can be observed that the filtering algorithm had a strong inhibition of random noise, which indicates that this filtering algorithm effectively reduced the noise interference and improved the quality and stability of the signal. The specific processing of this filtering algorithm was as follows:

(1) Recursive averaging filtering of the original tactile signal,

$$x_k = \frac{1}{N}\sum_{i=1}^{N-1} y_{k-i} \tag{9}$$

where y_{k-i} is the touch force measurement at moment $k-i$; x_k is the recursive average filter value at moment k; and N is the length of the selection queue.

(2) Kalman filter the obtained recursively averaged filtered signal with the state and measurement Equations, as in Equation (11), where x_{k-1} is the predicted value of the touch force at the moment $k-1$. u_{k-1} is the system input at moment $k-1$. w_{k-1} is the process

noise at moment $k-1$. x_k^- is the estimated value of the touch force at moment k, with the covariance Q. F, B are the system parameters.

$$\begin{cases} x_k^- = Fx_{k-1} + Bu_{k-1} + w_{k-1} \\ z_k = Hx_k + v_k \end{cases} \tag{10}$$

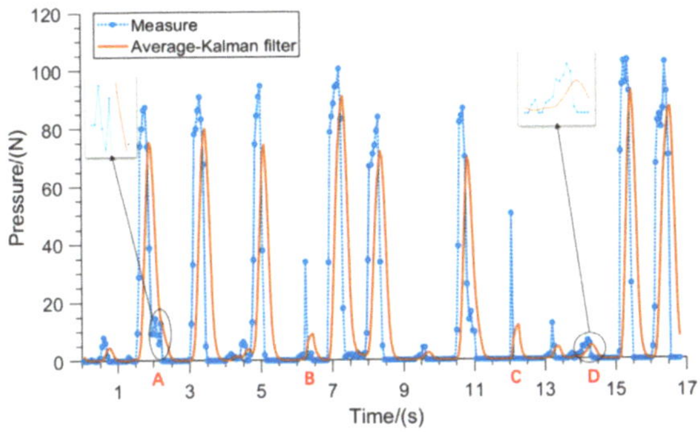

Figure 12. Comparison of the filtering of the contact force signal.

When the lesser toe consciously and actively presses the flexible pressure insole, the system will detect the waveform formed in the tactile force signal curve, and the disabled person will generate additional waveforms to form interference when he/she is in the movement state, but the amplitude of this waveform is obviously smaller compared with that formed by the lesser toe actively pressing the insole, and the interference waveform caused by walking can be shielded by setting the threshold, and at the same time, the threshold can also shield the toe from small mis-touches, which is also effective for the extraction of waveforms from the tactile force curve at rest. It is also effective in extracting the wave peaks of the touch force curve at rest. Therefore, choosing a reasonable threshold value can help distinguish whether the thumb actively makes a pressing action.

Through the above analysis, a pressing action can be identified by recognizing the crests of the tactile force curve and that the pressure is low in the absence of action. Therefore, judgment can be aided by increasing the threshold condition. As each person's weight and foot shape varies, the exact value of the threshold needs to be adjusted on an individual basis, which is determined by capturing the maximum value of the tactile force signal over some time when walking, standing, or sitting in a normal walking, standing, or sitting position. When the toe presses on the insole, the tactile force timing signal will generate peak signals that exceed the threshold value. Extracting these peak signals that exceed the threshold value can be used to determine whether the toe is consciously performing a pressing action. The specific peak extraction algorithm is as follows, the N-filtered tactile force data are recorded as $a_1, a_2, \ldots, a_N$, and the specific calculation formula is as follows:

$$\begin{cases} g(x) = \begin{cases} 1, x < 0 \\ 0, x \geq 0 \end{cases} \\ S = \sum\limits_{i=2}^{N} \dfrac{g((a_{i-1} - t)(a_i - t))}{2} \end{cases} \tag{11}$$

where $g(x)$ is the negative function defined. a_i are the ith filtered touch data. t is the touch threshold. S is the number of peaks above the threshold in the timing curve composed of N data. A set of data points satisfying $g((a_{i-1} - t)(a_i - t)) = 1$ represents a peak signal reflecting a single toe press action. Using the above algorithm, the number of peak signals that satisfy the condition can be obtained, thus identifying whether the toe was pressed or not during that period.

4.3. Combined Lidar and Camera Recognition of Object Size and Shape

According to the analysis in Section 2, this system uses a Lidar and monocular camera as the environment sensing module, the camera is used to identify the overall shape of the object, and the Lidar is used to identify the cross-sectional shape and size of the object. In Section 3, we used algorithms such as DTW to identify the cross-sectional shape and size of multiple objects in the environment, and in Section 4, we used the YOLOv8 algorithm to identify the shape of multiple objects in the environment. The system needs to obtain the independent shape and size of each object, so it was necessary to calibrate the objects in the two environments. In this section, a calibration algorithm was designed based on the traditional calibration algorithms of a Lidar camera.

There have been many model studies about camera coordinate transformation, and reference [23] described in detail the principle of camera imaging and the relationship between the camera coordinate system, image coordinate system, pixel coordinate system, and world coordinate system. The transformation relationship between the camera coordinate system and the pixel coordinate system is shown in Equation (12), where (X_C, Y_C, Z_C) is a point in the camera coordinate system,(u, v) are the coordinates of the point corresponding to the pixel coordinate system, f is the focal length of the camera, and f_u and f_v are the reciprocal of the pixel's object size in the x, y directions of the image plane, respectively. And (u_0, v_0) are the origin of the image coordinate system in the coordinates of the pixel coordinate system. The camera coordinate system and the lidar coordinate system are in the same three-dimensional space, they can be converted by Equation (13), where (X_L, Y_L, Z_L) is the point corresponding to (X_C, Y_C, Z_C) in the Lidar coordinate system, and R, T are the rotation and translation matrices of the two coordinate systems. Combine Equations (12) and (13) and let $Z_C A = \begin{bmatrix} f_u & 0 & u_0 \\ 0 & f_v & v_0 \\ 0 & 0 & 1 \end{bmatrix} \begin{bmatrix} f & 0 & 0 \\ 0 & f & 0 \\ 0 & 0 & 1 \end{bmatrix}$, $R' = AR$, $T' = AT$, the conversion equation between the laser coordinate system and the pixel coordinate system can be obtained, as shown in Equation (14), in which $R' = \begin{bmatrix} r_{11} & r_{12} & r_{13} \\ r_{21} & r_{22} & r_{23} \\ r_{31} & r_{32} & r_{33} \end{bmatrix}$, $T' = \begin{bmatrix} t_1 \\ t_2 \\ t_3 \end{bmatrix}$.

$$Z_C \begin{bmatrix} u \\ v \\ 1 \end{bmatrix} = \begin{bmatrix} f_u & 0 & u_0 \\ 0 & f_v & v_0 \\ 0 & 0 & 1 \end{bmatrix} \begin{bmatrix} f & 0 & 0 \\ 0 & f & 0 \\ 0 & 0 & 1 \end{bmatrix} \begin{bmatrix} X_C \\ Y_C \\ Z_C \end{bmatrix} \tag{12}$$

$$\begin{bmatrix} X_C \\ Y_C \\ Z_C \end{bmatrix} = R \begin{bmatrix} X_L \\ Y_L \\ Z_L \end{bmatrix} + T \tag{13}$$

$$\begin{bmatrix} u \\ v \\ 1 \end{bmatrix} = A \left(R \begin{bmatrix} X_L \\ Y_L \\ Z_L \end{bmatrix} + T \right) = R' \begin{bmatrix} X_L \\ Y_L \\ Z_L \end{bmatrix} + T' = \begin{bmatrix} r_{11} & r_{12} & r_{13} \\ r_{21} & r_{22} & r_{23} \\ r_{31} & r_{32} & r_{33} \end{bmatrix} \begin{bmatrix} X_L \\ Y_L \\ Z_L \end{bmatrix} + \begin{bmatrix} t_1 \\ t_2 \\ t_3 \end{bmatrix} \tag{14}$$

The essential task of the calibration of the Lidar and camera is to find R', T', and this paper used 2D Lidar, and uniformly set its Z-axis coordinate to 0. The solution algorithm of this paper was as follows: First of all, 15 groups of points were selected corresponding to (u, v), (X_L, Y_L) coordinates, in order to better find the corresponding points, this paper formed the object with a prism to select the marking points, as shown in Figure 13. Therefore, we can obtain the prism in pixel coordinates for the linear Equation $au + bv + c = 0$, a, b, c for the Equation coefficients, and the joint Formula (14) can be the Formula (15), so the solution to finding the R', T' can be transformed into solving for the values of $r_{11}, r_{12}, t_1, r_{21}, r_{22}, t_2$. Then, the collected coordinate data are transformed and brought into Equation (15) to construct a system of transcendental Equations, and R', T' can be obtained by solving the system of transcendental Equations using the SVD (singular value decomposition) method. As shown in Figure 13, after calibration using the above algorithm, the Lidar points (red points in the figure) were correctly projected onto the image.

$$\begin{cases} a(r11x + r12y + t1) + b(r21x + r22y + t2) + c = 0 \\ \begin{bmatrix} ax & ay & a & bx & by & b \end{bmatrix} \begin{bmatrix} r_{11} \\ r_{12} \\ t_1 \\ r_{21} \\ r_{22} \\ t_2 \end{bmatrix} = -c \end{cases} \tag{15}$$

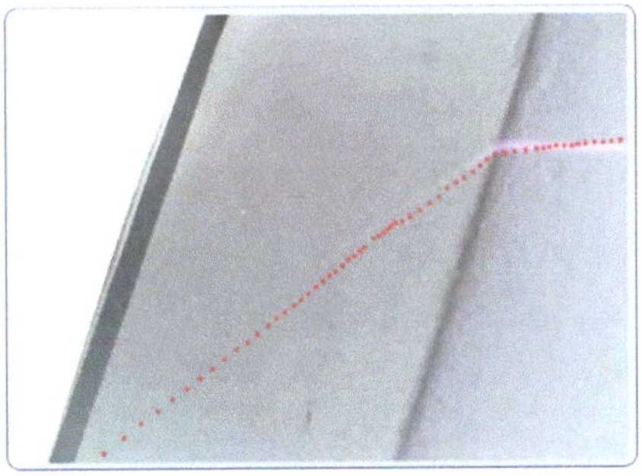

Figure 13. Two-dimensional Lidar–camera calibration algorithm.

4.4. Integrated Control Methods for Prosthetic Hands

According to the analysis in Section 2, the grasping posture of the prosthetic hand mainly depends on the shape and size of the object, and in this paper, we designed grasping postures for different kinds of objects. In Section 3, the objects were classified as columns, balls, and slices. As shown in Table 2, for columnar objects, if the cross-section size is small, we use the thumb, index finger, and middle finger to pinch and grasp the object. If the size is a little larger, we use five fingers to encircle the object. If the size is larger, we use the thumb and the rest of the fingers to pinch the object. For balls, if the size is small, we use the thumb and index finger to pinch and grasp the object. If the size is large, we use five fingers to try to encircle the object. For slice objects, if the cross-section size is small, we use three fingers to grasp the object in the direction of the cross-section. If the size is large, we use five fingers to pinch and grasp the object in the direction of thickness. For non-regular objects, a default action was designed to grasp the object, i.e., the hand is opened to its maximum extent and then closed.

Table 2. Object grasping posture design.

Cross-Section Size	0–20 mm	20–50 mm	50–150mm
Column			

Table 2. *Cont.*

Diameter	0–50 mm	50–150 mm
Ball	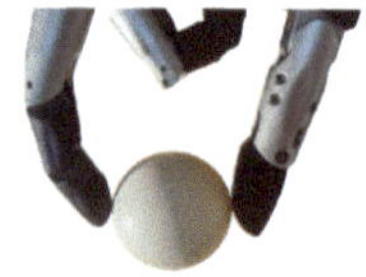	
Cross-section size	**0–50 mm**	**50–150 mm**
Slice		

Based on the above analyses of environment perception and grasping posture, this paper designed a novel control strategy for the prosthetic hand. When there are multiple objects in the environment, the prosthetic hand identifies the object located in the middle, so the subsequent gesture assignment is based on the information of the center object. The core formula of this control method is shown in Equation (16), where P is the result of the operation of the manipulator, $State_{plam}$ is the grasping gesture of the prosthetic hand, which is a function of the shape, size, and whether to press or not to identify the object, taking the value of one of the seven kinds of actions plus the default action in Table 2, which are $\{s_1, s_2, s_3, s_4, s_5, s_6, s_7, s_8\}$. And Pr takes the value of $\{0, 1\}$, which means no pressing and pressing insole, respectively; if pressing, then this means a need to release the grasping object, $State_{plam}$ takes a value s_8, if no pressing, then this means ready to grasp, according to the shape and size of the object corresponding to $s_1 - s_7$. I is the manipulation intention of the hand-helper system, which takes a value $\{0, 1\}$, $I = 0$ means manipulate manipulator, $I = 1$ means no intention.

To manipulate the manipulator, there are two ways to activate the manipulation intention of the prosthetic hand in this paper, i.e., the user actively presses the insole ($Pr = 1$) or the system automatically recognizes ($Pr = 0$). If the user presses the insole, $I = 1$, and if the system recognizes automatically, the expression of I is $I = \mu * G(X, Y)$, where μ takes the values of $\{0, 1\}$, which represent that the upper limb is in the motion or static state, respectively, and $G(X, Y)$ is the function of X and Y. X represents the position of the end of the hand, which takes the values $\left\{Pos_{init}, Pos_{front}, Pos_{near}\right\}$ and represents that the hand is in the initial state, and $G(X, Y)$ is the function of X and Y. X represents the position of the end of the hand, and the values $\left\{Pos_{init}, Pos_{front}, Pos_{near}\right\}$, which represent the hand in the initial state, hand in front of the torso, hand near the torso, respectively. Y represents the existence of an object in front or not, taking the value of {0,1}, and $G(X, Y)$ takes the value of $\{0, 1\}$, taking 1 when and only when $X = Pos_{front}$ and $Y = 1$, and 0 otherwise.

$$\begin{cases} P = State_{plam} * I \\ State_{plam} = \begin{cases} F(Size, Category)\ , Pr = 0 \\ s8\ , Pr = 1 \end{cases} \\ I = \begin{cases} \mu * G(X, Y)\ , Pr = 0 \\ 1\ , \ Pr = 1 \end{cases} \\ G(X, Y) = \begin{cases} 0\ , & REST \\ H(X, Y), & X = Pos_{front}\ and\ Y = 1 \end{cases} \end{cases} \tag{16}$$

5. Experiments and Discussion

5.1. Shape and Size Recognition of Object Sections

In this paper, Lidar is used to obtain the cross-sectional shape and size of objects. To test the recognition effectiveness of the method, this section used Lidar to sense complex environments with multiple objects, as shown in Figure 14. The experimental environment contained common fruit and vegetable stalls and supermarket shelves scenes.

Figure 14. Experimental environment.

In Section 3.2, cross-sectional shapes of common objects were classified into three categories: circular, straight-line segment, and L-shape. Circular cross-sections are found in objects like cylinders and spheres, straight-line segments in prisms (front view), and L-shapes in prisms (oblique view). Three types of circular cross-section objects (kiwi, radish, snacks, numbered 1–3), three types of straight-line segment objects (orthogonal food boxes, paper towel packets, numbered 4–6), and two types of L-shaped objects (oblique boxes, numbered 7–8) were recognized over 10 trials, with the results summarized in Table 3.

Table 3. Recognition results of eight kinds of object cross-sections.

Object Type	Object No.	Object Size (mm)	Positioning Successes	Number of Successful Shape Recognition	Average Recognized Size (mm)	Shape Recognition Rate	Dimension Tolerance (mm)
Circular	1	43	10	10	46	100%	3
	2	71	10	10	76	100%	5
	3	108	10	9	119	90%	11
Straight-line	4	11	10	10	12	100%	1
	5	34	10	10	36	100%	2
	6	71	10	10	73	100%	2
L-shape	7	70	10	9	63	90%	7
	8	95	10	9	84	90%	11
Average						96%	5

In Table 3, "Object Size" refers to the cross-sectional grasping size, "Successful Positioning" is the number of times the object was located, "Successful Shape Recognition" is the number of times the cross-sectional shape was correctly identified, "Average Recognition Size" is the mean size across 10 trials, "Shape Recognition Rate" is the ratio of successful recognitions, and "Dimensional Tolerance" is the absolute difference between the average recognition size and the actual object size. The results show that the method effectively identified objects of various types and sizes, with the dimensional tolerance within acceptable limits.

5.2. Integrated Control Experiment for Prosthetic Hand

To verify the applicability and feasibility of the integrated control method of the prosthetic hand, this experiment used the prosthetic hand to complete upper limb actions. To make the experimental results more reliable and universal, this section designed the experimental actions in Table 4 based on different scenarios and objects of different shapes and sizes.

Table 4. Experimental actions and action flow.

Action	Action Flow
Drink water	Upper limb moves towards the cup→recognizes the cup→grasps and moves to the mouth→retracts and releases
Fruit-eating	Move the upper limb towards the fruit stand→select and recognize a fruit→grasp and move to the mouth→release
Call	Upper limb moves towards the mobile phone→recognizes the mobile phone→grasps and moves to the ear→retracts and releases it
Shop	Upper limb moves towards the shelf→selects and recognizes the product→grasps and moves it to the shopping cart→releases it
Wear a hat	Upper limb moves to near the hat→toes press the insole→grasps the hat and moves it to the head→releases it

Drinking water, eating fruit, making a phone call, and shopping were the actions where the system automatically identified the object and assigned a grasping gesture. These actions roughly followed the following process: first, the upper limb moved to the vicinity of the object, the IMU detected the upper limb motion state. Second, the Lidar and the camera located and identified the object, then the prosthetic hand grasped the object according to the corresponding gesture. Finally, the prosthetic hand was released by pressing on the insole. For putting on a hat, the prosthetic hand was actively controlled by the user, with the following process. The process was as follows: The upper limb moved towards the hat, the IMU detected the upper limb motion state, the user took the initiative to press the insole to control the prosthetic hand to grasp the hat and put it on the head, finally pressing the insole to release the prosthetic hand.

In the experiment, 10 volunteers were invited to complete the above actions, and each volunteer wore the experimental equipment to repeat each action 20 times using the above action process, after receiving training. The experimental process and data were recorded. The equipment required for the experimental process is shown in Figure 1, where the volunteers wore flexible pressure sensing insoles on the soles of their feet, inertial sensing units on their upper limbs, and robotic hands, as well as a 2D Lidar and a monocular camera, to simulate the use of the environment for people with upper limb disabilities. Then, the experimental process was described in detail, limited by space, focusing on showing the flow of shopping actions as shown in Figure 15. The actions of the shopping process for supermarket fruits were used as an example to analyze the behaviors and data of the experimental process.

Shopping for fruits from a fruit pile is a common scene in daily life, volunteers wearing the prosthetic hand control device completed the action from the initial state to selecting and grasping the object to putting the object into a bag, recording the key steps of the experimental process as well as the data of multiple sensors. The experimental process of spherical fruit purchasing is shown in Figure 15a. The 2D Lidar recognized the point cloud data of the object and the recognition results are shown in Figure 16b. The results of the camera collecting the fruit pile object and recognizing the shape of the object are shown in Figure 17a, and the calibration between the monocular camera and the two-dimensional Lidar is shown in Figure 17b. The user completed the complete fruit object sourcing action, which can be decomposed into five main steps $A_1 - A_5$ as shown in Figure 15a. During the A_1 time, both the prosthetic hand system and the volunteer were ready and in the initial state, during the A_2 time, the upper limb drove the robotic hand to gradually move towards the fruit pile, the inertial sensing unit monitored the movement state of the upper limb in real time, and the upper limb was in the stationary state when detected. When the upper limb was in a stationary state and the palm of the hand was located in front of the body, the 2D Lidar started to detect whether there was an object near the palm of the hand, and when an object was detected, the 2D Lidar and the monocular camera were activated to identify the size and shape of the object, respectively, and the point cloud obtained from the scanning of the environmental objects by the 2D Lidar was synthesized from multiple frames, as shown in the above picture in Figure 16. The point cloud was pre-processed with noise reduction and DBSCAN clustering to obtain the four fruits shown in the following picture in Figure 16. The four fruits shown in the following picture are marked by different colors and the target object was located based on the rule that the target object is usually located in the center, i.e., the red circled object shown in Figure 16. The system recognized that the cross-section had a circular shape and its size was 78.34 mm, and the system obtained the calibration result shown in Figure 17b according to the 2D Lidar–monocular camera calibration algorithm. The system obtained the position of the target object in the image through coordinate mapping and used the object shape recognition model trained by YOLOv8 to identify the shape of the target object as a sphere, while the grasping strategy for larger-sized spherical objects was as shown in Table 2, i.e., using five fingers to try to encircle the object. During A_3 time, the system controlled the manipulator to execute the action according to the assigned grasping strategy. During A_4 time, the system completed the grasping of the object, and then the upper limb drove the manipulator towards the inside of the shopping bag. At time A_5, the manipulator placed the fruit inside the shopping bag, and then the system controlled the manipulator to release and return to the initial state, after recognizing the user's lesser toe pressing down on the flexible pressure-sensing insole. At this point, the complete process of the spherical fruit purchasing action was completed. During the whole process, the inertial sensing unit and the flexible pressure sensing insole accurately identified the gripping timing of the robot, the monocular camera, and the 2D Lidar successfully perceived the environmental information, located the target object, and identified the size and shape of the target object.

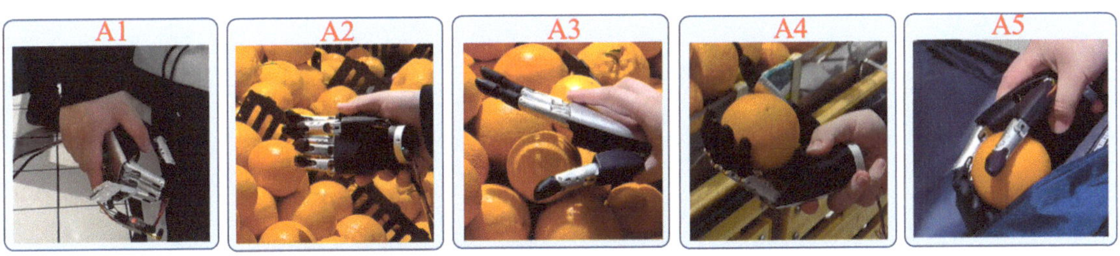

(**a**) Experimental process of spherical fruit procurement

(**b**) Experimental process for procurement of commodities from supermarket shelves

Figure 15. *Cont.*

(**c**) Experimental flow of columnar vegetable procurement

Figure 15. Shopping experiment process.

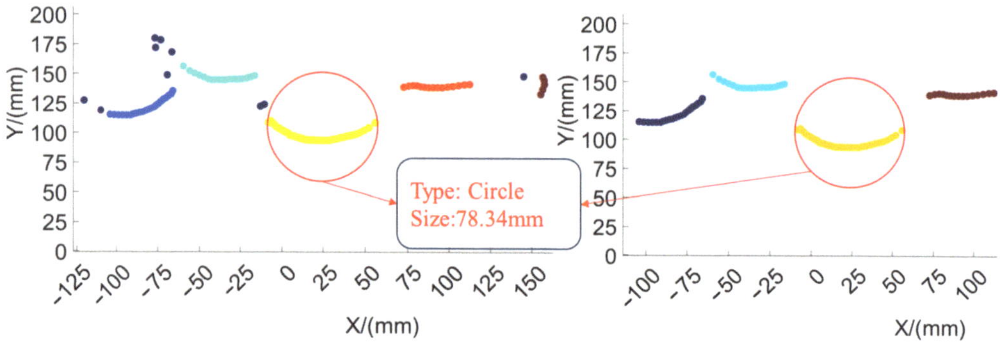

(**a**) Multi-frame composite point cloud (**b**) Pre-processed point cloud

Figure 16. Spherical fruit size recognition results.

(**a**) Fruit pile object shape recognition (**b**) Point cloud image calibration

Figure 17. Fruit object recognition and calibration.

The volunteers completed all the actions shown in Table 4, the experimental data and completion of each experiment were recorded, and the experimental results are displayed in Table 5. Here, Num indicates the number of times that the volunteers completed the corresponding actions in total, PRN indicates the number of times that the inertial sensing unit correctly recognized the position of the hand for all the times of the experiments, PRR indicates the ratio of the number of times that the upper limb position was recognized, TON indicated the number of times the Lidar and camera accurately located and identified the object in all the experiments, TOR is the ratio of the target object recognitions to the number of experiments, FPN indicates the number of times the flexible pressure sensing insole successfully identified the lesser toe pressing action, FPR is the ratio of the number of lesser toe pressing recognitions to the number of experiments, ACN indicates the number of times the corresponding action was completed for all times of the experiments, and SR represents the ratio of the number of times the action was completed to the number of experiments, and the average values of each recognition rate and completion rate are also calculated in the table.

Table 5. Experimental results of the integrated control of the prosthetic hand.

Action	Num	PRN	PRR	TON	TOR	FPN	FPR	ACN	SR
Drink water	200	189	94.50%	185	92.50%	198	99.00%	185	92.50%
Fruit-eating	200	188	94.00%	182	91.00%	199	99.50%	181	90.50%
Call	200	190	95.00%	184	92.00%	197	98.50%	184	92.00%
Shop	200	186	93.00%	183	91.50%	197	98.50%	183	91.50%
Average			94.13%		91.75%		98.88%		91.63%
Wear a hat	200					198		196	98.00%

The following conclusions can be drawn from the experimental results shown in Table 5.

(1) The position of the upper limb and the recognition of the motion state of the upper limb represented the action intention of the upper limb, and the average accuracy of its recognition was 94.13%, and the recognition rate of the different actions did not differ much, which met the requirements of the different user scenarios. This demonstrates that the method of recognition using the intention of the upper limb in the system has a high degree of feasibility and universality.

(2) Target recognition represents the environment perception ability of the system, which used 2D Lidar to recognize the size of the object and a monocular camera to recognize the shape of the object, and the average value of the target recognition rate was 91.75%, recognizing most of the scenes with good accuracy. The target recognition rate was similar for different scenes, indicating that the object recognition algorithm of the present system has good usability and generality.

(3) The recognition of the lesser toe pressing action represents the system's ability to recognize the user's active human–computer interaction intention. First, it had a high success rate for drinking water, eating fruits, making phone calls, shopping, and other scenarios, with an average recognition rate as high as 98.88%. Second, it also had a recognition rate as high as 98% in the hat wearing scenario, which relied on active human–computer interaction to control the prosthetic hand, which indicates that the user's lesser toe pressing on the flexible pressure sensing insole was capable of controlling the prosthetic hand system efficiently. This indicates that the user could control the prosthetic hand system efficiently by pressing the flexible pressure-sensing insole with his/her lesser toe.

(4) The success rate represents the rate at which each step in the whole action process was completed, and the average success rate was 91.63% for the scenario of automatic recognition and sensing by the system. The success rate was similar in the different scenarios, and the recognition rate was 98% for the scenario that relied on active human-computer interaction only, which demonstrates that the system designed in this paper has a high degree of feasibility and universal applicability.

(5) Due to the close cooperation of many sensors in the scene of automatic recognition and perception, the final success rate was not higher than the lowest success rate in each link, and at the same time, the target recognition was affected by the variety of complex environments, and the recognition rate was generally the lowest success rate in each link, which could be improved by adapting more point cloud scenes and increasing the number of picture training sets. This will be improved in future research.

6. Conclusions

In this study, we proposed a new method for recognizing the intention of upper limb movements: detecting the scale information of a grasped object based on the visual fusion of information from a miniature Lidar and pinhole camera, so as to automatically control the grasping posture of the manipulator; controlling the flow of grasping movements based on arm motion and toe haptic information. The hardware device in this paper is suitable for integrated mounting on the finger and also does not need to be in contact with the human skin, a feature that provides a great improvement in comfort and aesthetics compared to myoelectric- and electroencephalographic-controlled manipulators; the grasp and release commands of the manipulator used a human-in-the-loop decision-making method, which provides better interference immunity and reliability compared to manipulators controlled by signals from myoelectrics, EEGs, and speech. In our future research, we plan to more finely classify object shapes and grasping postures, and further delve into the stability and smoothness of the grasping behavior of the prosthetic hand. In addition, in the course of our research, we found that the comfort of the prosthetic hand was very important, all the current robotic hand for upper limb disabled people must be installed on the disabled person's residual limb skin through the prosthetic limb, which will cause discomfort such as sweating, skin inflammation, and will also cause pain in the skin due to the friction, and the implantable prosthesis of the upper limb may be a fundamental way to solve this problem.

Author Contributions: Conceptualization, J.C.; methodology, J.C. and B.Y.; software, B.Y.; validation, B.Y.; formal analysis, B.Y.; investigation, B.Y.; resources, B.Y.; data curation, B.Y.; writing—original draft preparation, B.Y.; writing—review and editing, J.C.; visualization, B.Y.; supervision, J.C.; project administration, J.C.; funding acquisition, J.C. All authors have read and agreed to the published version of the manuscript.

Funding: This research was funded by the National Natural Science Foundation of China, grant number 61873063.

Institutional Review Board Statement: All subjects gave their informed consent for inclusion before they participated in the study. The study was conducted in accordance with the Declaration of Helsinki, and the protocol was approved by the Ethics Committee of XDJK2020B029, SWU41015718, SWU20710953.

Informed Consent Statement: Informed consent was obtained from all subjects involved in the study. Written informed consent has been obtained from the patients to publish this paper.

Data Availability Statement: The data presented in this study are available on request from the corresponding author.

Conflicts of Interest: The authors declare no conflict of interest.

References

1. Huang, P. Study on the Key Technologies of Motion Intention Sensing and Recognition of Intelligent Lower Limb Prostheses. Ph.D. Thesis, The University of Chinese Academy of Sciences, Beijing, China, 2020.
2. Fleming, A.; Stafford, N.; Huang, S.; Hu, X.; Ferris, D.P.; Huang, H.H. Myoelectric control of robotic lower limb prostheses: A review of electromyography interfaces, control paradigms, challenges and future directions. *J. Neural Eng.* **2021**, *18*, 041004. [CrossRef] [PubMed]
3. Xu, B.; He, X.; Wei, Z.; Song, A.; Zhao, G. Research on continuous control system for Robot based on motor imagery EEG. *Chin. J. Sci. Instr.* **2018**, *39*, 10–19.
4. Nagai, K.; Goto, T.; Shimizu, T.; Dobashi, H.; Ito, K.; Hayashi, Y.; Loureiro, R.C.; Nasuto, S.J.; Harwin, W.S. Feasibility study on EEG driven robotic system to realize efficient stroke rehabilitation. In Proceedings of the 2015 IEEE International Conference on Rehabilitation Robotics (ICORR), Singapore, 11–14 August 2015; pp. 199–204.
5. Blana, D.; Van Den Bogert, A.J.; Murray, W.M.; Ganguly, A.; Krasoulis, A.; Nazarpour, K.; Chadwick, E.K. Model-based control of individual finger movements for prosthetic hand function. *IEEE Trans. Neural Syst. Rehabil. Eng.* **2020**, *28*, 612–620. [CrossRef] [PubMed]
6. Mishchenko, Y.; Kaya, M.; Ozbay, E.; Yanar, H. Developing a 3-to 6-state EEG-based brain-computer interface for a virtual robotic manipulator control. *IEEE Trans. Biomed. Eng.* **2019**, *66*, 977–987. [CrossRef] [PubMed]
7. Yang, B.; Peng-cheng, H.; Qing-hua, Y. Design of dexterous hand and its control system based on myoelectric control. *J. Mech. Electr. Eng.* **2016**, *33*, 388–394.
8. Tavakoli, M.; Benussi, C.; Lourenco, J.L. Single channel surface EMG control of advanced prosthetic hands: A simple, low cost and efficient approach. *Expert Syst. Appl.* **2017**, *79*, 322–332. [CrossRef]

9. Tigra, W.; Navarro, B.; Cherubini, A.; Gorron, X.; Gelis, A.; Fattal, C.; Guiraud, D.; Coste, C.A. A novel EMG interface for individuals with tetraplegia to pilot robot hand grasping. *IEEE Trans. Neural Syst. Rehabil. Eng.* **2016**, *26*, 291–298. [CrossRef] [PubMed]
10. Minsang, S.; Dukchan, Y.; Junghoon, K.; Youngjin, C. EMG-based Prosthetic Hand Control System Inspired by Missing-Hand Movement. In Proceedings of the 12th International Conference on Ubiquitous Robots and Ambient Intelligence (URAI 2015), Goyang, Republic of Korea, 28–30 October 2015; pp. 28–30.
11. Shima, K.; Fukuda, O.; Tsuji, T.; Otsuka, A.; Yoshizumi, M. EMG-based control for a feeding support robot using a probabilistic neural network. In Proceedings of the 2012 4th IEEE RAS & EMBS International Conference on Biomedical Robotics and Biomechatronics (BioRob), Rome, Italy, 24–27 June 2012; pp. 1788–1793.
12. Segil, J.L.; Huddle, S.A. Functional assessment of a myoelectric postural controller and multi-functional prosthetic hand by persons with trans-radial limb loss. *IEEE Trans. Neural Syst. Rehabil. Eng.* **2016**, *25*, 618–627. [CrossRef] [PubMed]
13. Shaoyan, D.; Yishuo, J. Research on MEMS sensor technology industry and the roadmap of China. *Inf. Commun. Technol. Policy* **2021**, *47*, 66.
14. Su, B.-Y.; Wang, J.; Liu, S.-Q.; Sheng, M.; Jiang, J.; Xiang, K. A CNN-based method for intent recognition using inertial measurement units and intelligent lower limb prosthesis. *IEEE Trans. Neural Syst. Rehabil. Eng.* **2019**, *27*, 1032–1042. [CrossRef] [PubMed]
15. Wilson, S.; Vaidyanathan, R. Upper-limb prosthetic control using wearable multichannel mechanomyography. In Proceedings of the 2017 International Conference on Rehabilitation Robotics (ICORR), London, UK, 17–20 July 2017; pp. 1293–1298.
16. Noccaro, A.; Cordella, F.; Zollo, L.; Di Pino, G.; Guglielmelli, E.; Formica, D. A teleoperated control approach for anthropomorphic manipulator using magneto-inertial sensors. In Proceedings of the 2017 26th IEEE International Symposium on Robot and Human Interactive Communication (RO-MAN), Lisbon, Portugal, 28 August–1 September 2017; pp. 156–161.
17. Merad, M.; Roby-Brami, A.; Jarrassé, N. Towards the implementation of natural prosthetic elbow motion using upper limb joint coordination. In Proceedings of the 2016 6th IEEE International Conference on Biomedical Robotics and Biomechatronics (BioRob), Singapore, 26–29 June 2016; pp. 821–826.
18. Cui, J.; Li, Z. Prediction of Upper Limb Action Intention Based on Long Short-Term Memory Neural Network. *Electronics* **2022**, *11*, 1320. [CrossRef]
19. Martin, H.; Donaw, J.; Kelly, R.; Jung, Y.; Kim, J.-H. A novel approach of prosthetic arm control using computer vision, biosignals, and motion capture. In Proceedings of the 2014 IEEE Symposium on Computational Intelligence in Robotic Rehabilitation and Assistive Technologies (CIR2AT), Orlando, FL, USA, 9–12 December 2014; pp. 26–30.
20. Ghazaei, G.; Alameer, A.; Degenaar, P.; Morgan, G.; Nazarpour, K. Deep learning-based artificial vision for grasp classification in myoelectric hands. *J. Neural Eng.* **2017**, *14*, 036025. [CrossRef] [PubMed]
21. Wu, Y.; Wang, Y.; Zhang, S.; Ogai, H. Deep 3D object detection networks using LiDAR data: A review. *IEEE Sens. J.* **2020**, *21*, 1152–1171. [CrossRef]
22. Li, Y.; Fan, Q.; Huang, H.; Han, Z.; Gu, Q. A Modified YOLOv8 Detection Network for UAV Aerial Image Recognition. *Drones* **2023**, *7*, 304. [CrossRef]
23. Ge, P.; Wang, Y.; Wang, B. Universal calibration for a ring camera array based on a rotational target. *Opt. Express* **2022**, *30*, 14538–14552. [CrossRef] [PubMed]

biomimetics

Article

Design and Optimization of a Custom-Made Six-Bar Exoskeleton for Pulp Pinch Grasp Rehabilitation in Stroke Patients

Javier Andrés-Esperanza *, José L. Iserte-Vilar and Víctor Roda-Casanova

Department of Mechanical Engineering and Construction, Universitat Jaume I, 12071 Castellón, Spain; jiserte@uji.es (J.L.I.-V.); vroda@uji.es (V.R.-C.)
* Correspondence: fandres@uji.es; Tel.: +34-964728143

Abstract: Stroke often causes neuromotor disabilities, impacting index finger function in daily activities. Due to the role of repetitive, even passive, finger movements in neuromuscular re-education and spasticity control, this study aims to design a rehabilitation exoskeleton based on the pulp pinch movement. The exoskeleton uses an underactuated RML topology with a single degree of mobility, customized from 3D scans of the patient's hand. It consists of eight links, incorporating two consecutive four-bar mechanisms and the third inversion of a crank–slider. A two-stage genetic optimization was applied, first to the location of the intermediate joint between the two four-bar mechanisms and later to the remaining dimensions. A targeted genetic optimization process monitored two quality metrics: average mechanical advantage from extension to flexion, and its variability. By analyzing the relationship between these metrics and key parameters at different synthesis stages, the population evaluated is reduced by up to 96.2%, compared to previous studies for the same problem. This custom-fit exoskeleton uses a small linear actuator to deliver a stable 12.45 N force to the fingertip with near-constant mechanical advantage during flexion. It enables repetitive pulp pinch movements in a flaccid finger, improving rehabilitation consistency and facilitating home-based therapy.

Keywords: exoskeleton; optimization; rehabilitation

check for updates

Citation: Andrés-Esperanza, J.; Iserte-Vilar, J.L.; Roda-Casanova, V. Design and Optimization of a Custom-Made Six-Bar Exoskeleton for Pulp Pinch Grasp Rehabilitation in Stroke Patients. *Biomimetics* **2024**, *9*, 616. https://doi.org/10.3390/biomimetics9100616

Academic Editor: Rafhael Milanezi de Andrade

Received: 15 September 2024
Revised: 2 October 2024
Accepted: 9 October 2024
Published: 11 October 2024

1. Introduction

In the last 20 years, the number of stroke cases among people aged 20 to 64 years has increased by 25%. Stroke often leads to neuromotor disabilities that can impair finger movement. Since finger movement is crucial for basic activities of daily life (ADL), there is significant motivation to prioritize finger rehabilitation following injury or stroke [1]. It is well established that repetitive flexion and extension movements of the fingers, even when performed passively, can promote neuromuscular re-education, help prevent spasticity, and manage pain associated with hand paralysis resulting from acquired brain damage [2,3].

In view of the consequences and to help with post-stroke recovery, autonomous solutions focused on the functional rehabilitation of the hand's prehensile capabilities are being developed, among which are hand exoskeletons. Exoskeletons may offer a more effective option than end-effector robot-assisted devices for treating finger motor impairment in stroke patients [4]. These exoskeletons are structures that totally or partially cover the affected hand, and according to the functions they can perform, a distinction can be made between augmentative or assistive (improving capabilities for the execution of specific tasks) for teleoperation (haptic devices oriented to virtual models) and hand exoskeletons for rehabilitation (RHEx) [5]. The RHEx try to recover the motor abilities of patients by forcing repetitive actions on the affected joints: the most common ones are opening and closing the fingers. In this way, it is possible to lengthen the sessions and reduce the physicians' permanent need for assistance. RHEx are not necessarily portable or lightweight; they can be fixed in a certain position. Even so, they are more versatile if they are portable because the patient can take them home to continue rehabilitation on their own. Moreover, the rapid progress of flexible electronics aligns with the growing demand for portable smart exoskeletons embedded with skin-interfacing flexible electronics (e-skin) [2,6,7] provide fine-grained control and feedback RHEx, allowing for continuous and real-time monitoring of physiological signals. This helps users regain movement and improve strength in affected extremities. Still, the lightness and pragmatism of the exoskeleton topology must be optimized.

The goal population largely conditions the suitability of an RHEx design. The movements allowed by the RHEx should follow patients' behavior naturally and without generating discomfort. The literature [5] compiles characteristics and criteria that should guide the design of RHEx. However, the evaluations presented in RHEx proposals only include partial aspects, on a few subjects, and rarely patients [2]. In this sense, the collaboration with rehabilitators in the context of the present research (DERAPPI project [8]) has made clear the characteristics of the patient that are suitable for therapy with RHEx:

- Plegic hand (MRC 0 [9,10] in all muscle groups of the hand).
- Flaccid hand (spasticity up to MAS 1 [11] would be admitted).
- Absence of rigidity in the fingers. In case of some rigidity, it cannot interfere with normal intelligent grasping movement.
- Skin is intact and there is no edema at the level of the hand.
- Cognitive status enough to follow simple instructions.
- Behaviorally stable.

For certain patients, it is essential to restrict the range of motion of specific joints to prevent articular degeneration. This goal is difficult to achieve with some RHEx designs inspired by soft robotics. Additionally, there is a lack of devices on the market that simultaneously target the mobilization of the index finger and are easy for either the physician or the patient to place on the hand [12]. Focusing on the recovery of the index finger is crucial since it plays a key role in many grasps and ADL. Regarding RHEx control, devices operated by a pushbutton activated by the healthy hand or programmed to function in a loop are highly suitable for stroke rehabilitation. Applying the concepts of robustness and simplicity makes it desirable to be able to use a simple control rather than having to command and coordinate several actions. The idea behind underactuation in robotics is to use an ingenious mechanical system that can adapt to the requirements of a grasp or a trajectory with one actuator [13]. This mechanical intelligence is commonly found in mechanical linkages where the different link lengths and joint types are determined at the design stage to follow a particular trajectory or adopt a particular posture. If this trajectory is entirely predetermined, then only one DOF suffices to follow it. Moreover, using a single actuator contributes to maintaining the lightweight and compact design of the device, which is desirable in a portable RHEx; therefore, the concept of using a single actuator for underactuation also merits consideration.

Several RHEx topologies can be discerned [12,14]:

(i) Those restraining each phalanx (Figure 1a–c), using mechanisms to match the axes of rotation with those of the phalanges. They strictly limit spasticity during rehabilitation. However, the complexity of the design increases as higher finger mobility is sought.

(ii) Those that hold only the fingertip (base-to-distal topology, Figure 1d), with the advantage that their design is simpler, as they do not have the restriction of matching the axes of rotation, so they can be lighter as well. Note that the three degrees of freedom that position and orient the distal phalange (DP) within the plane of flexion of the finger univocally condition the three degrees of freedom provided by the three joints: metacarpophalangeal (MCP), proximal, and distal interphalangeal (PIP, and DIP). However, the movement of the finger is less controlled, regarding the variety of exercises that can be performed by acting on the individual joints.

(iii) Those that fit over the entire finger, glove type (Figure 1e), using flexible and lightweight materials that adapt to the hand (soft robotics). They are less cumbersome and less expensive, but care must be taken in the forces applied.

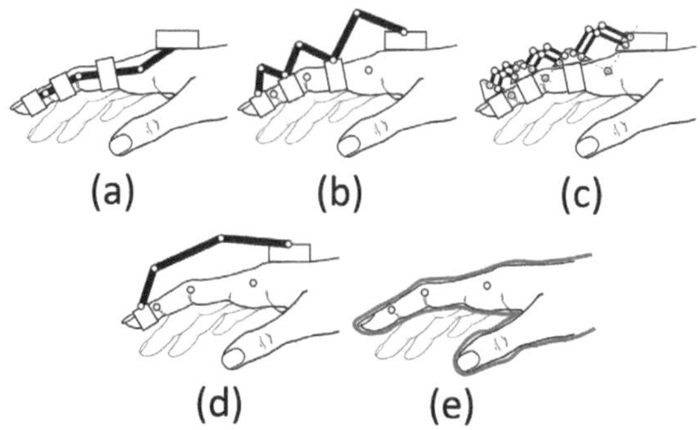

Figure 1. Schematics of various RHEx topologies for transmitting motion to the patient's fingers: (**a**) matched axes; (**b**) redundant linkage; (**c**) remote center of motion; (**d**) base-to-distal; and (**e**) compliant [12,14].

The choice between these approaches depends on the specific goals of the device and the patient's needs in terms of rehabilitation. In the case of soft robotics (iii), having no rigid joints to guide torques and forces, the user's skeleton becomes the guiding structure, and it may not be an appropriate choice in dealing with the patient's spasticity. The RHEx based on articulated mechanisms guide the movement based on a guiding structure (an articulated linkage that performs the function of guiding the movement) and a guiding chain (a system that transmits the actuation, usually a Bowden cable, electrical, pneumatic, or a system of additional links to the first ones [15]).

Control for RHEx can be relatively simple if they are aimed at task-specific training, allowing for passive control that overlooks the patient. For case (i), if independent control is desired for each degree of mobility of the hand, individual actuators will be required to adjust and control the movements of each finger independently. However, the design of the exoskeleton can be greatly simplified, and the cost of the device can be reduced by using sub-actuated designs, especially in case (ii), which makes them very economical and easy to transport. In this case, the RHEx may actuate on some degrees of freedom, leaving the others to move passively, i.e., allowing certain joints to freely adopt the position that best suits them, within their anatomical ranges of motion, to achieve the posture demanded by the exoskeleton.

In cases (i) and (ii), the location of the mechanism can be either dorsal (located on the dorsal part of the fingers, Figure 1b–d), palmar (on the palmar part of the fingers, hardly used because they interfere with interacting with objects [16]), or lateral (Figure 1a). Lateral devices leave the palm of the hand free and therefore allow objects to be grasped. In addition, they allow patients to see their hands, which is important for visualizing progress during rehabilitation. In any of the three arrangements, the mechanism is attached to the hand via Velcro®, flexible straps, or thimbles.

In recent years, different RHEx designs have been proposed in the literature [2,14,15,17], and there are already some commercial designs for diverse purposes [16,18]. There are very few RHEx reported in the literature that reach TRL level 9 [19,20], and, even though much research is being carried out on them, they are still far from being a practical solution. In a recent extensive literature search on the design and optimization of exoskeletons (over 722 studies from 2017 to 2023), it was shown that evolutionary computation (EC) methods are the most frequently used (genetic algorithms, particle swarm optimization, differential evolution, and evolutionary strategies) compared to other non-EC [21]. This outcome is anticipated given EC methods' numerous attributes, such as the ability to update multiple solutions simultaneously (population-based), their independence from gradient information, rapid convergence, and their capacity to tackle mixed optimization problems, including multimodal and multi-objective scenarios, to optimize different metrics (mainly force transmission (FT), workspace, weight, and size). However, this research found only two studies in which EC techniques were used for hand exoskeletons: one with an assistive application [22] and another for rehabilitation (RHEx) [23]. Close to these dates, we also found other four studies using EC for improving an RHEx [24–27]. Table 1 summarizes the studies related to the optimization of hand exoskeletons from 2014 to 2024.

Table 1. Hand exoskeletons optimized with evolutionary (EC) or non-evolutionary (nEC) computation techniques in terms of application (assistance (AHEx), rehabilitation (RHEx), or haptic (HHEx)), optimization metrics, and optimization method (2014–2023).

Authors	Application	Metrics [1]	Method
Li et al. [22]	AHEx	FT	Elitist Non-Dominated Sorting Genetic Algorithm (EC) [28,29]
Du et al. [23]	RHEx	FT	Single-Objective Genetic Algorithm (EC) [30]
Iqbal et al. [24]	RHEx	KM, CA, GI, FT	Weight-Based Genetic Algorithm (EC) [21]
Vanteddu et al. [25]	AHEx	AC	Weight-Based Genetic Algorithm (EC) [21]
Li et al. [26]	RHEx	M	Elitist Non-Dominated Sorting Genetic Algorithm (EC) [28,29]
Sarac et al. [27]	AHEx	FT	Single-Objective Genetic Algorithm (EC) [30]
Amirpour et al. [31]	HHEx	W, AC	Levenberg–Marquardt Algorithm (nEC) [32,33]
Bianchi et al. [34]	RHEx	FT, S	Levenberg–Marquardt Algorithm (nEC) [32,33]
Liang et al. [35]	AHEx, RHEx	W	Geometric differentiation (nEC) [36]
Xu et al. [37]	RHEx	W	Interior point algorithm (nEC) [38]
Qin et al. [39]	RHEx	AC	Goal attainment method (nEC) [40]
Secciani et al. [41]	AHEx	AC	Interior point algorithm (nEC) [38]

[1] FT: force transmission; W: workspace; S: size; AC: adjustability/calibration; KM: kinematic mapping; CA: collision avoidance; GI: global isotropy index; MI: manipulability index.

Du et al. [23], and Sarac et al. [27], used the Single-Objective Genetic Algorithm [30] to optimize the FT, whereas Iqbal et al. [24] used a Weight-Based Genetic Algorithm (WBGA) [21], the simplest classical method of solving multi-objective optimization problems, which considered factors like kinematic mapping, collision avoidance, and a global isotropy index, and FT. Vanteddu et al. [25] also used WBGA, the cost function being the weighted sum of errors for the end position of the grasp trajectory between the kinematic model and the HUST (Huazhong University of Science and Technology) dataset concerning the joint angles of a natural human finger. Li et al. [26] proposed a multi-parameter multi-objective optimization method (namely, the Elitist Non-Dominated Sorting Genetic Algorithm (NSGA–II) [28,29]) to enhance the three global manipulability measures simultaneously in a wearable index finger rehabilitation exoskeleton.

In those works, which used FT as the metric to observe in EC methods, the improvements achieved in either the method or the results are poorly described: Li et al. [22] worked with two functions, one to obtain the three contact forces via three phalanges to the object, distributed as evenly as possible, and another to maximize the sum of the forces exerted on the index finger phalanges by the proposed assistive exoskeleton. It is important to note that it only focuses on the objective in the final grasping, not on the entire finger travel as required in rehabilitation therapy. Their EC worked with a population of only 100 individuals, with a stall generation limit set to 500. Du et al. [23] focused on the transmission angle, a geometrical concept related to FT and mechanism geometry, but they do not detail the improvements achieved after 1000 generations (without detailing the population). Since the objective is to improve FT, it would be much more logical to work on concepts such as mechanical advantage. Iqbal et al. [24] do not describe their Overall Impact Factor used as a weighted optimization function, and how the FT is included. Sarac et al. [27] used an optimization function based on the sum of the squared torques applied to the MCP and PIP joints in their design. However, it is unclear how this function accounted for the full range of motion and the characteristics of the population studied. No studies were found using multi-parameter multi-objective optimization methods targeting metrics directly related to optimal FT for RHEx, in which the mechanical advantage is highly variable, and it is important to maintain a proper value for the entire range of motion, not only the final posture.

The goal of this work is to design a six-bar linkage intended for implementation in a hand-held RHEx, which can drive the DP from its initial to its final position, mirroring a pulp pinch grasp movement (aka. tip pinch [42]). This grasp has been documented as the most frequent in ADL [43,44]. The study focuses on six-bar mechanisms because this topology is of interest for the conception of RHEx to be placed either on the lateral or on the dorsal side of the index finger, partially mimicking the movement of the phalanges. Extending the design to other fingers would cause a single dorsal four-bar mechanism to interfere with the finger's dorsum. In the authors' previous works [45,46], a structured exploration of the design space for this linkage was performed to obtain good initial values for subsequent genetic optimization, following a methodology similar to that proposed in [47]. In those works, the design space for this linkage considered 10 free parameters. Three different values were selected for each of these variables, resulting in 59,049 different versions of the six-bar linkage. All these combinations were explored, and only 11,533 were regarded as valid. The maximum mechanical advantage was calculated for each valid solution, which was then used to sort and select the optimal design based on it.

In the present work, and for the optimization, we propose a targeted genetic scheme that reduces the number of generational calculations until an optimal solution is found. This scheme consists of two stages with two different targets, namely: first, the optimal location of a specific joint, and then the optimal dimensions of the linkage. Each one of the stages observed two quality metrics: averaged mechanical advantage along the travel, and its variability. These metrics are often conflicting, meaning improving one may worsen another. The solutions to multi-objective problems are frequently expressed as a set of Pareto optimal solutions, where no solution can be said to be better than another without considering trade-offs. The methodology here proposed lies in the field of ranking-based selection strategies with hierarchical objective prioritization. While it does not directly correspond to a widely recognized standard technique, it can be considered a variation of methods like NSGA-II or SPEA2 (Strength Pareto Evolutionary Algorithm 2) [48], where objectives are given relative importance instead of being treated entirely equally. According to our knowledge, it has never been applied to optimize the design of RHEx with such metrics, both the mechanical advantage and its variability, both related to FT.

The following section lists the design requirements from the scope of rehabilitation. Section 2, after meeting the specific requirements of the patient (dimensions and range of movement of the DP) to customize the proposed solution, delves into the mechanism topology selected for the design of the RHEx, as well as the methodology followed for its dimensional synthesis and optimization. Section 3 shows the results obtained in the current research. The final model is proposed at the end of Section 3. The last two sections are devoted to discussion and conclusions.

1.1. Design Requirements

In the context of the DERAPPI project [8], rehabilitators from the Brain Injury Unit of the Hospital La Magdalena (Castellón, Spain) made a detailed list of clinical needs for an RHEx to be integrated into the routine of clinical practice, for the rehabilitation of pulp pinch in subacute stroke patients:

- The thumb should be able to perform the opposition movement, or, if in its defect, keep it in opposition to the index finger, to be able to perform the bidigital pulp pinch.
- The RHEx should leave the palmar side of the hand and fingers clear so as not to interfere with the sensory stimuli generated when grasping an object and not to excessively favor the flexor pattern of the plegic hand.
- The RHEx should leave the wrist joint free to allow for the tenodesis effect in case it is present in the patient, namely, when finger movement involves passive wrist movement.
- Leave the pulp of the thumb and index finger free to allow tactile feedback during movement.
- The RHEx design must be accessible to different hand sizes.

For the RHEx to be safe in stroke patients, some biomechanical restrictions must also be verified in relation to the forces applied by the device: the use of linear actuators is justified compared to a more economical hydraulic system, since they allow for more detailed control of the movement, joint ranges, and pressure exerted in the pulp pinch movement.

From the observations made, it was concluded that the minimum force required to guide the distal phalanx of the index finger should be approximately (not less than) 5 N ($\approx$0.5 kgf) in a closing sequence of 1 s duration.

As listed above, thumb abduction is required to perform the pulp pinch. In the abducted position, the thumb can oppose the fingertips [42]. Since including this movement would add complexity and therefore weight and cost to the exoskeleton, it was decided that the exoskeleton structure itself would keep the thumb permanently in opposition to allow the pinch without any specific actuators for it, thus reducing the number of actuators needed without compromising the rehabilitation movement.

2. Materials and Methods

2.1. Characterization of the Two-Finger Gripper Motion

The present RHEx is intended for patients with reduced hand mobility. In relation to the functional range of motion, there are differences among patients, reaching up to 40 degrees of difference in some cases at the MCP, PIP, and DIP joints' flexion angles [49] due to the different anthropometry between subjects. It can be settled that the most appropriate approach is a customization of the dimensions of the mechanism to allow for a movement as close as possible to the natural movement of each patient. In this context, the use of contact measurement instruments (instrumented gloves, dimensional measurements) is not feasible for this type of patient [50]. Instead, 3D scanning has been shown to be the least invasive with the nature of the patient, as well as providing adequate detail of the anatomy and dimensions of the hand for subsequent adaptation of the mechanism, compared to other abovementioned methods.

Figure 2 shows the 3D-scanned hand of a stroke patient in the two boundary postures of a pulp pinch grasp movement, with the starting posture being the functional resting position [51]. Thumb motion during the pulp pinch has been measured by videogrammetry in numerous studies, indicating that healthy subjects maintain the thumb in opposition throughout the pinch [51,52]. Jahn et al. [53] identified three distinct phases that the joints progress through during the execution of the pulp pinch: the initiation phase, the preshaping phase, and the pinch phase. In the initiation phase, little to no movement occurs. It is important to note that most rehabilitation exercises are designed to begin from the functional resting position [51], as is assumed in this work (Figure 2): the wrist is positioned between 10° and 30° of extension, the thumb is in opposition and abduction, and the finger joints are in semiflexion [54]. This posture is often referenced by splints designed to support the hand in such impairment conditions [55].

The preshaping phase accounts for the majority of the movement and begins with the first noticeable increase in motion. The pinch phase starts when the slope decreases as the movement comes to a stop. Thus, the preshaping phase is the focus of gesture rehabilitation as it predisposes and defines the intended grasp. According to the physicians involved in this study, with the exoskeleton's goal being rehabilitation rather than assistance with ADLs, it was observed that the thumb's range of motion from the functional resting position is minimal. In fact, the index is the first finger to have all joints actively engaged during the preshaping phase, followed by the thumb. The preshaping phase ends in a proximal-to-distal pattern for the index finger, while in the thumb, the last joint to become actively involved is the MCP, which is primarily predisposed when in the functional resting position [53]. Therefore, in this context, it is considered appropriate for rehabilitation to focus initially on the movement of the index finger, with the thumb showing minimal motion or being stabilized within a splint.

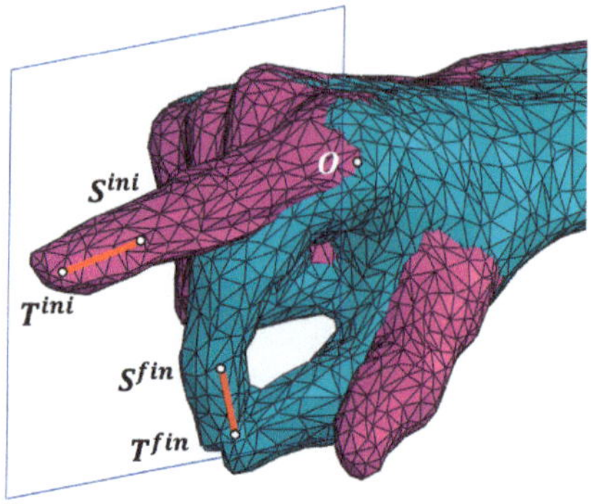

Figure 2. Overlapped 3D-scanned hand in two different postures: extended index (pink) and pulp pinch grasp (green).

The initial position of the DP of the index finger is defined by points S^{ini} (DIP joint) and T^{ini} (tip of the finger), and the final position is defined by points S^{fin} and T^{fin}. Point O is set at the MCP joint on the index as the connection of the first mobile link of the RHEx with the palm (ground bar). Finally, a working plane is set as parallel to the one fitted to $\{S^{ini}, T^{ini}, S^{fin}, T^{fin}\}$ data, minimizing the sum of the squared distances between the points and the plane through a least squares approach [56], at a distance so that it does not interfere with the hand. In this plane, the RHEx will perform its motion. To proceed with the design, we obtain orthogonal projections on this plane of the distal segment (ST) at the initial and final positions and of the O point.

2.2. Linkage Topology

Multiple four-bar linkages, made up of a series of concatenated crossed four-bar mechanisms, are complex assemblies where the motion is transferred through a sequence of interconnected four-bar linkages, enabling the design of intricate mechanical systems with sophisticated movement capabilities. This versatility has led to significant innovation and a variety of designs. Our research group has experience in applying different topologies for the design of prosthetic devices, such as the Toronto/Bloorview/MacMillan (TBM) [57,58], consisting of two coupled four-bar mechanisms. The proximal one is coupled with a crank–slider mechanism in its first inversion (cylinder being fixed), with the crank being part of link 4 and the coupler being part of link 1 (see Figure 3a). This results in having a linear actuator as input, and the proximal four-bar mechanism involves links 1-4-6-7. This inversion of the crank–slider implies having the linear actuator integrated within the ground bar (links 1 and 2 at the dorsum of the hand). The RML (Robotics and Mechatronics Lab of the Virginia Tech University, Blacksburg, Virginia, United States) [59] has a similar topology, although the crank–slider providing a linear input is in its third inversion (the coupler link (1) of the crank–slider is fixed). In this mechanism, the proximal four-bar mechanism also involves links 1-4-6-7, although link 1 is the ground at the dorsum of the hand. The third inversion of the crank–slider gives the designer more freedom to optimize the location of the actuator for improved performance (see Section 2.3.3). In addition, the crank is an extension of link 6 (Figure 3c) instead of link 4 (Figure 3b), which becomes a quaternary link. This last variation has the advantage of leaving more spare space at the level of the hand, preventing any interaction and allowing the user better visual feedback. For all these reasons, RML topology was used for the forthcoming RHEx design. All these topologies have one degree of mobility for the 8 links involved.

As stated earlier, the RML, as the topology here studied, has been used for different proposals of RHEx [37,59–61], although no optimization has been conducted under the requirement of an optimal force transmission [25].

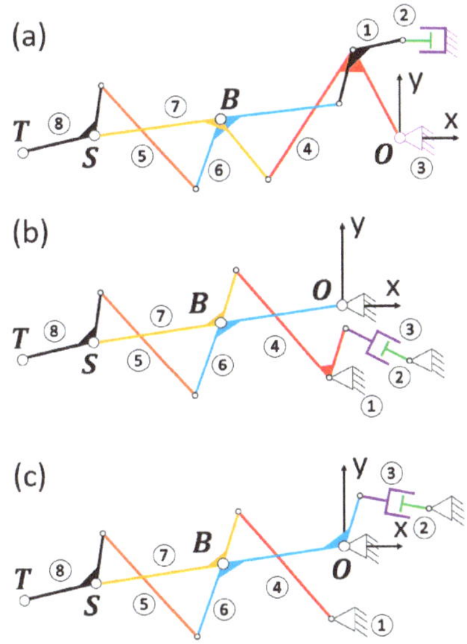

Figure 3. TBM and RML mechanisms compared: (**a**) TBM consisting of two four-bar mechanisms and a crank–slider in the first inversion; (**b**) change of the crank–slider to its third inversion; (**c**) resulting RML mechanism having selected link 4 as the coupler of the crank–slider.

2.3. RHEx Design

2.3.1. Dimensional Synthesis

In the process of synthesizing the two four-bar mechanisms within the linkage, let us consider a recurrent nomenclature for both, the proximal and distal ones concerning the patient's hand (see Figure 4). The segments of each four-bar mechanism are named from L_1 to L_4, indicating with the subscript p or d whether they belong to the proximal or distal mechanism, respectively. Segments L_{2p} (belonging to link 6) and L_{2d} (belonging to link 7) are articulated in B, where both four-bar mechanisms are concatenated through the relative movement of segments L_{1d} (also belonging to link 6) and L_{3p} (also belonging to link 7).

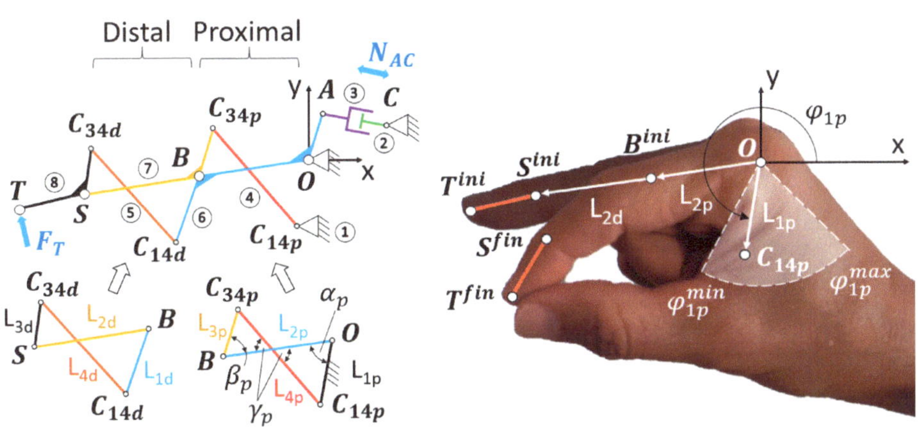

Figure 4. (**Left**) nomenclature of the segments that form the concatenated four-bar mechanisms; (**Right**) region where to locate point C_{14p}, delimited by the maximum and minimum values of φ_{1p} and L_{1p}.

The synthesis to be performed involves two positions (initial and final) of the ST segment corresponding to the pulp pinch grasp (Figure 4). As has been mentioned before, let us designate S^{ini} and T^{ini} as the points that make up this segment in the initial position. Analogously, S^{fin} and T^{fin} correspond to the final posture.

By the inherent nature of the flexion of the finger, the distance between S^{ini} and O is longer than the distance between S^{fin} and O. Therefore, any point B^{ini} that divides the line $S^{ini}O$ into the two segments L_{2p} and L_{2d}, for the construction of the proximal and distal four-bar mechanisms, guarantees the existence of two possible solutions for the flexed finger posture, namely, B^{fin}_{high} and B^{fin}_{low} (see Figure 5a). Of these two, for mimicry with the two-finger pincer movement, B^{fin}_{high} will be considered to continue the synthesis procedure (for the sake of brevity, B^{fin}).

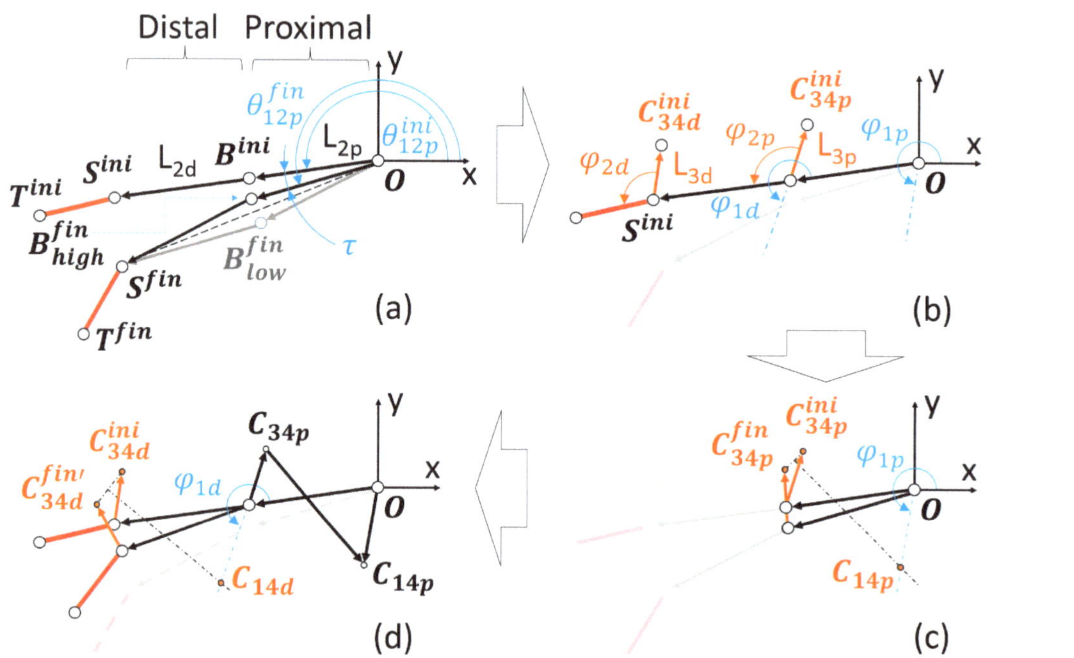

Figure 5. (**a**) Location of B^{ini} and determination of the lengths L_{2p} and L_{2d} and of B^{fin}. (**b**) Selection of the design parameters: L_{3p}, φ_{2p}, φ_{1p}, L_{3d}, φ_{2d}, φ_{1d}. (**c**) Determination of C_{14p} and therefore of L_{1p} and L_{4p}. (**d**) Superimposition of L_{2p} bars in both postures to obtain C_{14d}, L_{1d}, and L_{4d} in an analogous procedure.

Without loss of generality and for further optimization, we started considering a B^{ini} corresponding to the midpoint of the segment $S^{ini}O$. With the lengths L_{2p} and L_{2d} known, values can be proposed for the following design parameters: L_{3p}, φ_{2p}, φ_{1p}, L_{3d}, φ_{2d}, φ_{1d} (see Figure 5b). Due to the functional purpose of the RHEx, maximum and minimum values have been established for these parameters, as shown in Table 2. The first three correspond to one length and two angles of the proximal four-bar mechanism: length L_{3p} and angle φ_{2p} (the latter measured with respect to L_{2d}) allow for the locating of the C_{34p} joint between bars L_{3p} and L_{4p} (in both positions corresponding to extended finger, C^{ini}_{34p}, and flexed, C^{fin}_{34p}). The angle φ_{1p} with respect to the ground bar stipulates the line on which the joint C_{14p}, between bars L_{1p} and L_{4p}, will be located. Figure 4 (right) shows the region where point C_{14p} should be located, delimited by the maximum and minimum values of

φ_{1p} and L_{1p} chosen after observing the anatomical dimensions of the patient's hand. More specifically, the limit values of φ_{1p} are specified in Table 2, while any value obtained after the synthesis for L_{1p}, with a maximum value of 50 mm, is considered valid. This value is based on the subjectivity of not exceeding the margins of the first dorsal interosseous of the patient's hand in the pulp pinch posture.

Table 2. Parameters on which to perform variations across the different tests.

	L_{3p}	φ_{2p}	φ_{1p}	L_{3d}	φ_{2d}	φ_{1d}
Min.	10 mm	90°	230°	4 mm	90°	230°
Max.	25 mm	170°	340°	8 mm	170°	340°

In a graphical synthesis process, C_{14p} is determined by drawing the bisector of the segment that joins the representation joints C_{34p}^{ini} and C_{34p}^{fin} (see Figure 5c). Consequently, the lengths of L_{1p} and L_{4p} are also determined.

We proceed in an analogous way to obtain the distal four-bar mechanism, being φ_{2d} defined with respect to the distal segment (ST). Previously, it was necessary to make a rotation of τ to superimpose the L_{2p} bars of the mechanism in both postures (initial and final) and thus work with $C_{34d}^{fin\prime}$ (see Figure 5a,d). Results of L_{1d} with a maximum value of 50 mm were considered valid. Again, this value is subjectively eligible to avoid invading the palmar space in excess during any grasp.

2.3.2. Optimization Algorithm

The synthesis procedure described in the previous section was programmed in Matlab® R2018b. This same platform was maintained for the subsequent programming of the optimization process, which requires recurrent repetitions of the synthesis process as well as the selection of solutions based on the metrics described below. No Matlab® optimization library was used in order to have full control of the design of operations.

A genetic scheme was adopted for the present study, in two stages (thus having two generations) with two different targets. The first step was to provide discretion to the optimization in order to reduce the study population. To this end, the relevance of the location of the B-joint (B^{ini}, Figure 5a) between the proximal and distal four-bar mechanisms was studied, resulting in a pre-selection. Next, the influence of the variation in the values (dimensions) of the different lengths and angles set for the synthesis of the mechanism was investigated. Figure 6 summarizes the two generations in which, in the first instance, the B^{ini} localization was optimized by studying a population of 24 mechanisms, on which a first selection was made. In the second generation, the mechanisms resulting from varying certain design parameters in the selected mechanisms were studied.

As objective functions, the mean value of the mechanical advantage for the entire range of motion of the exoskeleton, as well as its level of variation with respect to this mean value, were observed. Heuristic multi-objective approaches with dynamic objective prioritization refer to optimization techniques where multiple objective functions are optimized, not treating all objective functions equally (as in typical Pareto-based methods). Instead, the algorithm prioritizes certain objectives over others (hierarchical preference or dynamic prioritization), often in stages, or dynamically (where the objective importance changes based on some conditions or measures). These hierarchical approaches have the advantages of (i) flexibility, as they allow different treatment of objectives based on the stage of optimization or problem-specific priorities; (ii) adaptability, as they help balance exploration and exploitation in complex sceneries; and (iii) simplicity, as they provide an intuitive way based on heuristics [62,63]. Both generations and their optimization details reflected in Figure 6 are explained in detail hereunder.

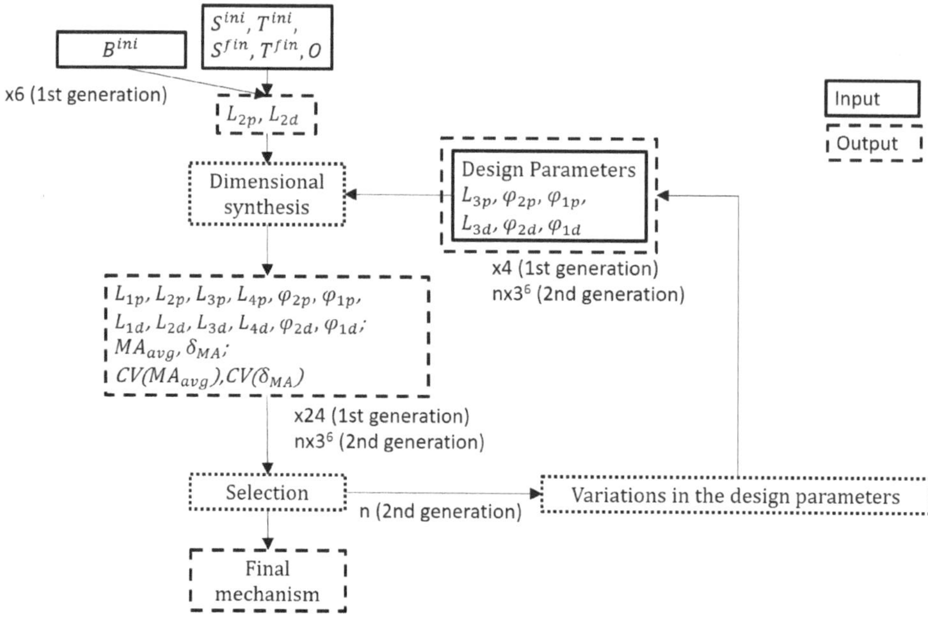

Figure 6. Two generations were studied, for a total of 2211 mechanisms.

1. First Generation: Optimization of the location of the joint B

In the first stage, the selection of point B^{ini} was reconsidered. For this, and in the posture corresponding to the extended finger (Figure 4-right), we could determine a series of possible relocations of point B^{ini} in polar coordinates, namely, describing a circumference around the first B^{ini}-hypothesis, at a radius $R = 5$ mm, and every 45°. In any of these other relocations of B^{ini}, the initial posture is no longer undetermined, and it is easy to observe that those options for point B^{ini} that result in a configuration above the line $S^{ini}O$ maintain that configuration throughout the movement of the mechanism. The same is true for relocations of B^{ini} that result in a configuration below the line $S^{ini}O$. These solutions are discarded because they offer final configurations (flexed finger) that could interfere with some of the rehabilitation activities (see $B^{fin}_{225°}$, $B^{fin}_{270°}$, and $B^{fin}_{315°}$ in Figure 7a).

The initial synthesis process, explained in Section 2.3.1, is valid for any of these relocations of B^{ini}. Given the complexity of the mechanism, it was likely that any attempt to select values between the limits of Table 2 would result in a non-existing solution in either of the four-bar mechanisms. Instead, it was decided to generate random values between these limits, guaranteeing the existence of solutions through the stages of the synthesis (proximal four-bar, followed by the distal one) by performing iterative calculations until a valid solution was found. Four simulations were run for each hypothesized B^{ini} point to have a sufficiently diverse population (see Figure 7b).

To assess the performance of each design, two quality metrics were evaluated: the averaged mechanical advantage (MA_{avg}) over the entire RHEx travel, and the degree of irregularity of the mechanical advantage (δ_{MA}) over the same travel, defined as

$$MA_{avg} = \frac{1}{\theta_{12p}^{fin} - \theta_{12p}^{ini}} \cdot \int_{\theta_{12}^{ini}}^{\theta_{12}^{fin}} MA(\theta_{12p}) \cdot d\theta_{12p}, \quad (1)$$

$$\delta_{MA}(\%) = \frac{MA_{max} - MA_{min}}{MA_{avg}} \cdot 100. \quad (2)$$

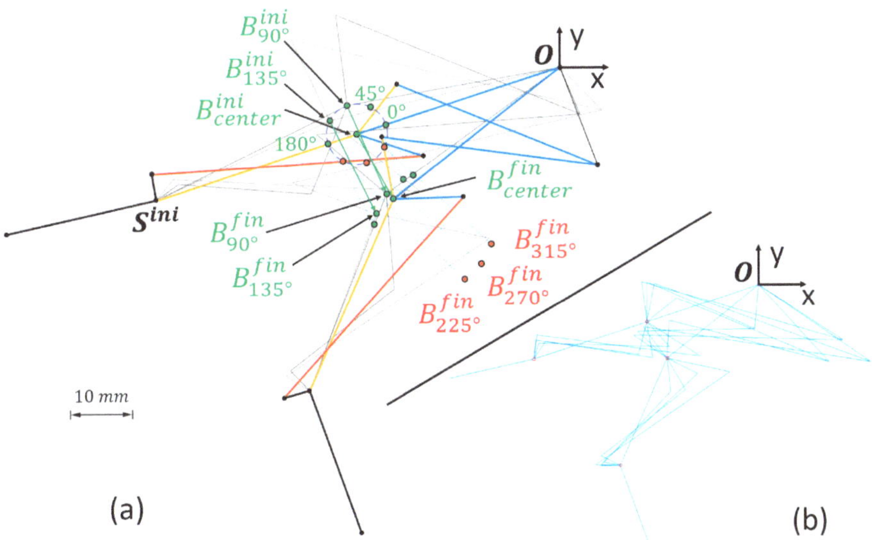

Figure 7. (**a**) The figure illustrates how some initial re-locations of B^{ini} around the initial hypothesis (B^{ini}_{center}) lead to configurations where B^{fin} ends up above (accepted solutions, in green) or below (rejected solutions, in red) the line $S^{ini}O$. In grey in the background, two additional mechanisms generated under the $B^{ini}_{90°}$ and $B^{ini}_{135°}$ hypotheses are exemplified. (**b**) Four proposals generated after dimensional synthesis for the same B^{ini}_{center}, with random values for the parameters of Table 2.

For each mechanism, and for the calculation of the MA_{avg}, the mechanical advantage (MA) was evaluated at four equidistant positions along the θ_{12p} travel (see Figure 5a), including the initial and final postures. This allowed for the interpolating of a third-degree polynomial function $MA(\theta_{12p})$ for each mechanism. The evaluation of the MA at each of the four postures was performed using approximate methods, namely (see Figure 4-left): (i) without loss of generality, point A was considered at a distance of 10 mm from O; (ii) concerning the posture being studied at a θ_{12p} value, displacements of point A (Δs_A) and point T (Δs_T) were evaluated (these displacements were measured from the posture resulting from a decrease $\Delta\theta_{12p}$ to the posture resulting from an increase $\Delta\theta_{12p}$ ($\Delta\theta_{12p} = 10^{-5}$ rad)); and (iii) MA for the referred posture was calculated as

$$F_T \cdot \dot{s_T} = N_{AC} \cdot \dot{s_A} \rightarrow MA = \frac{F_T}{N_{AC}} \approx \frac{\Delta s_A}{\Delta s_T}. \tag{3}$$

For the calculation of Equation (3), we assume an ideal mechanism so the power developed at point A by the linear actuator (N_{AC}) must coincide with the power developed at point T.

When evaluating the mechanical advantage, it should be noted that all the dimensions of the mechanism were already known, after the synthesis of the previous section. Therefore, the displacement Δs_T was estimated from the variation undergone by the concatenated four-bar mechanisms, caused by the rotation of A around O, here approximated to Δs_A. Without loss of generality, for the proximal mechanism depicted in Figure 4-left, it follows that

$$\vec{OB} + \vec{OC_{34p}} + \vec{C_{34p}C_{14p}} - \vec{OC_{14p}} = \vec{0}. \tag{4}$$

Projecting the vectors of the loop into a local coordinate system with the abscissa axis along L_{2p} we have

$$L_{4p} \cdot cos\gamma_p = L_{2p} - L_{3p} \cdot cos\beta_p - L_{1p} \cdot cos\alpha_p, \quad (5)$$

$$L_{4p} \cdot sin\gamma_p = L_{3p} \cdot sin\beta_p + L_{1p} \cdot sin\alpha_p. \quad (6)$$

By squaring the Equations (5) and (6), adding them, and using the fundamental trigonometric identities, the dependence on the angle γ_p is removed. This allows obtaining the value of β_p as a function of α_p. The distal four-bar mechanism can be solved analogously. Thus, for each value of θ_{12p} and its variation ($\Delta\theta_{12p}$), the values of α_p, β_p and their variations ($\Delta\alpha_p = -\Delta\theta_{12p}$, $\Delta\beta_p$) are known univocally. It allowed us to find the displacement required in Equation (3).

With all this, in the first generation, a total of 24 simulations were performed, which comprised the random synthesis of four RHEx proposals for each point B^{ini}.

From the number of resulting proposals, the coefficient of variation (CV) of the above quality metrics, Equations (1) and (2), were studied. The CV shows the extent of variability concerning the mean of the population. It is defined as the ratio of the standard deviation (σ) to the mean ($\overline{x}$), namely:

$$CV(\%) = \frac{\sigma}{\overline{x}} \cdot 100, \quad (7)$$

where x denotes the values of the metrics whose CV is evaluated. From the metric with the higher CV, i.e., the higher relative variability in that metric, the first quartile of the population with better values (lower δ_{MA} or higher MA_{avg}) was pre-selected. Half of the pre-selected cases with the best values for the other evaluated metric was selected for further optimization.

2. Second Generation: Optimization of the Design Parameters

To assess the influence that the parameters in Table 2 have on the design, the influence of an incremental, null, or decremental variation of each of them was considered ($\pm$2 mm for the lengths, $\pm 5^\circ$ for the angles, in case there was variation). Considering the 6 parameters, it gave a total of $3^6 = 729$ possible combinations for each of the mechanisms selected in the previous section. The dimensional synthesis of each of them was performed as described in Section 2.3.1.

From the total amount of resulting proposals, we observed again the metric with higher CV to pre-select the first quartile of the population with better values (lower δ_{MA} or higher MA_{avg}). Finally, the one with better value in the remaining quality metric has been chosen for further development of the RHEx.

In summary, this algorithm can be seen as a hierarchical multi-objective selection approach, where objectives (δ_{MA} or MA_{avg}) are prioritized based on their relative variability (CV), followed by selection based on the other objective. This step involves prioritizing one metric over the other based on its variability, introducing a hierarchy in the selection process. The algorithm reinforces the idea that both metrics are important, but with a differential emphasis depending on the variability. Not being a standard method, it can be classified as a heuristic multi-objective approach with dynamic objective prioritization.

2.3.3. Linear Actuator Positioning

Point A depicted in Figure 4-left, where the linear actuator is anchored observes the following design conditions: (i) it performs a rotation around point O; (ii) the rotation covers a circumferential arc whose central angle is the same as the angle τ traveled by the bar L_{2p} in the movement of the RHEx from extension to flexion (Figure 5a); and (iii) the radius of the circumference (OA) is such that the length of the chord (L_{chord}) joining the two endpoints of the arc matches with the stroke length of the linear actuator. At the prolongation of this chord, the attachment of the other end of the linear actuator (point C in Figure 4-left) must be found. Also, the chord and the radii joined to the two endpoints of the described arc constitute an isosceles triangle from the resolution of which the value of the radius is determined univocally, namely:

$$OA = \frac{L_{chord}}{2 \cdot sin(\tau/2)}. \quad (8)$$

By observing these three conditions, it is guaranteed that the transmission angle [64] remains as close to the optimum as possible, i.e., moderately close to perpendicular to the radius OA, during the entire stroke. The solutions can still be infinite, and it is necessary to decide the solution to suit the design of the dorsal support of the hand on which the other end of the actuator will articulate.

3. Results

3.1. Preliminary Designs from the Optimization of the Location of the Joint B

Figure 8 shows the wide spectrum of dimensions, within the ranges of the problem (Table 2), that has been addressed by generating and simulating 24 proposals of mechanisms for the different possible B^{ini} locations. They are the result of simulating four mechanisms of varied dimensions based on each of the six possible B^{ini} locations announced in Figure 7a.

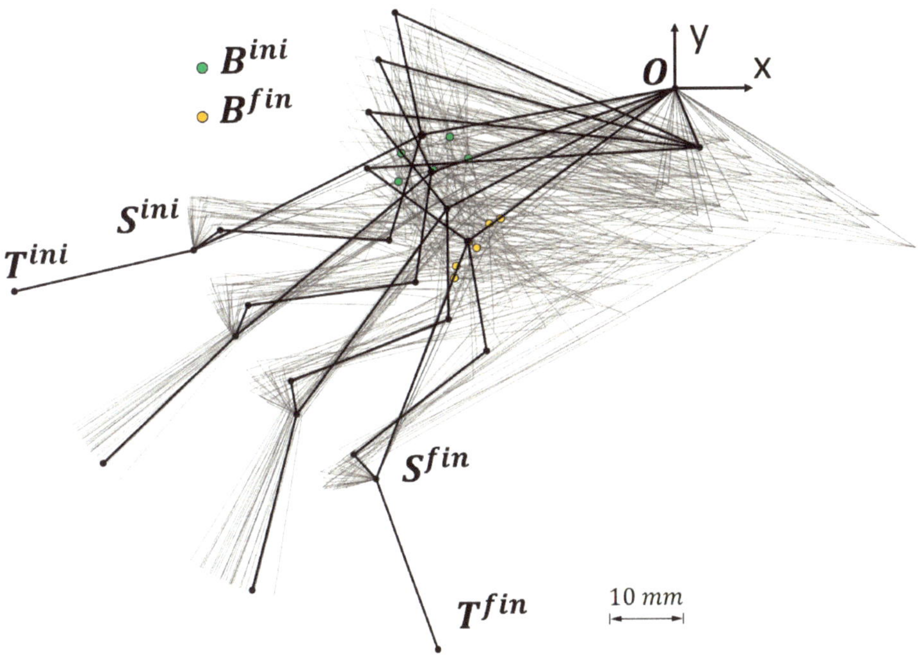

Figure 8. In black, the sequence of motion of the RHEx from extension ($S^{ini}T^{ini}$) to flexion ($S^{fin}T^{fin}$) is shown. The grey background represents the wide spectrum of possible linkages studied, achieved by generating and simulating four mechanisms with varied dimensions for each of the six possible B^{ini} locations detailed in Figure 7a.

The CV of the δ_{MA} had a value of 52.8%, higher than the CV of the MA_{avg} (12.6%). Therefore, according to the procedure explained in Section 2.3.2-(1), the quartile with the lowest δ_{MA} amongst the 24 cases was considered first. It corresponds to cases represented in Figure 9, namely, one $B^{ini}_{90^\circ}$ subcase, one $B^{ini}_{180^\circ}$, and two subcases of $B^{ini}_{45^\circ}$ and $B^{ini}_{135^\circ}$ (see Figure 7a). Figure 10 depicts, from the abovementioned cases, the three with the highest

MA_{avg}. In the figures, these variations are grouped after their B^{ini} for the sake of brevity, the subcases from the same B^{ini} being tagged with *a* and *b* sub-indexes.

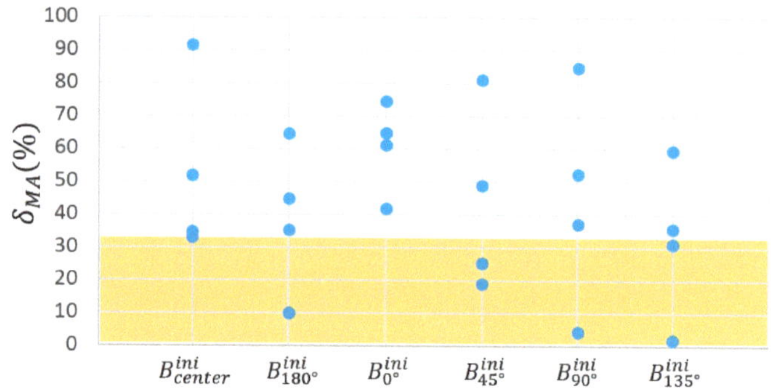

Figure 9. Preselection for the B^{ini} locations, as explained in Section 2.3.2-(1), was conducted based on δ_{MA} as it exhibits a higher coefficient of variation (CV) compared to MA_{avg} across the 24 proposed cases. The colored region highlights the quartile of the population with the lowest δ_{MA}, narrowing down this preselection to six possible RHEx proposals.

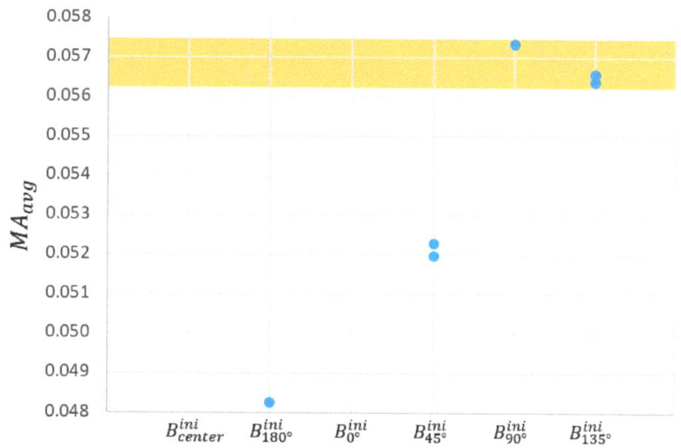

Figure 10. Selection for the optimal B^{ini} locations, as explained in Section 2.3.2-(1), was conducted based on MA_{avg} over the previous pre-selection of 6 cases. The colored region highlights the half of the pre-selected cases with the best MA_{avg}, that is, the three cases selected for further optimization.

Table 3 shows in detail the results of the dimensional synthesis for the six pre-selected cases and the three finally selected for further study, named $B^{ini}_{135°a}$, $B^{ini}_{135°b}$, and $B^{ini}_{90°}$.

Table 3. Dimensions and values of $\delta_{MA}(\%)$ and MA_{avg} for the six preselected cases from the optimization of the location of the joint B^{ini}. Lengths (L) in mm, angles (φ) in *deg*, δ_{MA} in %.

	L_{1p}	L_{2p}	L_{3p}	L_{4p}	φ_{1p}	φ_{2p}	L_{1d}	L_{2d}	L_{3d}	L_{4d}	φ_{1d}	φ_{2d}	δ_{MA}	MA_{avg}
$B^{ini}_{135°a}$	8.995	38.290	17.755	42.741	250.575	98.641	21.534	31.256	4.018	45.738	334.882	134.750	1.881	0.056
$B^{ini}_{90°}$	8.406	34.950	15.141	42.827	292.691	109.698	15.831	34.950	4.537	21.696	247.841	156.550	4.292	0.057
$B^{ini}_{180°}$	13.169	39.591	11.140	35.899	266.028	122.145	6.838	29.591	4.736	30.169	285.022	99.866	10.005	0.048
$B^{ini}_{45°a}$	8.620	31.256	15.084	37.617	273.232	104.319	14.883	38.290	5.563	30.328	248.439	98.123	18.783	0.052
$B^{ini}_{45°b}$	8.076	31.256	14.555	33.955	236.341	94.383	39.015	38.290	6.881	70.426	334.430	105.638	25.161	0.052
$B^{ini}_{135°b}$	9.819	38.290	14.746	51.646	313.820	90.096	18.820	31.256	4.493	40.292	321.311	141.643	30.965	0.057

3.2. Designs from the Subsequent Optimization of the Design Parameters

Figure 11 summarizes the three best mechanisms over the different locations of B^{ini} and the $3 \cdot 729 = 2187$ mechanisms resulting from varying the parameters of Table 2, as explained in Section 2.3.2-(2). The CV observed across this whole population showed very uneven results for the δ_{MA} (71.2%) and the MA_{avg} (0.8%).

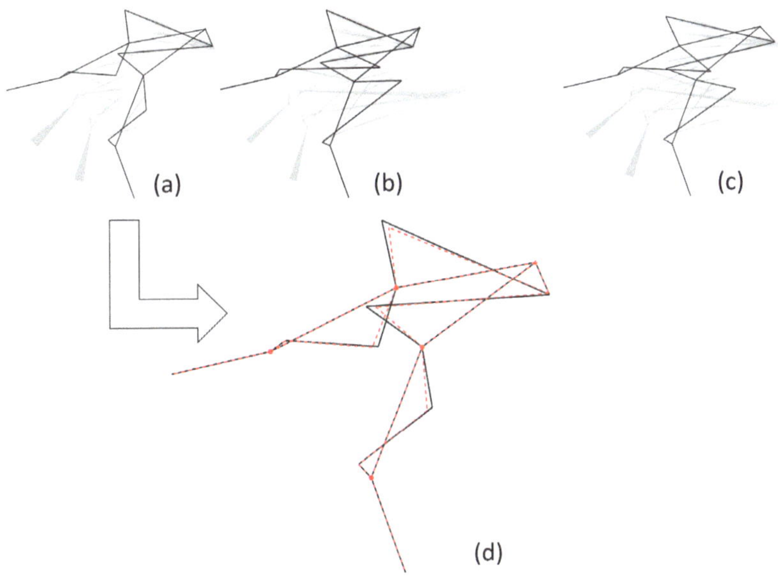

Figure 11. (**a–c**) In black, the representation of the three mechanisms selected from Table 3 (after optimization for the location of joint B^{ini}, following Section 2.3.2-(1)). The corresponding sets of mechanisms resulting from parameter variations are shown in grey in the background. (**d**) In black, the selected best mechanism after the optimization of the design parameters (following Section 2.3.2-(2)), overlapping with the preliminary design (in red).

Figure 12 summarizes the dispersion of the different δ_{MA}. The region containing the first quartile of the population arranged in increasing order of δ_{MA} is highlighted, i.e., 547 versions of the mechanism with the lower δ_{MA}. In the figure, these populations are grouped after their B^{ini} for the sake of brevity. Basically, they include versions of the original mechanisms documented in Table 2 as $B^{ini}_{90°}$ and $B^{ini}_{135°a}$.

Among those pre-selected and ordered by their MA_{avg}, the $B^{ini}_{90°}$ versions ranked the best 218 proposals. It is noteworthy that the very low CV shown by the MA_{avg} implied an improvement of only $8.5 \cdot 10^{-5}$ in that metric when choosing the mechanism with better MA_{avg} instead the one with better δ_{MA}. However, this change would imply a worsening of δ_{MA} of 6.7%. For that reason, the version of $B^{ini}_{90°}$ which ranked better at the pre-selection based on the δ_{MA} was the one selected to achieve the RHEx (Figure 11d). Values for this mechanism are shown in Table 4.

Table 4. Best mechanism after the optimization of the design parameters. The values shown for the version of $B^{ini}_{90°}$ were taken for the design of the RHEx. Lengths (**L**) in mm, angles (φ) in *deg*, δ_{MA} in %.

	L_{1p}	L_{2p}	L_{3p}	L_{4p}	φ_{1p}	φ_{2p}	L_{1d}	L_{2d}	L_{3d}	L_{4d}	φ_{1d}	φ_{2d}	δ_{MA}	MA_{avg}
$B^{ini}_{90°}$	8.796	34.950	17.141	45.379	292.691	104.698	15.247	34.950	4.537	23.158	252.841	156.550	0.884	0.057

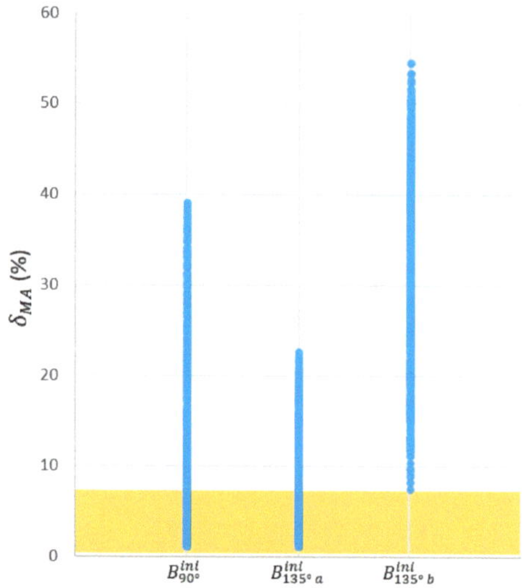

Figure 12. Dispersion of the different δ_{MA} of the $3 \cdot 3^6 = 2187$ mechanisms resulting from an incremental, null, or decremental variation of the six parameters of Table 2 in those three selected cases which resulted from the optimization of the location of the joint B^{ini} (resumed in Table 3). The colored region highlights the quartile of the population with the lowest δ_{MA}, resulting in 547 proposals of RHEx.

3.3. *Study of Correlations*

3.3.1. Study of Correlations in the Location of the Joint B

In Section 2.3.2-(1), a first location of B^{ini} was proposed as the midpoint of the segment $S^{ini}O$ and then repositioned to five different locations, shown in Figure 7a. For each hypothesis, four mechanisms were generated. In total, 20 mechanisms were available at the various B^{ini} locations, whose quality metrics (MA_{avg} and δ_{MA}) could be contrasted with those of the 4 mechanisms at the first B^{ini} location. This resulted in 80 possible combinations on which to perform a correlation study between the variation of the quality metrics and the relocation of B^{ini}.

Overall, no correlation was observed between B^{ini} location and the variation in the value of δ_{MA} (−0.125 correlation coefficient). However, it is presumed that the location of B^{ini} can be correlated with the variation in the value of MA_{avg} (0.732 correlation coefficient).

The average observed variation of the MA_{avg}, for each possible direction for the relocation of B^{ini}, was as described in the Table 5, with the largest increases for the 135° and 90° directions. The results are consistent with the preselection made in Section 3.1.

Table 5. Average variation in the MA_{avg} for each of the directions of the displacement of B^{ini} depicted in Figure 7.

Dir. Displacement B^{ini}	ΔMA_{avg}
0°	−0.0046
45°	0.0079
90°	0.0131
135°	0.0117
180°	0.0038

3.3.2. Study of Correlations in the Design Parameters

With the 2187 mechanisms explained in Section 3.2., a correlation study was carried out between each of the variables in Table 2 and the quality metrics used, MA_{avg} and

δ_{MA}. Specifically, we studied the correlation between the variation in each of the varied parameters with respect to those in the original mechanism (resulting from the calculations in Section 2.3.2-(1)) and the variations in the metrics concerning the metric values in that same mechanism.

Positive values, in general, might indicate a direct relationship, i.e., when the parameter increases, the metric also tends to increase. This was the case with L_{3p} concerning MA_{avg} and δ_{MA} but the relationship was negligible. Also, L_{3d} had a weak positive correlation with δ_{MA}. However, L_{3d} had a weak negative correlation with MA_{avg}. Regarding the angular parameters, φ_{2p} had a very weak negative correlation with MA_{avg} and with δ_{MA}. On the other hand, φ_{2d} had correlations very close to 0 with both metrics, thus dismissing any linear relationship between φ_{2d} and these metrics.

In summary, all correlations summarized in Table 6 were weak, suggesting that there was no linear relationship between the parameters L_{3p}, φ_{2p}, L_{3d}, and φ_{2d} with the quality metrics MA_{avg} and δ_{MA}.

Table 6. Correlation coefficients between the design parameters varied and the quality metrics.

	L_{3p}	φ_{2p}	L_{3d}	φ_{2d}
MA_{avg}	0.0634	−0.0155	−0.1893	0.0024
δ_{MA}	0.0808	−0.0137	0.1650	−0.0148

3.4. Final Model

Taking into account that the MA_{avg} values did show a correlation with the relocation of B^{ini} in the 90° direction (see Figure 7a), a new simulation was performed in which the distance (radius) with respect to the first location of B^{ini} was increased to 10 mm. We simulated not only the mechanism resulting from maintaining the parameters L_{3p}, φ_{2p}, φ_{1p}, L_{3d}, φ_{2d}, and φ_{1d} of the mechanism selected in Section 3.2 (values shown in Table 4) but also the 729 mechanisms resulting from variations of these variables. The mechanism directly obtained by displacing the B^{ini} point had a MA_{avg} of 0.0634 ($\delta_{MA} = 4.919$). For the 729 variations, the MA_{avg} had a mean value of 0.102 (0.063 min, 0.898 max, $CV = 123.11$) and the δ_{MA} had a mean value of 109.94 (1.554 min, 741.190 max, $CV = 171.895$). Note that the best MA_{avg} had the highest δ_{MA} values associated with them. In the end, after a screening performed according to the procedure described in Section 3.2, the dimensions of the mechanism described in Table 7 were adopted for the design of the RHEx.

Table 7. Dimensions taken for the final design of the RHEx. Lengths (L) in mm, angles (φ) in *deg*, δ_{MA} in %.

	L_{1p}	L_{2p}	L_{3p}	L_{4p}	φ_{1p}	φ_{2p}	L_{1d}	L_{2d}	L_{3d}	L_{4d}	φ_{1d}	φ_{2d}	δ_{MA}	MA_{avg}
$B^{ini}_{90°}$	3.650	36.007	15.141	46.176	286.665	99.698	25.122	36.007	4.537	19.513	252.522	156.550	1.554	0.063

After the dimensional synthesis of the RHEx adapted to the patient's anatomy, there is still room for the choice and location of the actuator. The basis for its installation has already been discussed in Section 2.3.3. With the assumption of distance $OA = 10$ mm with which the mechanical advantage has been calculated, observing Equation (8), with $\tau = 29.308°$ in the final design, it is found that the stroke length required in the linear actuator would be $L_{chord} = 5.059$ mm. These calculations must be confronted with the reality of the actuators available on the market. For this purpose, the duration of the pulp pinch closing sequence was first estimated to be 1 s, applying a force of 5 N (see Section 1.1). Approximating the distance traveled by T to be about 40 mm, measured as the distance between the patient's index fingertip and the end of the thumb, this implied a power requirement (H_T) of

$$H_T = 5\,\text{N} \cdot \frac{40 \cdot 10^{-3}\,\text{m}}{1\,\text{s}} = 0.2\,\text{W}. \tag{9}$$

The actuator selected should not provide less power than that of Equation (9): the model chosen for the development was the Actuonix® PQ12 (maximum load of 50 N, 20 mm stroke length (i.e., available L_{chord}), 10 mm/s max speed [65]).

Furthermore, recalling Equation (3), it is observed that if the distance OA is multiplied by a factor k, the overall mechanical advantage of the mechanism varies by the same factor:

$$\Delta s'_A = k \cdot \Delta s_A \rightarrow MA' \approx \frac{k \cdot \Delta s_A}{\Delta s_T} = k \cdot MA. \tag{10}$$

Recalling Equation (8), the radius OA taking benefit from the stroke length of the above-mentioned actuator is $OA = 39.528$ mm, and, following Equation (10) with $k \approx 39.528/10$, the averaged mechanical advantage for the final design of RHEx is $MA'_{avg} = 0.249$. All requirements were met:

$$H_A = 50\ \text{N} \cdot 0.010\ \frac{m}{s} = 0.5\ \text{W} > 0.2\ \text{W}, \tag{11}$$

$$F_T = N_{AC} \cdot MA'_{avg} = 50\ \text{N} \cdot 0.249 = 12.45\ \text{N} > 5\ \text{N}. \tag{12}$$

In Equation (12), the value of MA has been considered as that of MA_{avg}, given the low value of δ_{MA}. Figure 13 shows the final RHEx design with the dimensions resulting from the optimization process and with the selected linear actuator. The RHEx was modeled in SolidWorks® 2025 with the dimensions obtained after optimization in Matlab®R2018b, and later, it was printed in Acrylonitrile Butadiene Styrene (ABS) via Fused Deposition Modeling (FDM) on a 3D printer, namely, a CoLiDo® mod. X3045 (CoLiDo Ibérica, Valencia, Spain) with Repetier-Host v2.3.2 (www.repetier.com accessed on 10 June 2024) software. The proposed assembly is grounded on the dorsum of the palm, and it avoids any interference with it during the movement.

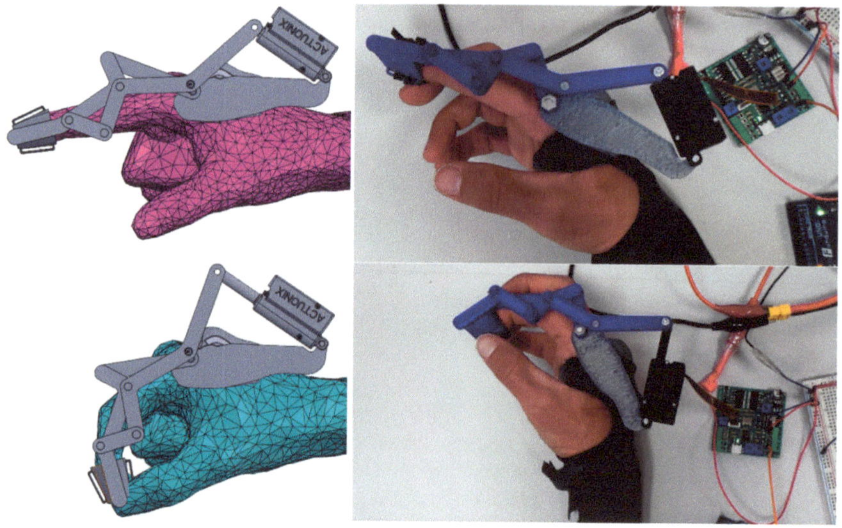

Figure 13. Final design and testing of the RHEx.

4. Discussion

The focus on improving the quality of life and rehabilitation prospects for post-stroke patients has become increasingly significant in recent years. Hand exoskeletons, including assistive ones, should be as compact and lightweight as possible to enhance wearability. Underactuated designs, like the one examined in this study, are well-suited for this goal, as they aim for compact actuator sizes while delivering high output forces (relative to the actuator's size) and ensuring efficient power transmission through the links. Despite progress, research on exoskeletons with enhanced mechanical advantages and ergonomics still requires revisions and refinements to meet these specific needs. The scanning process, along with the mechanism optimization presented here, are essential tools for developing custom-made, optimal mechanisms tailored to each patient.

The results of this research have focused on a 6-bar topology which is versatile for the design of exoskeletons both for the whole hand (in the case of a dorsal layout) and for index finger rehabilitation, as has been the case (with a more compact lateral layout).

The integration of design and optimization techniques raises an important question: how do engineers determine the most suitable method or algorithm for optimizing different design criteria. To address these challenges, alternative optimization approaches have been developed, including nature-inspired methods from evolutionary computation (EC). The strategy followed in this research provides design procedures applicable to similar exoskeletons. First, the search for the optimal location for the concatenation joint of the four-bar mechanisms (B-joint), and second, the search for correlations on the design variables (the location coordinates of the B-joint included). We have worked with quality metrics that are of interest for the performance of the task, such as MA_{avg} and δ_{MA}. It should be noted that the lack of correlations between the variation of the design parameters and the quality metrics evidenced the high non-linearity of the system. Only the coordinates of the B-joint showed some significance. All things considered, these metrics have allowed us to obtain a mechanism with a quasi-constant MA value across its travel, with an average value (MA_{avg}) higher than the result of our previous studies [45,46] mentioned earlier (this former design had values of $MA = 0.064$ at the final posture and $MA = 0.051$ at the initial posture, calculated according to Equation (3), with $\delta_{MA} = 22.6\%$).

By targeting genetic optimization schemes, we reduced the number of iterative or generational calculations required to reach an optimal solution. In this case, compared to a previous optimization for the same problem [45,46], we only evaluated 2211 different mechanisms, instead of 59,049. This signifies the elimination of 96.2% of the cases that were randomly generated in those studies.

5. Conclusions

The selection and design of the best topologies for the design of collaborative or rehabilitative exoskeletons is a growing field because their applications in modern life have been boosted in parallel to the explosion of control techniques and skin-interfacing flexible electronics (e-skin). Moreover, the integration of artificial intelligence in these systems could further enhance their functionality, enabling predictive diagnostics and personalized rehabilitation strategies based on real-time data analysis. This would make rehabilitation more efficient and accessible to patients both in clinical settings and at home. However, the optimization of the topologies to be used in the kinematic chain is crucial; therefore, the choice of good optimization functions must ensure that they evaluate the mechanism throughout its functional range.

The targeted genetic optimization scheme here exposed, by observing the MA_{avg} and δ_{MA} and their relationship with specific parameters identified in the different stages of the mechanism synthesis, allows us to arrive at designs that improve the overall behavior of the RHEx. On the one hand, this has made it possible to reduce the computational cost which can be abused with current means but would not justify a good selection process. On the other hand, this optimization can be easily applied to any other exoskeleton after studying its functionality.

The following conclusions can be made on the resulting prototype of RHEx:

- The movement of the RHEx is custom-made and adequate for the patient's requirements. Being an underactuated mechanism, it can be actuated by one small linear actuator. The power supply of the prototype is a low-voltage battery (up to 12 Volts) which poses no risk to the user.

- It is verified that the RHEx has a low weight and that the fastening system to the hand is not uncomfortable.
- Regarding the safety of the prototype and future work, the following can be taken into account:
- To leave the fingertip free, designs that adhere the final segment of the RHEx to the nail and sides of the patient's distal phalanx may be considered.
- The same procedure can be adapted and replicated for an RHEx design allowing flexion of the thumb, which now remains fixed in opposition.

The RHEx design protocol here presented, from conception to optimization, is adaptable to any hand size, ensuring that institutions can effectively serve a wide range of patients during rehabilitation. By utilizing 3D printing technology based on a precise scan of the patient's hand, it becomes possible to create exoskeletons that are perfectly tailored to individual hand dimensions. This level of customization enhances both fit and comfort, which is challenging to achieve with conventional, one-size-fits-all devices [66].

Additionally, 3D printing allows healthcare facilities to produce these devices in-house at a very low cost, using materials like PLA or ABS, which are both affordable and recyclable [67,68]. This approach not only ensures that the developed technology is accessible to patients with different hand sizes but also reduces costs and delivery times by enabling on-site production.

Author Contributions: Conceptualization, J.A.-E., J.L.I.-V. and V.R.-C.; methodology J.A.-E. and V.R.-C.; software, J.A.-E. and J.L.I.-V.; validation, J.A.-E. and J.L.I.-V.; formal analysis, J.A.-E.; investigation, J.A.-E., J.L.I.-V. and V.R.-C.; resources, J.A.-E., J.L.I.-V. and V.R.-C.; data curation, J.A.-E.; writing—original draft preparation, J.A.-E.; writing—review and editing, J.A.-E. and J.L.I.-V.; supervision, J.A.-E. and J.L.I.-V.; project administration, J.L.I.-V.; funding acquisition, J.L.I.-V. All authors have read and agreed to the published version of the manuscript.

Funding: This research was supported by grant PID2020-118021RB-I00 funded by MICIN/AEI/10.13039/501100011033; and by grant number 015 DERAPPI funded by the Fundació per al Foment de la Investigació Sanitària i Biomèdica de la Comunitat Valenciana (FISABIO) and Universitat Jaume I (UJI).

Institutional Review Board Statement: The study was conducted in accordance with the Declaration of Helsinki and approved by the Ethics Committee of the Universitat Jaume I (record number CD/006/2019, 6 March 2019).

Data Availability Statement: The biometrics and the programming code in Matlab® that supported the findings of this research are located in controlled access data storage at the Biomechanics and Ergonomics repositories of the Universitat Jaume I. The data used in this study are available on request from the corresponding author due to sensitivity reasons.

Acknowledgments: Brain Injury Unit of the Hospital La Magdalena (Castellón, Spain) for the specifications in the rehabilitation process. The authors thank J. Fuentes-Ballesteros for contributing to 3D printing and assembly.

Conflicts of Interest: The authors declare no conflicts of interest. The funders had no role in the design of the study; in the collection, analyses, or interpretation of data; in the writing of the manuscript; or in the decision to publish the results.

References

1. Scivoletto, G.; Morganti, B.; Molinari, M. Early versus Delayed Inpatient Spinal Cord Injury Rehabilitation: An Italian Study. *Arch. Phys. Med. Rehabil.* **2005**, *86*, 512–516. [CrossRef] [PubMed]
2. Tran, P.; Jeong, S.; Herrin, K.R.; Desai, J.P. Review: Hand Exoskeleton Systems, Clinical Rehabilitation Practices, and Future Prospects. *IEEE Trans. Med. Robot. Bionics* **2021**, *3*, 606–622. [CrossRef]
3. Yue, Z.; Zhang, X.; Wang, J. Hand Rehabilitation Robotics on Poststroke Motor Recovery. *Behav. Neurol.* **2017**, *2017*, 3908135. [CrossRef] [PubMed]
4. Moggio, L.; de Sire, A.; Marotta, N.; Demeco, A.; Ammendolia, A. Exoskeleton versus End-Effector Robot-Assisted Therapy for Finger-Hand Motor Recovery in Stroke Survivors: Systematic Review and Meta-Analysis. *Top. Stroke Rehabil.* **2022**, *29*, 539–550. [CrossRef] [PubMed]

5. Davila-Vilchis, J.M.; Avila-Vilchis, J.C.; Vilchis-Gonzalez, A.H. Laz-Aviles Design Criteria of Soft Exogloves for Hand Rehabilitation-Assistance Tasks. *Appl. Bionics Biomech.* **2020**, *2020*, 2724783. [CrossRef] [PubMed]
6. Chen, Y.; Gao, Z.; Zhang, F.; Wen, Z.; Sun, X. Recent Progress in Self-Powered Multifunctional e-Skin for Advanced Applications. *Exploration* **2022**, *2*, 20210112. [CrossRef]
7. Hu, X.; Ma, Z.; Zhao, F.; Guo, S. Recent Advances in Self-Powered Wearable Flexible Sensors for Human Gaits Analysis. *Nanomaterials* **2024**, *14*, 1173. [CrossRef]
8. Iserte-Vilar, J.L. Development of an Exoskeleton for Pulp Pinch Grasping Rehabilitation in Stroke Patients (DERAPPI). Available online: https://be.uji.es/en/proyectos/proyecto_derappi.htm (accessed on 1 May 2024).
9. Kleyweg, R.P.; Van der Meché, F.G.; Schmitz, P.I.M. Interobserver Agreement in the Assessment of Muscle Strength Guillain-Barre Syndrome. *Muscle Nerve* **1991**, *14*, 1103–1109. [CrossRef]
10. Medical Research Council MRC Muscle Scale. Available online: https://www.ukri.org/councils/mrc/facilities-and-resources/find-an-mrc-facility-or-resource/mrc-muscle-scale/ (accessed on 4 July 2024).
11. Harb, A.; Kishner, S. Modified Ashworth Scale. Available online: www.ncbi.nlm.nih.gov/books/NBK554572/ (accessed on 1 July 2024).
12. Birouaș, F.; Avram, F.; Nilgesz, A.; Mihalca, V.O. A Review Regarding Hand Exoskeleton Technologies for Rehabilitation. *Recent Innov. Mechatron.* **1970**, *5*, 1–5. [CrossRef]
13. Birglen, L.; Laliberté, T.; Gosselin, C. *Underactuated Robot. Hands*; Springer: Berlin/Heidelberg, Germany, 2008; Volume 40, ISBN 978-3-540-77458-7.
14. Ferguson, P.W.; Shen, Y.; Rosen, J. Hand Exoskeleton Systems-Overview. In *Wearable Robotics: Systems and Applications*, 1st ed.; Academic Press: Cambridge, MA, USA, 2019; pp. 149–170, ISBN 9780128146590.
15. du Plessis, T.; Djouani, K.; Oosthuizen, C. A Review of Active Hand Exoskeletons for Rehabilitation and Assistance. *Robotics* **2021**, *10*, 40. [CrossRef]
16. Sarac, M.; Solazzi, M.; Frisoli, A. Design Requirements of Generic Hand Exoskeletons and Survey of Hand Exoskeletons for Rehabilitation, Assistive, or Haptic Use. *IEEE Trans. Haptics* **2019**, *12*, 400–413. [CrossRef]
17. Huamanchahua, D.; Toledo-Garcia, P.; Aguirre, J.; Huacre, S. Hand Exoskeletons for Rehabilitation: A Systematic Review. In Proceedings of the 2022 IEEE International IOT, Electronics and Mechatronics Conference (IEMTRONICS), Toronto, ON, Canada, 1–4 June 2022; pp. 1–6. [CrossRef]
18. Kabir, R.; Sunny, M.S.H.; Ahmed, H.U.; Rahman, M.H. Hand Rehabilitation Devices: A Comprehensive Systematic Review. *Micromachines* **2022**, *13*, 1033. [CrossRef]
19. Noronha, B.; Accoto, D. Exoskeletal Devices for Hand Assistance and Rehabilitation: A Comprehensive Analysis of State-of-the-Art Technologies. *IEEE Trans. Med. Robot. Bionics* **2021**, *3*, 525–538. [CrossRef]
20. Barreda-Galvez, T.; Terrazas-Rodas, D. Novel Technologies of Exoskeleton Systems Applied to Rehabilitation for Hand Therapies: A Technological Review. In Proceedings of the 2023 International Seminar on Intelligent Technology and Its Applications (ISITIA), Surabaya, Indonesia, 26–27 July 2023; pp. 328–335. [CrossRef]
21. Stroppa, F.; Soylemez, A.; Yuksel, H.T.; Akbas, B.; Sarac, M. Optimizing Exoskeleton Design with Evolutionary Computation: An Intensive Survey. *Robotics* **2023**, *12*, 106. [CrossRef]
22. Li, H.; Cheng, L.; Sun, N.; Cao, R. Design and Control of an Underactuated Finger Exoskeleton for Assisting Activities of Daily Living. *IEEE/ASME Trans. Mechatron.* **2022**, *27*, 2699–2709. [CrossRef]
23. Du, J.; Tian, Y.; Zhang, D.; Wang, H.; Zhang, Y.; Cheng, B.; Niu, J. Mechanism Design and Performance Analysis of a Wearable Hand Rehabilitation Robot. *Machines* **2022**, *10*, 1211. [CrossRef]
24. Iqbal, J.; Khan, H.; Tsagarakis, N.G.; Caldwell, D.G. A Novel Exoskeleton Robotic System for Hand Rehabilitation—Conceptualization to Prototyping. *Biocybern. Biomed. Eng.* **2014**, *34*, 79–89. [CrossRef]
25. Vanteddu, T.; Sebastian, B.; Ben-Tzvi, P. Design Optimization of RML Glove for Improved Grasp Performance. In Proceedings of the ASME 2018 Dynamic Systems and Control Conference, Atlanta, GA, USA, 30 September–3 October 2018; Volume 1, pp. 1–8. [CrossRef]
26. Li, G.; Cheng, L.; Sun, N. Design, Manipulability Analysis and Optimization of an Index Finger Exoskeleton for Stroke Rehabilitation. *Mech. Mach. Theory* **2022**, *167*, 104526. [CrossRef]
27. Sarac, M.; Solazzi, M.; Sotgiu, E.; Bergamasco, M.; Frisoli, A. Design and Kinematic Optimization of a Novel Underactuated Robotic Hand Exoskeleton. *Meccanica* **2017**, *52*, 749–761. [CrossRef]
28. Fonseca, C.M.; Fleming, P.J. Genetic Algorithms for Multiobjective Optimization: Formulation Discussion and Generalization. In Proceedings of the International Conference on Genetic Algorithms, Urbana-Champaign, IL, USA, 17–21 July 1993; Volume 93, pp. 416–423.
29. Maulik, U.; Bandyopadhyay, S.; Mukhopadhyay, A. Genetic Algorithms and Multiobjective Optimization. In *Multiobjective Genetic Algorithms for Clustering*; Springer: Berlin/Heidelberg, Germany, 2011; pp. 25–50, ISBN 978-3-642-16615-0.
30. Goldberg, D.E.; Holland, J.H. Genetic Algorithms and Machine Learning. *Mach. Learn.* **1988**, *3*, 95–99. [CrossRef]
31. Amirpour, E.; Savabi, M.; Saboukhi, A.; Gorii, M.R.; Ghafarirad, H.; Fesharakifard, R.; Rezaei, S.M. Design and Optimization of a Multi-DOF Hand Exoskeleton for Haptic Applications. In Proceedings of the 2019 7th International Conference on Robotics and Mechatronics (ICRoM), Tehran, Iran, 20–21 November 2019; pp. 270–275. [CrossRef]

32. Levenberg, K. A Method for the Solution of Certain Non-Linear Problems in Least Squares. *Q. Appl. Math.* **1944**, *2*, 164–168. [CrossRef]
33. Marquardt, D.W. An Algorithm for Least-Squares Estimation of Nonlinear Parameters. *J. Soc. Ind. Appl. Math.* **1963**, *11*, 431–441. Available online: https://www.jstor.org/stable/2098941 (accessed on 7 June 2024). [CrossRef]
34. Bianchi, M.; Cempini, M.; Conti, R.; Meli, E.; Ridolfi, A.; Vitiello, N.; Allotta, B. Design of a Series Elastic Transmission for Hand Exoskeletons. *Mechatronics* **2018**, *51*, 8–18. [CrossRef]
35. Liang, R.; Xu, G.; Teng, Z.; Li, M.; Zhang, S.; Zheng, X.; Zhang, K.; He, B. A General Arthropod Joint Model and Its Applications in Modeling Human Robotic Joints. *IEEE Access* **2021**, *9*, 7814–7822. [CrossRef]
36. Porteous, I.R. *Geometric Differentiation: For the Intelligence of Curves and Surfaces*, 2nd ed.; Cambridge University Press: Cambridge, UK, 2001; ISBN 9780521810401.
37. Xu, W.; Pradhan, S.; Guo, Y.; Bravo, C.; Ben-Tzvi, P. A Novel Design of a Robotic Glove System for Patients with Brachial Plexus Injuries. *Proc. ASME Des. Eng. Tech. Conf.* **2020**, *10*, V010T10A042. [CrossRef]
38. Wright, M.H. The Interior-Point Revolution in Optimization: History, Recent Developments, and Lasting Consequences. *Bull. Am. Math. Soc.* **2005**, *42*, 39–56. [CrossRef]
39. Qin, C.; Li, P.; Yang, X.; Li, B. A Design of Hand Rehabilitation Exoskeleton Mechanism Adapted to Different Finger Lengths. In Proceedings of the 2019 IEEE International Conference on Robotics and Biomimetics (ROBIO), Dali, China, 6–8 December 2019; pp. 1621–1626. [CrossRef]
40. Gembicki, F.W.; Haimes, Y.Y. Approach to Performance and Sensitivity Multiobjective Optimization: The Goal Attainment Method. *IEEE Trans. Automat. Contr.* **1975**, *20*, 769–771. [CrossRef]
41. Secciani, N.; Bianchi, M.; Ridolfi, A.; Vannetti, F.; Volpe, Y.; Governi, L.; Bianchini, M.; Allotta, B. Tailor-Made Hand Exoskeletons at the University of Florence: From Kinematics to Mechatronic Design. *Machines* **2019**, *7*, 22. [CrossRef]
42. Feix, T.; Romero, J.; Schmiedmayer, H.B.; Dollar, A.M.; Kragic, D. The GRASP Taxonomy of Human Grasp Types. *IEEE Trans. Hum. Mach. Syst.* **2016**, *46*, 66–77. [CrossRef]
43. Vergara, M.; Sancho-Bru, J.L.; Gracia-Ibáñez, V.; Pérez-González, A. An Introductory Study of Common Grasps Used by Adults during Performance of Activities of Daily Living. *J. Hand Ther.* **2014**, *27*, 225–234. [CrossRef]
44. Gracia-Ibáñez, V.; Sancho-Bru, J.L.; Vergara, M. Relevance of Grasp Types to Assess Functionality for Personal Autonomy. *J. Hand Ther.* **2018**, *31*, 102–110. [CrossRef]
45. Gabarda López, M. *Analysis and Optimization of a Mechanism for a Pincer-Gripper Exoskeleton*; Universitat Jaume I: Castelló de la Plana, Spain, 2023.
46. Gabarda-Lopez, M.; Roda-Casanova, V.; Andres-Esperanza, J.; Iserte, J.L.; Fuentes-Ballesteros, J.F.; Jarque-Bou, N.J. Design and Optimization of a Six-Bar Linkage to Assist in the Rehabilitation of the Pulp Pinch Grip in Stroke Patients. In Proceedings of the 28th Congress of the European Society of Biomechanics, European Society of Biomechanics, Maastricht, The Netherlands, 9–12 July 2023.
47. Sanchez-Marin, F.; Roda-Casanova, V. An Approach for the Global Search for Top-Quality Six-Bar Dwell Linkages. *Mech. Mach. Theory* **2022**, *176*, 104974. [CrossRef]
48. Zitzler, E.; Laumanns, M.; Thiele, L. *SPEA2: Improving the Strength Pareto Evolutionary Algorithm*; ETH Zurich: Zurich, Switzerland, 2001.
49. Hume, M.C.; Gellman, H.; McKellop, H.; Brumfield, R. Functional Motion of the Joints of the Hand. *Am. Acad. Orthop. Surg.* **1999**, *15*, 240–243.
50. Roda-Sales, A.; Sancho-Bru, J.L.; Vergara, M. Problems Using Data Gloves with Strain Gauges to Measure Distal Interphalangeal Joints' Kinematics. *Sensors* **2022**, *22*, 3757. [CrossRef]
51. Kapandji, A.I. *Physiology of the Joints. Vol. 1. The Upper Limb*, 6th ed.; Elsevier: Amsterdam, The Netherlands, 2007; ISBN 9780443103506.
52. Roach, S.S.; Short, W.H.; Werner, F.W.; Fortino, M.D. Biomechanical Evaluation of Thumb Opposition Transfer Insertion Sites. *J. Hand Surg. Am.* **2001**, *26*, 354–361. [CrossRef]
53. Jahn, J.; Janes, W.E.; Saheb-Al-Zamani, M.; Burbank, C.M.; Brown, J.M.; Engsberg, J.R. Identification of Three Movement Phases of the Hand during Lateral and Pulp Pinches Using Video Motion Capture. *Hand* **2013**, *8*, 123–131. [CrossRef]
54. Lannin, N.A.; Horsley, S.A.; Herbert, R.; McCluskey, A.; Cusick, A. Splinting the Hand in the Functional Position after Brain Impairment: A Randomized, Controlled Trial. *Arch. Phys. Med. Rehabil.* **2003**, *84*, 297–302. [CrossRef]
55. Videler, A.; Eijffinger, E.; Nollet, F.; Beelen, A. A Thumb Opposition Splint to Improve Manual Dexterity and Upper-Limb Functioning in Charcot-Marie-Tooth Disease. *J. Rehabil. Med.* **2012**, *44*, 249–253. [CrossRef]
56. Schmidt, V. Planefit. Available online: www.mathworks.com/matlabcentral/fileexchange/36353-planefit (accessed on 6 June 2024).
57. Andrés, F.J.; Pérez-González, A.; Rubert, C.; Fuentes, J.; Sospedra, B. Comparison of Grasping Performance of Tendon and Linkage Transmission Systems in an Electric-Powered Low-Cost Hand Prosthesis. *J. Mech. Robot.* **2018**, *11*, 011018. [CrossRef]
58. Dechev, N.; Cleghorn, W.L.; Naumann, S. Multiple Finger, Passive Adaptive Grasp Prosthetic Hand. *Mech. Mach. Theory* **2001**, *36*, 1157–1173. [CrossRef]
59. Refour, E.; Sebastian, B.; Ben-Tzvi, P. Two-Digit Robotic Exoskeleton Glove Mechanism: Design and Integration. *J. Mech. Robot.* **2018**, *10*, 025002. [CrossRef]

60. Refour, E.; Sebastian, B.; Ben-Tzvi, P. Design and Integration of a Two-Digit Exoskeleton Glove. In Proceedings of the ASME 2017 International Design Engineering Technical Conferences and Computers and Information in Engineering Conference, Cleveland, OH, USA, 6–9 August 2017; Volume 5A-2017, pp. 1–8. [CrossRef]
61. Guo, Y.; Xu, W.; Pradhan, S.; Bravo, C.; Ben-Tzvi, P. Personalized Voice Activated Grasping System for a Robotic Exoskeleton Glove. *Mechatronics* **2022**, *83*, 102745. [CrossRef] [PubMed]
62. Ishibuchi, H.; Tsukamoto, N.; Nojima, Y. Evolutionary Many-Objective Optimization: A Short Review. In Proceedings of the IEEE Congress on Evolutionary Computation (CEC 2008), Hong Kong, China, 1–6 June 2008; pp. 2419–2426.
63. Debenham, J.; Wagner, R.R. *Multiobjective Optimization-Interactive and Evolutionary Approaches*; Branke, J., Deb, K., Kaisa Miettinen, R.S., Eds.; Springer: Berlin/Heidelberg, Germany, 2008; Volume 3588, ISBN 9783540889076.
64. Balli, S.S.; Chand, S. Transmission Angle in Mechanisms (Triangle in Mech). *Mech. Mach. Theory* **2002**, *37*, 175–195. [CrossRef]
65. Actuonix Motion Devices Actuator Selector Tool. Available online: https://www.actuonix.com/actuator-selector-tool?sn=166599480&e=cor&ajx=1&nsrt=CID_N1&ontd=2 (accessed on 10 July 2024).
66. Buonamici, F.; Furferi, R.; Governi, L.; Lazzeri, S.; McGreevy, K.S.; Servi, M.; Talanti, E.; Uccheddu, F.; Volpe, Y. A Practical Methodology for Computer-Aided Design of Custom 3D Printable Casts for Wrist Fractures. *Vis. Comput.* **2020**, *36*, 375–390. [CrossRef]
67. Rojek, I.; Dorożyński, J.; Mikołajewski, D.; Kotlarz, P. Overview of 3D Printed Exoskeleton Materials and Opportunities for Their AI-Based Optimization. *Appl. Sci.* **2023**, *13*, 8384. [CrossRef]
68. European Commission In-House Medical Devices. Available online: https://health.ec.europa.eu/medical-devices-topics-interest/house-medical-devices_en (accessed on 27 September 2024).

biomimetics

Article

Gender Differences in Performing an Overhead Drilling Task Using an Exoskeleton—A Cross-Sectional Study

Bettina Wollesen [1,2,*], Julia Gräf [1], Sander De Bock [3], Eligia Alfio [3,4], María Alejandra Díaz [3,4] and Kevin De Pauw [3]

1 Department of Human Movement Science, Universität Hamburg, 20148 Hamburg, Germany; julia.graef@uni-hamburg.de
2 Institute of Movement Therapy and Movement-Oriented Prevention and Rehabilitation, German Sports University Cologne, 50933 Cologne, Germany
3 Human Physiology and Sports Physiotherapy Research Group, Vrije Universiteit Brussel, 1050 Brussels, Belgium; sander.de.bock@vub.be (S.D.B.); eligia.alfio@vub.be (E.A.); ma.diaz@vub.be (M.A.D.); kevin.de.pauw@vub.be (K.D.P.)
4 Human Robotics Research Center, 1050 Brussels, Belgium
* Correspondence: b.wollesen@dshs-koeln.de; Tel.: +49-221-4982-4790

Abstract: (1) Exoskeletons offer potential benefits for overhead working tasks, but gender effects or differences are unclear. This study aimed to compare the performance as well as subjective body strain and comfort of men and women using an upper-body exoskeleton. (2) n = 20 female and n = 16 male participants performed an overhead drilling task with and without a passive upper-body exoskeleton in a randomized cross-over study. The task performance of different movement phases, perceived exertion, and ease of use were measured to compare gender differences. One- and two-way analyses were used to compare genders in the different conditions. The body mass index (BMI) was included as a covariate. (3) Gender differences in task performance were found for error integrals ($p < 0.001$) with higher values in male participants. Moreover, there was a significant interaction effect for gender x exoskeleton use. While females showed performance decrements in aiming with exoskeleton use, the males' performance increased ($p = 0.025$). No other gender differences were observed. (4) Gender differences in task performance using an upper-body industrial exoskeleton were less detectable than expected, indicating that body composition and anthropometrics might be valuable indicators for performance including assisting devices. Moreover, future studies should also integrate the examination of muscle activity to gain more insights into potential gender movement control patterns.

Keywords: supportive systems; gender comparison; over-shoulder working task

Citation: Wollesen, B.; Gräf, J.; De Bock, S.; Alfio, E.; Díaz, M.A.; De Pauw, K. Gender Differences in Performing an Overhead Drilling Task Using an Exoskeleton—A Cross-Sectional Study. *Biomimetics* **2024**, *9*, 601. https://doi.org/10.3390/biomimetics9100601

Academic Editor: Rafhael Milanezi de Andrade

Received: 29 August 2024
Revised: 3 October 2024
Accepted: 4 October 2024
Published: 7 October 2024

1. Introduction

Exoskeletons have emerged as innovative tools to augment human physical capabilities, offering promising applications across various fields, including rehabilitation and industrial work [1]. Among these, upper-body shoulder exoskeletons are designed to assist with tasks that involve repetitive lifting, holding, or overhead activities, aiming to reduce the physical strain and the risk of musculoskeletal disorders [2,3].

Despite the growing body of research on exoskeleton efficacy, there remains a critical gap in understanding how these devices affect diverse user groups, particularly between genders. Men and women often exhibit physiological and biomechanical differences, which can influence the performance and effectiveness of assistive devices like exoskeletons [4]. Factors such as muscle strength or range of motion according to body dimensions vary between genders and may lead to differing interactions with the exoskeleton [5].

Men generally have higher absolute muscle strength, particularly in the upper body, compared to women [6–8]. This can affect how men utilize the exoskeleton's support in tasks requiring high-force output [9]. Women may have comparable muscle endurance, but their lower muscle mass can influence how the exoskeleton's torque distribution and load bearing are perceived, potentially affecting long-term use and fatigue rates [8,10,11].

Men and women differ in body proportions, including differences in shoulder width, arm length, and torso dimensions. Dissimilarities in body composition, such as a higher percentage of body fat in women and greater muscle mass in men [12], can affect the anthropometric fit and therefore the functionality of exoskeletons. These differences may influence how body weight and the additional weight of the exoskeleton are distributed and how the exoskeleton stabilizes the body during use. In addition, women might adopt different postural adjustments due to their wider pelvis and different center of gravity, which can further impact the exoskeleton's fit and function [13].

Moreover, women tend to have a greater range of motion in their shoulder joints compared to men [14], which could affect how they interact with the exoskeleton. This increased flexibility might allow for a greater range of overhead motion but could also lead to other requirements from the exoskeleton regarding stabilization and support [15]. Women may also adopt distinct postural adjustments and kinematic strategies when using an exoskeleton for overhead tasks and might therefore show distinct movement patterns compared to men, which could be influenced by the exoskeleton's assistance [16]. Additionally, women might excel in tasks requiring fine motor control and precision [17], which could interact uniquely with the exoskeleton's assistance. This might result in different task performance metrics such as speed, accuracy, and coordination compared to men [16–19].

The study by Gräf et al. [19] has shown that when wearing an upper-body exoskeleton while executing overhead tasks, errors in cognitive tasks, especially under fatigue conditions, were significantly reduced. Furthermore, while muscle fatigue led to a general decrease in movement time to finish a task, it did not alter the overall speed–accuracy trade-off, indicating the motor system's ability to compensate and adapt.

All these variations can impact the fit and comfort of the exoskeleton, potentially affecting performance and ease of use. The poor fit of the exoskeleton caused by these differences can lead to discomfort or suboptimal support during tasks, influencing motor performance [20]. Therefore, women might report higher levels of discomfort or perceived exertion when using an exoskeleton designed with a male anthropometry in mind. Females might experience discomfort or suboptimal support if the exoskeleton is not designed to accommodate their body proportions [21], potentially affecting precision and work productivity [22–24].

Understanding the gender differences in wearing an upper-body exoskeleton is crucial for the development of designs that are inclusive and effective for all users. Addressing these differences involves considering physiological, biomechanical, and ergonomic factors to optimize fit, comfort, and performance.

While general trends suggest that motor precision could differ between males and females when using an exoskeleton, the extent and nature of these differences are highly dependent on individual physiological and biomechanical factors, as well as the design and adjustability of the exoskeleton itself. Tailoring exoskeletons to accommodate gender-specific needs can help optimize assistance, work productivity, and precision motor precision for both males and females.

This study aims to compare the effects of wearing an upper-body shoulder exoskeleton between women and men by assessing task performance of an overhead nailing task including perceived exertion and comfort. We seek to determine if there are significant gender-based differences in the use of the device.

Since the exoskeleton was designed and constructed for male users, we hypothesize that women will perform worse during the drilling task and have a higher rating of perceived exertion and lower usability score when using the exoskeleton in comparison to male users.

The findings of this study could inform the design and development of more inclusive exoskeletons, ensuring optimal performance and safety for all users, regardless of gender.

2. Materials and Methods

2.1. Study Design

This study followed a 2 × 2 randomized balanced cross-over study design (cf., Figure 1).

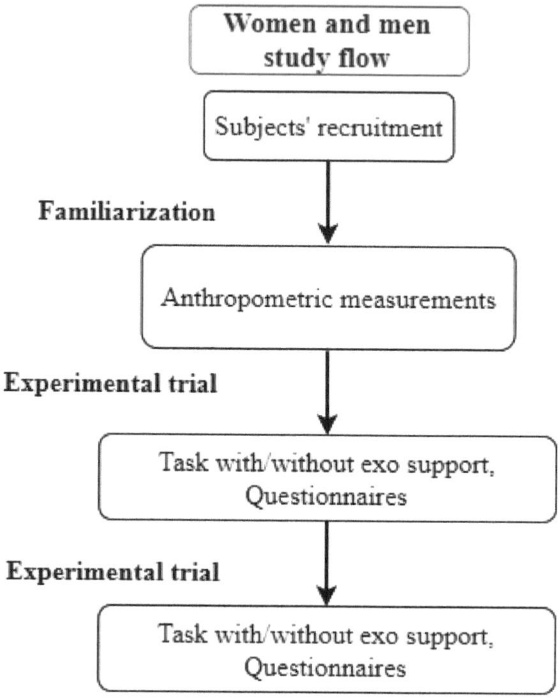

Figure 1. Study flow. Note: exo support = passive support.

2.2. Participants

2.2.1. Sample Size Calculation

To compare male and female participants with and without exoskeleton support, a power calculation with G*Power revealed a total sample size of 41 subjects (ANOVA; fixed effects, main effects and interactions, ß-Power of 0.8, estimated effect size of 0.45, and alpha of 0.05).

2.2.2. Inclusion and Exclusion Criteria

Biological females and males aged 18 and older, right-handed, and without a history or current manifestation of musculoskeletal disorders or injuries in the upper extremity within the past 12 months were included.

Participants were excluded if they had any injuries or health-related complaints. Only right-handed participants were included.

2.2.3. Participant Description

Participants were recruited on or in the surroundings of the VUB campus in two rounds. First, a number of n = 16 males were examined. Afterwards, a total number of n = 21 females were incorporated into the experiment. n = 1 female needed to be excluded due to poor data quality.

The mean age of the sample was 31.2 ± 14.5 years for the females and 31.9 ± 9.1 years for the males.

The anthropometrics naturally differed between female and male participants with higher values for males (cf., Table 1). As the BMI also differed significantly, the performance data had to be controlled for BMI.

Table 1. Overview of the participant data.

Anthropometrics	Female n = 20 (MW ± SD)	Male n = 16 (MW ± SD)	F (1,34) *p*
Body height (m)	1.65 ± 0.04	1.81 ± 0.06	93.592 **<0.001**
Body mass (kg)	61.3 ± 8.5	81.4 ± 7.6	54.495 **<0.001**
BMI	22.5 ± 2.7	24.8 ± 2.4	7.042 **0.012**
Fingertip height (m)	2.09 ± 6.9	2.35 ± 8.3	99.773 **<0.001**
Fingertip height (m) with elbow at 90 degrees	1.73 ± 5.4	1.92 ± 4.1	126.463 **<0.001**

2.2.4. Restrictions and Prohibitions for the Subjects

Subjects were asked not to drink caffeine-containing beverages or alcohol and to refrain from vigorous training 24 h before testing to avoid confounding influences on cognitive and physical performance.

2.3. *Measurements*

2.3.1. Anthropometric Data

To unify the working situation, the height of the working position needed to be adjusted to body height including the arm length and the hand/elbow position during the drilling task. Therefore, next to common anthropometrics like body height and body mass, the fingertip height and the fingertip height with the elbow at shoulder height with 90-degree flexion were also measured.

2.3.2. Drilling Performance

To quantify the working performance and facilitate segmentation of the acquired signals, force sensors and accelerometers were attached to the overhead working set-up and the weights that would be lifted. To determine the proper overhead height, the method of Sood et al. [25] was used (hand height with the shoulder and elbow in a 90-degree angle + 0.4 × (hand height with the arm in full extension—hand height with the shoulder and elbow in a 90-degree angle). To evaluate the precision performance in overhead work, a custom task with high to moderate test–retest reliability (adapted from Kim et al. [26]) was developed. In this task, participants used an electric screwdriver (Black & Decker, New Britain, CT, USA, 1.14 kg) to tighten 20 pre-inserted bolts in an aluminum plate positioned at overhead height. We determined that a task was successfully completed until all 20 bolts were tightened (mean task duration). The bolts were partially screwed into the plate using self-fixing nuts, standardizing the distance between the bolt heads and the plate. A plexiglass mask was placed over the bolts to interfere with the screwdriver tip if the screwing trajectory deviated by more than ±2° from perpendicular to the plate [26]. Participants pressed a push button at pelvic crest height, tightened a bolt at overhead height, and pressed the button again (cf., Figure 2), repeating this process for all bolts. They were instructed to perform the task quickly and accurately, minimizing errors. Sensorized aluminum and plexiglass plates tracked the contact between the screwdriver bit and the bolt, as well as screwing errors. Pressing the push button marked the start of each movement.

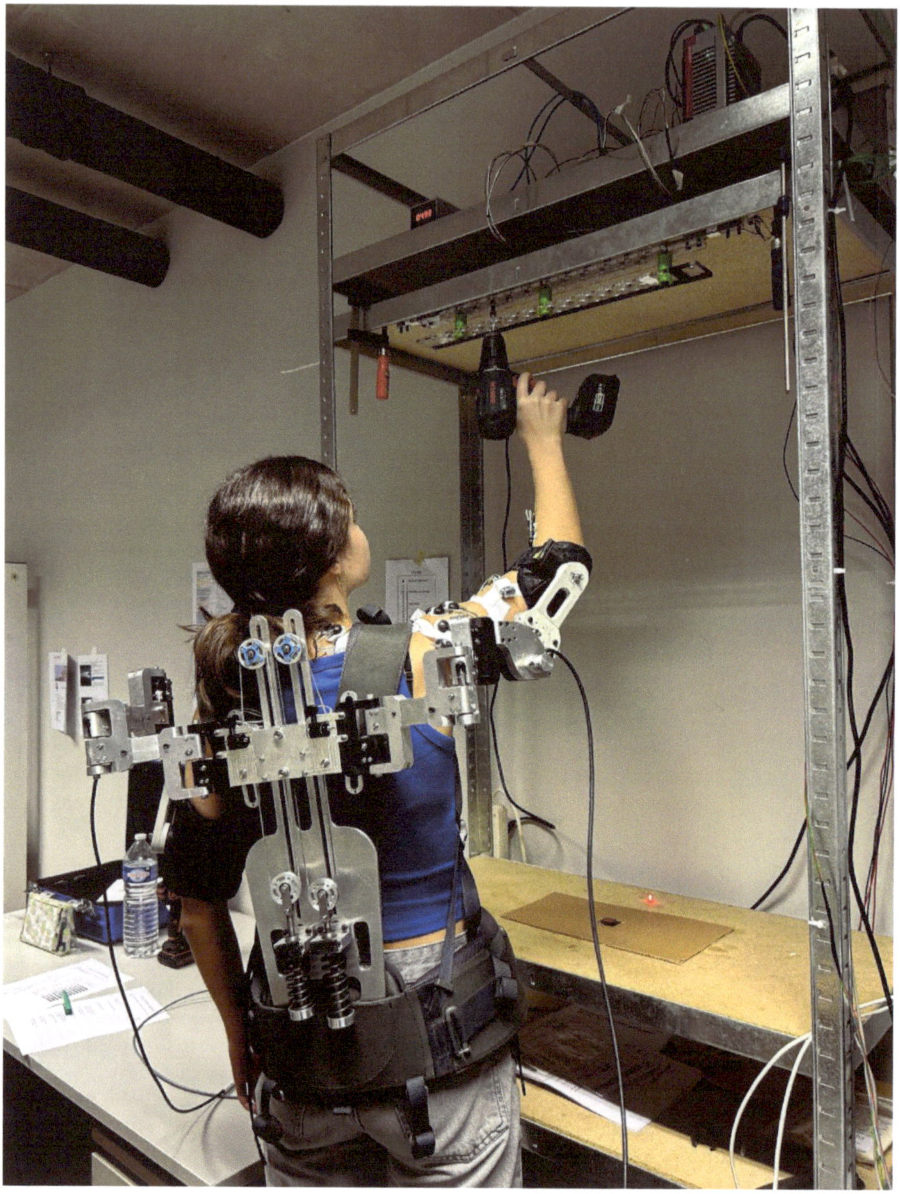

Figure 2. Laboratory setting with the execution of the industrial overhead task.

During the overhead drilling task, the exoskeleton's encoder data and load cell data of the sensorized overhead work bench were acquired at a frequency of 1000 Hz, using Beckhoff components (Beckhoff, Verl, Germany) interconnected via an Ethercat communication protocol. Pressing the push button at hip-level indicated movement initiation. The load cells connected to the bolts allowed tracking when the screwdriver bit was touching a bolt. The encoder data were used to differentiate between lifting the arm from hip level towards the bolts, and the slower aiming process before the bolts were reached. These data allowed all acquired data to be segmented into a lifting and aiming phase, and operative phase. In the latter phase, the bolts were tightened. Data from the load cells connected to the plexiglass surrounding the bolts documented the error. In order to encompass the duration of the error, the error score was computed as the time-integrated output from the force sensors connected to the plexiglass plates.

The Exo4Work's right shoulder joint, fitted with an RM22 Rotary Magnetic Encoder, tracked shoulder elevation angle (1 kHz; RLS, Komenda, Slovenia). All measurements were recorded on the right side, so tasks were performed right-handed, with the left hand at rest.

2.3.3. Perceived Exertion BORG Scale

For the male participants in this study, the 10-point Borg scale was used [27,28].

For the second part of this study, including female participants, the subjective assessment of perceived exertion was carried out using the 100-point Borg scale [27–30], which is more precise in accounting for the RPE. This scale allows participants to subjectively assess the intensity of exertion after each drilling trial, with 0 representing no exertion and 100 indicating maximal effort. The assessment refers to the overall perceived exertion and specific body regions, such as the neck, shoulders, arms, upper and lower back, and legs. A chart was shown for this purpose. The rating of the perceived exertion was expressed verbally by the participants.

The familiarization to the RPE scales was performed by verbal instructions in the initial session. Familiarizing participants with the scale involved presenting the scale and asking for their ratings during the experimental trial. This process helped participants become accustomed to the task and measurement tools being used.

To compare both CR scales, we downscaled the female 100 CR scale to the 10-point scale.

2.3.4. System Usability Scale (SUS)

The System Usability Scale (SUS) provides a reliable tool for measuring usability, comprising ten questions with response options ranging from "Strongly Agree" to "Strongly Disagree" [31]. This tool facilitates the assessment of diverse products and services, encompassing hardware, software, mobile devices, websites, and applications.

2.3.5. Passive Shoulder Exoskeleton Characteristics

The upper-body exoskeleton (Exo4Work) used in this study was developed by the research group of the Vrije Universiteit Brussel and KU Leuven. This industrial exoskeleton weighs 3.8 kg and features 6 degrees of freedom to ensure compatibility with the glenohumeral joint. The support offered by the exoskeleton can be modulated by adjusting the pretension of the spring, an element of the passive remote actuation mechanism. The exoskeleton can provide upper-body assistance of 0.5 Nm within a shoulder flexion range of 0–35° with the support peaking at 3 Nm at 105°. The set-up including pretension of the spring was standardized and kept the same for each subject. The exoskeleton can be worn on the body like a backpack with a hip belt, which ensures the transfer of the weight from the back to the legs. The device was attached to the upper arms with Velcro straps, similar to commercially available devices. For more information about the exoskeleton (kinematics, torque distribution, and hysteresis), readers should refer to Rossini et al., 2021 [32].

2.4. Procedures

All tests took place at the laboratory of the Human Physiology and Sports Physiotherapy Research Group (MFYS, U-Residence, campus Etterbeek). This study was conducted by the standards of the Declaration of Helsinki and the local medical ethics committee (Vrije Universiteit Brussel and Universitair Ziekenhuis Brussel, B.U.N.: 143201941463). Upon initial arrival, the subject was informed about the protocol and signed an informed consent.

The experiment consisted of a total of three laboratory visits in which general participant characteristics (e.g., body height and weight, measurements of body parts) were initially recorded (visit 1) and participants were introduced to the experimental protocol, the laboratory environment, and the Exo4Work exoskeleton (familiarization)—a passive upper-body exoskeleton (PSE). Over the following two visits, the participants completed the experimental protocol, which followed a 2 × 2 randomized counterbalanced cross-over approach (with and without an exoskeleton). Each trial lasted approximately 1.5 h. In between the first and the second visit, at least 48 h was scheduled. In between the second and the third laboratory visit, 6 to 9 days was foreseen.

Familiarization

The first visit to the lab included familiarization with the study protocol and the exoskeleton to get to know the routine and reduce learning effects throughout the experimental trials. This involved the execution of the overhead precision task with the Exo4Work with a duration up to, 12 times including with a 3 min break in between each repetition. The session was terminated when the participants became physically very tired.

After completing the task, participants rated their physical fatigue on the 10-point or 100-point Borg Scale and completed the SUS questionnaire to assess system usability.

2.5. Statistics

To analyze the data, one-way ANOVA was performed to compare demographic variables (body weight and height, BMI, fingertip height, and fingertip height with elbow at 90 degrees) to detect gender differences. Moreover, a two-way ANOVA was conducted to compare male and female data in terms of performance, perceived exertion, and scoring of the SUS. The BMI was included as a covariate to control for its potential influence on the outcomes. A Bonferroni correction was applied to adjust the alpha level for multiple comparisons. For all statistical tests, the significance level was set at $\alpha = 0.05$. Statistical analysis was conducted with SPSS 29.0.

3. Results

3.1. Drilling Performance

As shown in Table 2, there were no gender differences for the mean task duration; however, the task duration changed with exoskeleton use and increased for females and decreased for males ($F(1,33) = 0.954$; $p = 0.327$; $eta^2 = 0.030$). Nevertheless, this difference failed to be significant if the data were controlled for BMI (cf., Table 2 and Figure 3).

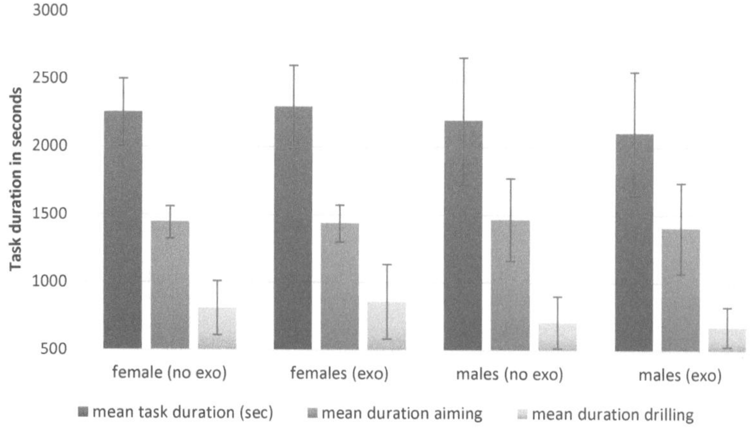

Figure 3. Results of the task duration for the different conditions.

Table 2. Task performance data of both genders (with and without exoskeleton use).

Performance	Female noExo (M ± SD)	Female Exo (M ± SD)	Male noExo (M ± SD)	Male Exo (M ± SD)	Gender Differences F p eta^2	Interactions no Exo/Exo and Gender F p eta^2
mean task duration (s)	2253.7 ± 246	2295.1 ± 307.2	2194.9 ± 467.1	2104 ± 455.3	0.954 0.327 0.030	3.492 0.071 0.098
mean duration aiming	1443 ± 120.7	1435.5 ± 138.7	1465 ± 305.8	1405 ± 336.8	0.142 0.709 0.005	1.257 0.271 0.038
mean duration drilling	810.5 ± 200.8	859.6 ± 275.8	708.6 ± 191.2	678.5 ± 146.5	2.149 0.152 0.063	1.609 0.214 0.048
mean error integral aiming	0.133 ± 0.031	0.165± 0.043	0.313 ± 0.137	0.279 ± 0.120	**25.153** **<0.001** **0.440**	**5.509** **0.025** **0.147**
mean error integral drilling	0.102 ± 0.034	0.144 ± 0.163	0.274 ± 0.151	0.264 ± 0.133	**16.766** **<0.001** **0.344**	1.717 0.199 0.051

Note: M = mean, SD = standard deviation.

However, there were gender differences for the mean error integral for aiming ($F(1,33) = 25.153$; $p < 0.001$; eta$^2 = 0.440$), even when controlling for BMI ($F(1,33) = 5.509$; $p < 0.001$; eta$^2 = 0.147$). Furthermore, there were differences in the mean error integral for drilling with overall higher errors for the males ($F(1,33) = 16.766$; $p < 0.001$; eta$^2 = 0.344$). Moreover, for the mean errors of aiming, the error rate increased significantly with exoskeleton use for the aiming phase, while the errors of males decreased (cf., Table 2 and Figure 4).

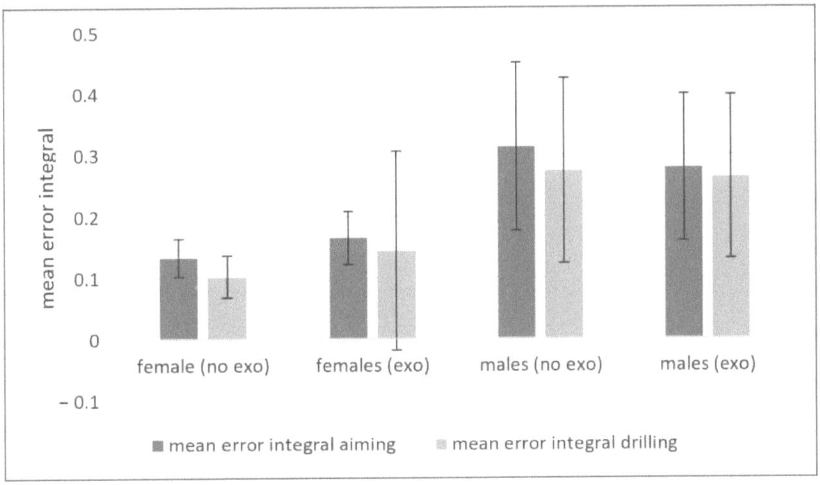

Figure 4. Results of the mean error integrals for the different conditions.

3.2. Perceived Exertion

The RPE analysis did not show any differences in the main effect of condition (with and without exoskeleton) and gender. Both genders had a score of 4.0 ± 1 for working without an exoskeleton and 3.8 ± 1.5 for working with an exoskeleton ($F(1,33) = 0.472$;

$p = 0.397$, $eta^2 = 0.014$). There was a between-subject effect for the BMI ($F = 11.985$; $p = 0.033$; $eta^2 = 0.131$).

3.3. System Usability

The evaluation of the system usability did not lead to significant gender differences for the SUS score (female score: 31.3 ± 16.7; male score 22.8 ± 11.1; $F(1,30) = 2.825$; $p = 0.103$; $eta^2 = 0.086$).

4. Discussion

Motor precision, or the ability to execute movements with accuracy and control, varies between males and females when using occupational exoskeletons, influenced by a multitude of factors. While specific studies on motor precision differences in gender, regarding exoskeleton use, are limited, general trends in motor control and biomechanical differences may provide valuable insights. Females tend to have superior fine motor control and precision in certain tasks including overhead work, which could influence how they use an exoskeleton for precision-dependent activities. Males have greater gross motor control, which could impact tasks requiring strength and broad movements more than those requiring fine precision. The movement patterns are determined by the muscle activation patterns [17,33]. These patterns can influence how the exoskeleton assists with precision tasks, potentially making it more effective for one gender over the other depending on the design. Therefore, this study aimed to compare the performance of an overhead precision drilling task between males and females with and without exoskeleton use. We hypothesized that females would exhibit slower task performance, especially when using the exoskeleton, potentially attributed to greater discomfort.

4.1. Performance

Surprisingly, the results of the mean duration of the task performance during the drilling task showed no significant differences for females and males. We supposed this in line with previous research, suggesting that females generally take more time in tasks requiring fine motor control due to differences in muscle strength and endurance [34]. However, controlling the analysis by the covariate BMI, these expected differences were not observed.

Next to this, for the main duration of the task performance as well as for the drilling part, the values worsened for the females and improved for the males (cf., Table 2), but this difference failed to be significant ($p = 0.071$). These findings were also not expected. We assumed that the use of an upper-body exoskeleton might mitigate potential differences, reducing the task duration for both genders, and that exoskeletons can be effective in equalizing performance between genders by compensating for physiological differences, particularly in tasks requiring sustained muscle engagement [35]. However, the exoskeleton support did not alter drilling times across genders, implying that drilling efficiency is less influenced by exoskeleton support and more dependent on individual skill or strength, which aligns with previous findings that highlight gender differences in tasks requiring upper-body strength [6]. Moreover, women typically have a greater range of motion in the shoulder joints, which can affect how they interact with the exoskeleton [33]. This increased flexibility might require adjustments in the exoskeleton's design to accommodate a wider range of movement without causing discomfort or reducing effectiveness. It has to be noted that these results were observed including the BMI as a control variable in the measurements following the results of a systematic review by Cote [17]. As performance differences in movement patterns might also be explained by strength and endurance, and not by gender itself, these aspects should be integrated into future study designs to avoid a potential bias in the data [17].

Regarding the errors, the mean error integral for aiming and drilling was higher in males, which might be attributed to a more aggressive or less precise approach often observed in male participants in similar tasks [36]. Men and women often exhibit different movement patterns and strategies for performing tasks. Men rely more on strength, while women might utilize more coordinated and precise movements. These differences can influence how effectively each gender uses the exoskeleton to enhance performance and reduce strain. Interestingly, the use of an exoskeleton increased the error rate for aiming in females, while it decreased for males. This could indicate that while the exoskeleton provides stabilization that benefitted males, it could have imposed constraints on females, potentially due to differences in body mechanics and anthropometrics or how each gender interacts with assistive devices [37,38]. The observed increased errors for females suggest that exoskeletons might need to be tailored differently for each gender to optimize performance, aligning with previous research indicating that assistive technologies often need gender-specific adjustments [39].

4.2. Perceived Exertion

Contrary to the performance metrics, the analysis of perceived exertion (RPE) revealed no significant differences between genders, either with or without the exoskeleton. Both genders reported similar exertion levels, suggesting that the exoskeleton's support did not differentially impact the subjective workload of the task. This finding contrasts with studies that have reported higher perceived exertion in females during physically demanding tasks, possibly due to different muscle fatigue rates [34,40]. However, the absence of a significant difference in this study may be due to the relatively moderate intensity of the task.

The notable effect of BMI on perceived exertion (with a higher BMI associated with increased exertion) underscores the importance of considering individual physical characteristics in the design and use of exoskeletons. This finding aligns with the literature that highlights the influence of body composition on task performance and perceived exertion, especially in tasks requiring sustained upper-body engagement [35].

4.3. System Usability

The evaluation of system usability while executing tasks with the upper-body exoskeleton, using the System Usability Scale (SUS), revealed no significant gender differences, though the scores were marginally higher for females. Despite the observed gender-specific differences in performance, both genders reported similar levels of usability for the exoskeleton, suggesting that the device is generally well suited for both males and females in terms of usability. The lack of significant differences in SUS scores could imply that the perceived usability of the exoskeleton was not strongly affected by the actual performance outcomes. This is consistent with studies suggesting that usability perceptions are often more influenced by the subjective experience of ease of use rather than objective performance metrics [31,41,42]. Moreover, there is some evidence that exoskeleton acceptance combined with training on self-efficacy to use the device was associated with the willingness to continue using the device [43].

The slightly higher SUS scores for females, though not statistically significant, might indicate a tendency for females to rate the usability of assistive devices more positively. However, given the small effect size, this observation warrants further investigation with larger sample sizes or diverse usability measures and setting options to determine if there is a meaningful difference in usability perceptions between genders. This could provide a more comprehensive understanding of any subtle gender-related perceptions or preferences, e.g., adaptations to physical requirements, especially for female users regarding system usability.

4.4. Strength and Limitations

One strength of this study is the age- and gender-balanced cohort of this study, securing comparability of the task performance. Moreover, the experiments were conducted in a highly standardized environment mimicking a real industrial environment and controlling for external factors that could have influenced performance.

However, there are also some potential limitations. We found a significant influence of BMI on the performance data. We suppose that the individual upper-extremity maximal strength might have gained more insights into the performance differences between the participants. Therefore, we suggest controlling for maximum strength and strength endurance for future gender comparisons.

Next to this aspect, the experimental set-up was designed and restricted for right-handed persons. Therefore, the results do not represent potential differences in the performance of left-handed individuals and could not be generalized for left-handed persons.

We have to address the fact that the individuals′ shoulder angle might differ from the exoskeleton shoulder support angle. We only measured the angle of the exoskeleton shoulder joint.

Moreover, the participants were not recruited in industrial settings and were therefore not familiar with the task performance. On the other hand, as our analysis integrated in-between subject comparisons for the task conditions with and without exoskeletons as well as a randomized order within the experiment, this disadvantage of the participant group might be relatively small.

Additionally, the use of a specific drilling platform in this study might limit replication in future studies. Nevertheless, future studies should also use real-world industrial scenarios to confirm whether there are differences in lab versus real-world studies.

In summary, further research is needed to develop exoskeletons that can be easily adjusted to accommodate a wide range of body types and movement patterns, ensuring that both men and women can benefit equally from these advanced assistive technologies.

5. Conclusions

The analysis showed significant gender-related variations in task performance, particularly in the error rates during aiming and drilling.

The effectiveness of the exoskeleton in assisting with specific tasks vary between genders due to differences in strength, endurance, and movement patterns. Women might benefit more from exoskeletons that enhance precision and fine motor control, while men might see greater benefits from those that augment brute strength and load-bearing capacity. However, some of these effects are reduced if the performance data are controlled for BMI.

Nevertheless, these differences did not translate into changes in perceived exertion or system usability, indicating that while performance metrics may vary, the subjective experience of using the exoskeleton remains similar across genders.

Ensuring that the exoskeleton is anthropometrically designed to fit both male and female body types is crucial. An ill-fitting exoskeleton can lead to decreased performance, discomfort, and even injury. Adjustability in key areas, such as shoulder straps, torso supports, and arm cuffs, can help accommodate different body shapes and sizes.

Author Contributions: Conceptualization, S.D.B. and K.D.P.; methodology, S.D.B., K.D.P., B.W. and J.G.; software, S.D.B.; validation, S.D.B., K.D.P. and B.W.; formal analysis, S.D.B., J.G., B.W., M.A.D. and E.A.; investigation, S.D.B., M.A.D. and B.W.; resources, K.D.P.; data curation, S.D.B., M.A.D., E.A., B.W. and J.G.; writing—original draft preparation, B.W. and J.G.; writing—review and editing, B.W., J.G., S.D.B., M.A.D., E.A. and K.D.P.; visualization, J.G., B.W. and S.D.B.; supervision, B.W. and K.D.P.; project administration, S.D.B. and K.D.P.; funding acquisition, K.D.P. and B.W. All authors have read and agreed to the published version of the manuscript.

Funding: This work was supported by the Brussels Institute of Advanced Studies (Grant: BrIAS2024). Maria Alejandra Diaz and Eligia Alfio are funded by the Federal Public Service for Policy and Support (AidWear project).

Institutional Review Board Statement: This study was conducted in accordance with the Declaration of Helsinki and approved by the Medical Ethics Committee of the VUB and UZ Brussel (B.U.N. 143201941463).

Data Availability Statement: The data set can be requested via the corresponding author.

Acknowledgments: We would like to thank Nina Heijens and Alessandra Clara Preckher for their assistance in gathering data.

Conflicts of Interest: The authors declare no conflicts of interest.

References

1. Reid, C.R.; Nussbaum, M.A.; Gregorczyk, K.; Harris-Adamson, C.; Kyte, K.; Lowe, B.; Smets, M.; Zmijewski, R. Industrial exoskeletons: Are we ready for prime time yet? *Proc. Hum. Factors Ergon. Soc. Annu. Meet.* **2017**, *61*, 1000–1004. [CrossRef]
2. McFarland, T.; Fischer, S. Considerations for industrial use: A systematic review of the impact of active and passive upper limb exoskeletons on physical exposures. *IISE Trans. Occup. Ergon. Hum. Factors* **2019**, *7*, 322–347. [CrossRef]
3. Theurel, J.; Desbrosses, K. Occupational Exoskeletons: Overview of Their Benefits and Limitations in Preventing Work-Related Musculoskeletal Disorders. *IISE Trans. Occup. Ergon. Hum. Factors* **2019**, *7*, 264–280. [CrossRef]
4. Crea, S.; Beckerle, P.; de Looze, M.; de Pauw, K.; Grazi, L.; Kermavnar, T.; Masood, J.; O'Sullivan, L.W.; Pacifico, I.; Rodriguez-Guerrero, C. Occupational exoskeletons: A roadmap toward large-scale adoption. Methodology and challenges of bringing exoskeletons to workplaces. *Wearable Technol.* **2021**, *2*, e11. [CrossRef] [PubMed]
5. Leibman, D.; Choi, H. Individual Differences in Body Mass, Biological Sex, and Physical Fitness Affecting Human-Exoskeleton Interactions. *Proc. Hum. Factors Ergon. Soc. Annu. Meet.* **2023**, *67*, 1794–1800. [CrossRef]
6. Miller, A.E.J.; MacDougall, J.D.; Tarnopolsky, M.A.; Sale, D.G. Gender differences in strength and muscle fiber characteristics. *Eur. J. Appl. Physiol. Occup. Physiol.* **1993**, *66*, 254–262. [CrossRef] [PubMed]
7. Maughan, R.J.; Watson, J.S.; Weir, J. Strength and cross-sectional area of human skeletal muscle. *J. Physiol.* **1983**, *338*, 37–49. [CrossRef]
8. Ryman Augustsson, S.; Bersås, E.; Magnusson Thomas, E.; Sahlberg, M.; Augustsson, J.; Svantesson, U. Gender differences and reliability of selected physical performance tests in young women and men. *Adv. Physiother.* **2009**, *11*, 64–70. [CrossRef]
9. Ojelade, A.; Morris, W.; Kim, S.; Kelson, D.; Srinivasan, D.; Smets, M.; Nussbaum, M.A. Three passive arm-support exoskeletons have inconsistent effects on muscle activity, posture, and perceived exertion during diverse simulated pseudo-static overhead nutrunning tasks. *Appl. Ergon.* **2023**, *110*, 104015. [CrossRef]
10. de Bock, S.; Ampe, T.; Rossini, M.; Tassignon, B.; Lefeber, D.; Rodriguez-Guerrero, C.; Roelands, B.; Geeroms, J.; Meeusen, R.; de Pauw, K. Passive shoulder exoskeleton support partially mitigates fatigue-induced effects in overhead work. *Appl. Ergon.* **2023**, *106*, 103903. [CrossRef]
11. Srinivasan, D.; Sinden, K.E.; Mathiassen, S.E.; Côté, J.N. Gender differences in fatigability and muscle activity responses to a short-cycle repetitive task. *Eur. J. Appl. Physiol.* **2016**, *116*, 2357–2365. [CrossRef] [PubMed]
12. Janssen, I.; Heymsfield, S.B.; Wang, Z.; Ross, R. Skeletal muscle mass and distribution in 468 men and women aged 18–88 yr. *J. Appl. Physiol.* **2000**, *89*, 81–88. [CrossRef] [PubMed]
13. Lamers, E.P.; Soltys, J.C.; Scherpereel, K.L.; Yang, A.J.; Zelik, K.E. Low-profile elastic exosuit reduces back muscle fatigue. *Sci. Rep.* **2020**, *10*, 15958. [CrossRef] [PubMed]
14. Barnes, C.J.; Van Steyn, S.J.; Fischer, R.A. The effects of age, sex, and shoulder dominance on range of motion of the shoulder. *J. Shoulder Elb. Surg.* **2001**, *10*, 242–246. [CrossRef]
15. Kibler, W.B.; Sciascia, A.D.; Uhl, T.L.; Tambay, N.; Cunningham, T. Electromyographic analysis of specific exercises for scapular control in early phases of shoulder rehabilitation. *Am. J. Sports Med.* **2008**, *36*, 1789–1798. [CrossRef]
16. Schmalz, T.; Schändlinger, J.; Schuler, M.; Bornmann, J.; Schirrmeister, B.; Kannenberg, A.; Ernst, M. Biomechanical and metabolic effectiveness of an industrial exoskeleton for overhead work. *Int. J. Environ. Res. Public Health* **2019**, *16*, 4792. [CrossRef]
17. Côté, J.N. A critical review on physical factors and functional characteristics that may explain a sex/gender difference in work-related neck/shoulder disorders. *Ergonomics* **2012**, *55*, 173–182. [CrossRef]
18. Ashta, G.; Finco, S.; Battini, D.; Persona, A. Passive Exoskeletons to Enhance Workforce Sustainability: Literature Review and Future Research Agenda. *Sustainability* **2023**, *15*, 7339. [CrossRef]
19. Gräf, J.; Grospretre, S.; Argubi-Wollesen, A.; Wollesen, B. Impact of a passive upper-body exoskeleton on muscular activity and precision in overhead single and dual tasks: An explorative randomized crossover study. *Front. Neurol.* **2024**, *15*, 1405473. [CrossRef]
20. Kranenborg, S.E.; Greve, C.; Reneman, M.F.; Roossien, C.C. Side-effects and adverse events of a shoulder-and back-support exoskeleton in workers: A systematic review. *Appl. Ergon.* **2023**, *111*, 104042. [CrossRef]
21. Gutierrez, N.; Ojelade, A.; Kim, S.; Barr, A.; Akanmu, A.; Nussbaum, M.A.; Harris-Adamson, C. Perceived benefits, barriers, perceptions, and readiness to use exoskeletons in the construction industry: Differences by demographic characteristics. *Appl. Ergon.* **2024**, *116*, 104199. [CrossRef] [PubMed]
22. Cardoso, A.; Colim, A.; Sousa, N. The effects of a passive exoskeleton on muscle activity and discomfort in industrial tasks. In *Occupational and Environmental Safety and Health II*; Springer: Cham, Switzerland, 2020; pp. 237–245.
23. Luger, T.; Bosch, T.; Hoozemans, M.; de Looze, M.; Veeger, D. Task variation during simulated, repetitive, low-intensity work–influence on manifestation of shoulder muscle fatigue, perceived discomfort and upper-body postures. *Ergonomics* **2015**, *58*, 1851–1867. [CrossRef] [PubMed]

24. Luger, T.; Bär, M.; Seibt, R.; Rieger, M.A.; Steinhilber, B. Using a back exoskeleton during industrial and functional tasks—Effects on muscle activity, posture, performance, usability, and wearer discomfort in a laboratory trial. *Hum. Factors* **2023**, *65*, 5–21. [CrossRef]
25. Sood, D.; Nussbaum, M.A.; Hager, K. Fatigue during prolonged intermittent overhead work: Reliability of measures and effects of working height. *Ergonomics* **2007**, *50*, 497–513. [CrossRef]
26. Kim, S.; Nussbaum, M.A.; Esfahani, M.I.M.; Alemi, M.M.; Alabdulkarim, S.; Rashedi, E. Assessing the influence of a passive, upper extremity exoskeletal vest for tasks requiring arm elevation: Part I–"Expected" effects on discomfort, shoulder muscle activity, and work task performance. *Appl. Ergon.* **2018**, *70*, 315–322. [CrossRef] [PubMed]
27. Borg, G. *Borg's Perceived Exertion and Pain Scales*; Human Kinetics: Champaign, IL, USA, 1998.
28. Borg, G.A.V. Psychophysical bases of perceived exertion. *Med. Sci. Sports Exerc.* **1982**, *14*, 377–381. [CrossRef] [PubMed]
29. Borg, E.; Borg, G. A comparison of AME and CR100 for scaling perceived exertion. *Acta Psychol.* **2002**, *109*, 157–175. [CrossRef]
30. Borg, G.; Borg, E. A new generation of scaling methods: Level-anchored ratio scaling. *Psychologica* **2001**, *28*, 15–45.
31. Brooke, J. SUS: A "quick and dirty" Usability Scale. *Usability Eval. Ind./Taylor Fr.* **1996**, *189*, 4–7.
32. Rossini, M.; De Bock, S.; van der Have, A.; Flynn, L.; Rodriguez-Cianca, D.; De Pauw, K.; Lefeber, D.; Geeroms, J.; Rodriguez-Guerrero, C. Design and evaluation of a passive cable-driven occupational shoulder exoskeleton. *IEEE Trans. Med. Robot. Bionics* **2021**, *3*, 1020–1031. [CrossRef]
33. Renda, E.; Lamanuzzi, S.; Dal Maso, F.; Côté, J.N. The effects of hand dominance, fatigue, and sex on muscle activation during a repetitive overhead fatiguing task. *Hum. Mov. Sci.* **2023**, *92*, 103149. [CrossRef] [PubMed]
34. Hunter, S.K. Sex differences in human fatigability: Mechanisms and insight to physiological responses. *Acta Physiol.* **2014**, *210*, 768–789. [CrossRef] [PubMed]
35. Schwartz, M.; Desbrosses, K.; Theurel, J.; Mornieux, G. Biomechanical consequences of using passive and active back-support exoskeletons during different manual handling tasks. *Int. J. Environ. Res. Public Health* **2023**, *20*, 6468. [CrossRef] [PubMed]
36. Liutsko, L.; Muiños, R.; Tous Ral, J.M.; Contreras, M.J. Fine motor precision tasks: Sex differences in performance with and without visual guidance across different age groups. *Behav. Sci.* **2020**, *10*, 36. [CrossRef]
37. Raghuraman, R.N.; Barbieri, D.F.; Aviles, J.; Srinivasan, D. Age and gender differences in the perception and use of soft vs. rigid exoskeletons for manual material handling. *Ergonomics* **2024**, 1–18. [CrossRef]
38. Roda-Sales, A.; Vergara, M.; Sancho-Bru, J.L.; Gracia-Ibáñez, V.; Jarque-Bou, N.J. Effect of assistive devices on hand and arm posture during activities of daily living. *Appl. Ergon.* **2019**, *76*, 64–72. [CrossRef]
39. Perry, J.C.; Rosen, J.; Burns, S. Upper-limb powered exoskeleton design. *IEEE/ASME Trans. Mechatron.* **2007**, *12*, 408–417. [CrossRef]
40. Hunter, S.K. Sex differences and mechanisms of task-specific muscle fatigue. *Exerc. Sport Sci. Rev.* **2009**, *37*, 113–122. [CrossRef] [PubMed]
41. Sauer, J.; Sonderegger, A.; Schmutz, S. Usability, user experience and accessibility: Towards an integrative model. *Ergonomics* **2020**, *63*, 1207–1220. [CrossRef]
42. Sonderegger, A.; Sauer, J. The influence of design aesthetics in usability testing: Effects on user performance and perceived usability. *Appl. Ergon.* **2010**, *41*, 403–410. [CrossRef]
43. Siedl, S.M.; Mara, M. Exoskeleton acceptance and its relationship to self-efficacy enhancement, perceived usefulness, and physical relief: A field study among logistics workers. *Wearable Technol.* **2021**, *2*, e10. [CrossRef] [PubMed]

biomimetics

MDPI

Article

The Design of the Dummy Arm: A Verification Tool for Arm Exoskeleton Development

Suzanne J. Filius [1,*,†], **Bas J. van der Burgh** [2,†] and **Jaap Harlaar** [1,3]

1 Department of BioMechanical Engineering, Delft University of Technology, 2628 CD Delft, The Netherlands; j.harlaar@tudelft.nl
2 Department of Precision and Microsystems Engineering, Delft University of Technology, 2628 CD Delft, The Netherlands
3 Department of Orthopedics and Sports Medicine, Erasmus MC University Medical Center, 3015 GD Rotterdam, The Netherlands
* Correspondence: s.j.filius@tudelft.nl
† These authors contributed equally to this work.

Abstract: Motorised arm supports for individuals with severe arm muscle weakness require precise compensation for arm weight and elevated passive joint impedance (e.g., joint stiffness as a result of muscle atrophy and fibrosis). Estimating these parameters in vivo, along with the arm's centre of mass, is challenging, and human evaluations of assistance can be subjective. To address this, a dummy arm was designed to replicate the human arm's anthropometrics, degrees of freedom, adjustable segment masses, and passive elbow joint impedance (eJimp). This study presents the design, anthropometrics, and verification of the dummy arm. It successfully mimics the human arm's range of motion, mass, and centre of mass. The dummy arm also demonstrates the ability to replicate various eJimp torque-angle profiles. Additionally, it allows for the tuning of the segment masses, centres of mass, and eJimp to match a representative desired target population. This simple, cost-effective tool has proven valuable for the development and verification of the Duchenne ARm ORthosis (DAROR), a motorised arm support, or 'exoskeleton'. This study includes recommendations for practical applications and provides insights into optimising design specifications based on the final design. It supplements the CAD design, enhancing the dummy arm's application for future arm-assistive devices.

Keywords: dummy limb; mechanical phantom limb; exoskeleton evaluation; arm replica; upper extremity; joint impedance

Citation: Filius, S.J.; van der Burgh, B.J.; Harlaar, J. The Design of the Dummy Arm: A Verification Tool for Arm Exoskeleton Development. *Biomimetics* **2024**, *9*, 579. https://doi.org/10.3390/biomimetics9100579

Academic Editors: Rafhael Milanezi de Andrade and Stanislav N. Gorb

Received: 30 July 2024
Revised: 4 September 2024
Accepted: 20 September 2024
Published: 24 September 2024

1. Introduction

Assistive wearable technologies, such as motorised orthosis or exoskeletons, work in parallel to the human skeleton to support or augment human motion in terms of strength, endurance, or function [1–4]. The application of such devices varies [5], from industrial [4,6] to military [7] to medical (e.g., rehabilitative [8], or daily assistive [9]) use. Developing such devices requires extensive human testing and evaluation, which brings several challenges.

The first challenge with human testing relates to safety, especially at the beginning of the development cycle, as it is difficult to predict how new sensors or software systems behave when interacting with humans. For instance, failure in safety limits or unexpected behaviour might result in fast, unexpected movements of the limb or even extend the limb beyond its natural joint limits, potentially resulting in dangerous situations. Moreover, when developing exoskeletons for a vulnerable population, such as people with severe muscle weakness or children, obtaining (medical) ethical approval and recruiting subjects that are willing to be exposed to the potential risks are time-consuming processes and are therefore not feasible for every phase of development [1]. In addition to safety issues, developing an exoskeleton is time-consuming, labour-intensive, and, therefore, costly.

Furthermore, for exoskeletons that aim to compensate for the weight of the human limb, it is essential to accurately know the properties of the limb to verify the controller's performance. However, the properties of human limb, such as their weight and centre of mass (CoM), can only be estimated indirectly (e.g., using tables from the literature [10,11]). Especially in the case of populations with neuromuscular disorders, the anthropometrics might deviate from the available tables due to, for instance, muscle atrophy, bone density reductions, joint contractures, or deformations.

Moreover, it is difficult for humans to objectively quantify the amount of assistance provided or distinguish between small changes in controller settings, leading to subjective or even biased evaluations of the controller's performance.

During the development of the Duchenne ARm ORthosis (DAROR), as described in [12], these challenges led to the need for a mechanical phantom. Such a phantom, like a Dummy Arm, with similar human-limb characteristics, can serve as a tool for verification for various compensation strategies, and enhances the fast iteration of exoskeleton control development [1]. The DAROR is an investigational motorised arm exoskeleton that enhances arm function. It is intended to provide assistance in daily activities for people with neuromuscular diseases such as Duchenne Muscular Dystrophy (DMD) that cope with severe muscle weakness and elevated joint stiffness, or, more accurately, 'passive joint impedance' (pJimp). The impedance controller of the DAROR aims to compensate for the weight of the user's arm and the elbow pJimp (denoted as eJimp for the remainder of this work). The term pJimp describes the passive mechanical impedance against motion [13] resulting from the passive tissue around the joint and limb inertia. For instance, in people with DMD, this pJimp is elevated, presumably resulting from muscle atrophy, higher levels of connective tissue, and the development of joint contractures [14].

In the literature, few examples of dummy limbs were found, describing mannequins with a primary focus on mimicking the outer dimensions of the human limb [15–17]. Others, described rather complex and costly designs which equipped with sensors and actuators or soft tissues [1,5,18,19]. For the development of the DAROR, an upper-extremity phantom limb that is cost-effective, simple in design, and mimics the characteristics of the human arm, including eJimp, was required to serve as a verification tool for the compensation models.

This work presents the design and verification of a relatively simple, reproducible, cost-effective dummy arm that mimics human arm characteristics. The novelty of this design is that it allows us to replicate (elevated) eJimp and is easily adjustable in mass and CoM to fit various exoskeleton target populations. By sharing our design, we promote the development of future arm exoskeletons and facilitate the rapid iteration and objective evaluation of exoskeleton controllers.

2. Materials and Methods

2.1. Dummy Arm Design Requirements

The dummy arm was required to have the following properties:

- A representative and adjustable segment mass and CoM [10,11,20–22];
- A representative forearm anthropometric to fit the DAROR interface sleeve(s);
- A similar DAROR range of motion; see Table 1;
- An attachment at the base of the frame at the shoulder joint;
- An adjustable (linear) eJimp;
- An interface to attach additional objects at the wrist location.

Table 1. DAROR shoulder range of motion.

Joint Rotation	Target
GH elevation	11° to 137°
GH horizontal	−46° to 138°
GH axial	−113° to 65°
El flexion/extension	2° to 120°

Abbreviation: GH, glenohumeral joint; El, elbow.

2.2. Dummy Arm Design

The design of the dummy arm can be divided into several sub-design problems: the shoulder joint, the elbow joint, the elbow joint impedance, the connection frame, and the mass of the upper and lower arm. The design of the dummy arm, with its components together with the DAROR, is shown in Figure 1.

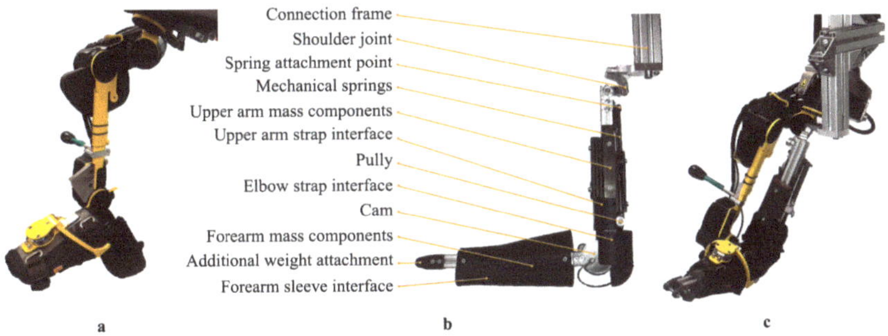

Figure 1. Picture of the (**a**) DAROR exoskeleton, the (**b**) dummy arm design with its components, and (**c**) the dummy arm placed inside the DAROR exoskeleton.

2.2.1. Shoulder Joint

The shoulder joint is a complex joint due to its high number of degrees of freedom (e.g., glenohumeral elevation and horizontal and axial rotation, according to the ISB recommendations [23–25]), as well as its large range of motion. A serial chain of three revolute joints was used to mimic the three degrees of freedom of the shoulder. However, with three revolute joints, several configurations are possible. Each has its advantages and disadvantages with respect to its range of motion and the possible occurrence of gimbal locking. As the range of motion is primarily restricted by the exoskeleton, which consists of three actuators mounted in series at an inclination of approximately 60° with respect to each other [12], a similar configuration for the dummy arm is used. Each of the three revolute joints consists of a set of two ball bearings (MR95-2Z) to minimise friction and provide support to possible bending moments. The joints are connected through 3D-printed or milled aluminium brackets (for the final design milled aluminium brackets were used).

2.2.2. Elbow Joint

The elbow joint can be considered a hinge joint. Therefore, the design of the elbow joint is relatively straightforward, consisting of two ball bearings (625ZZ) mounted to the upper arm frame on either side of an axis and attached to the forearm frame.

2.2.3. Passive Elbow Joint Impedance

The impedance of the dummy arm elbow joint is created using a wrapping cam mechanism. This mechanism consists of a spring attached to a belt, which wraps around a cam attached to the forearm. Upon rotation of the forearm, the belt wraps/unwraps around the cam, stretching/relaxing the spring. This is shown schematically in Figure 2. A cam is attached on either side of the elbow with the belt either wrapping the cam in a clockwise or counterclockwise direction to simulate impedance during both the flexion and extension of the elbow. The current design aimed to mimic the linearised torque profile of the passive eJimp similar to [26], which is achieved using a circular cam. However, different torque profiles can be generated depending on the shape of the cam. The torque-angle profile can be further tuned by changing the spring stiffness (k_s) or the spring's pretension (x_0). An additional advantage of the wrapping cam mechanism is that the springs can be mounted parallel to the upper arm frame, resulting in a compact design.

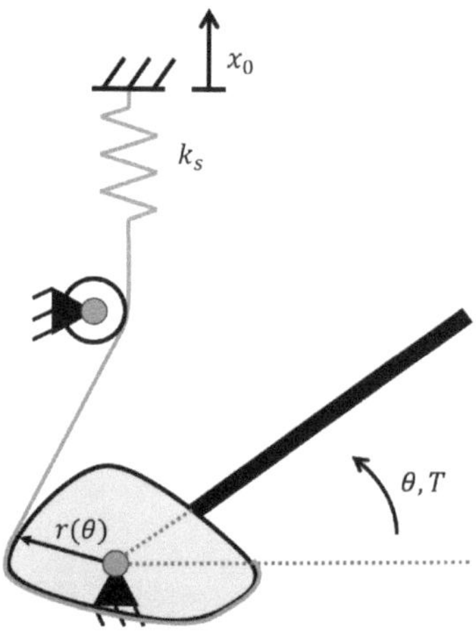

Figure 2. Schematic of the wrapping cam mechanism to generate the desired joint stiffness. Here, k_s is the spring stiffness, θ the angle of the elbow, r is a function of θ, defining the cam shape (for the current prototype, r is constant, resulting in a circular cam), and x_0 is the pretension applied to the spring. The torque T can be expressed as $T = T(\theta, k_s, x_0)$. A detailed discussion on deriving the cam shape for a given torque profile can be found in [27,28].

2.2.4. Connection Frame

To connect the different components (joints, springs, and weights), a skeleton for the upper and lower arm is made of aluminium extrusion beams (20 × 20 mm), which allows the various components to be attached freely along the beam's length. As a result, the system is modular, allowing for easy adjustment of different parts. Additionally, the dummy arm can mimic different human arm sizes using various lengths of beams. A rigid 3D-printed PLA shell is attached to the frame, which mimics the forearm [29] for interfacing with the exoskeleton sleeve.

2.2.5. Mass and Centre of Mass of the Upper and Lower Arm

Steel plates are mounted to the aluminium beams to mimic the overall mass and the CoM of the human forearm and upper arm. The location for the mounting of the steel plates is based on a CAD model, which is used to estimate the CoM. The dimensions of the dummy arm are based on the biometric data from [10], resulting in an approximate length of 250 mm for the lower arm and 300 mm for the upper arm. The segment mass parameters are based on the work of [10,11,20,22], with a mass of approximately 1.6 to 2 kg for the upper arm and 0.9–1.1 kg for the forearm. The mass of the hand (ca. 400 g) is not included, but it can be attached separately alongside additional weights to mimic lifted objects in the hand if deemed necessary. The advantage of using a modular system consisting of aluminium extrusion profiles is that properties such as its mass and CoM can be easily adjusted to the desired application by removing or adding weights and shifting their position. As such, the dummy arm can be adjusted to simulate different arms.

2.2.6. Verification of Design

The verification of the dummy arm consists of two parts. First, the direct measurement of the segment properties of the dummy arm is performed by weighing the segment mass and estimating the CoM in the direction of the limb with respect to the proximal joint. This latter property was measured by hanging the upper and forearm individually to a rod with two ropes connected to the endpoints of the limb segments. The location on the rod where it needed to be held fixed to keep the rod in balance (level) was taken as the CoM location. The second part verifies the realised eJimp with four different spring-type configurations. This is done using the DAROR set-up. The DAROR moves the dummy arm elbow joint (position-controlled) through a range-of-motion cycle (either ca. 10–109° or 40–116°) with approximately 8 static intervals, while the shoulder joint remained in a neutral position (ca. GH elevation: 10°, horizontal 10°, axial 0°). A video showing this procedure is available in the Supplementary Materials. The joint torque is measured by the deflection of the series elastic element in the actuators (torque accuracy 0.5 Nm, torque precision 0.4 Nm, ref. [12]) in the elbow actuator. The range-of-motion cycle is repeated with and without springs attached to the dummy arm elbow joint. First, these torque-angle profiles are fitted to a 5th-order polynomial (polyfit, polyval) using MATLAB (version 2021b). Then, these torque-angle profiles measured with and without the attached springs were subtracted to receive the measured eJimp of the dummy arm. Additionally, a first-order fit of this eJimp was created to compare the realised Jimp slope, offset, and equilibrium (zero-crossing) to the average human eJimp measured in twelve non-disabled male individuals in the study of Filius et al. [26].

3. Results

The dummy arm was developed and fabricated based on the design requirements. Table 2 summarizes the realised characteristics of the dummy arm in terms of its mass, CoM, and eJimp characteristics, and compares it to the intended human model and the CAD design parameters. The table shows that the mass and CoM of the realised dummy arm is comparable to the human arm model, but is tunable by adjusting the position and number of steel plates, as mentioned in Section 2.2.5.

Table 2. Comparison of the mass and centre of mass (CoM) based on CAD with real-life measurements.

	Human Model		Dummy Arm		CAD	
	Mass [a] kg	CoM [b] mm	Mass kg	CoM mm	Mass kg	CoM [c] mm
UA	1.6–2	137	1.8	140	1.7	$\begin{bmatrix} 2.5 \\ -140.2 \\ -2.0 \end{bmatrix}$
FA	0.9–1.1	106	1.0	110	1.0	$\begin{bmatrix} -0.2 \\ -112.7 \\ 2.6 \end{bmatrix}$

Abbreviations: UA: Upper arm; FA: Forearm; CoM, centre of mass; eJimp, passive elbow joint impedance; z, zero-crossing (equilibrium point); M, mean; SD, standard deviation. All CoM values are expressed with respect to the proximal joint centre. [a] Retrieved from [10,11]. [b] Based on average locations of COM as ratio of segment length of dummy arm, retrieved from Table 4 of [10]. [c] The second element (y-axis) is aligned with the limb in a proximal direction, and the first element (x-axis) points laterally.

3.1. *Joint Impedance Realisation*

The realised eJimp in the dummy arm is compared to the measured human eJimp from our previous work [26] and presented in Figure 3. This previous work used an active elbow support set-up to identify the human eJimp in twelve non-disabled male individuals. The actuator placed at the elbow joint rotated the participant's forearm over the elbow's range of motion in the horizontal plane. In this plane, the gravitational component of the passive forces exerted on the elbow joint remains constant. A six-degrees-of-freedom force/torque sensor on the forearm interface sleeve measured the elbow joint moments, representing the human eJimp [26]. Figure 3 shows the human eJimp model as a group average and standard error of the mean of the identified human eJimp and compares this to the realised eJimp of the four spring types.

Depending on the selected spring type and pre-tension of the springs, different eJimp characteristics were realised; see Table 3. Spring type 4 has the most representative equilibrium point (i.e., where the torque-angle curve crosses the zero-line) compared to the human model. The slope of type 2 is more representative of the slope of the non-disabled eJimp, whereas the slope of spring type 4 could represent an elevated eJimp.

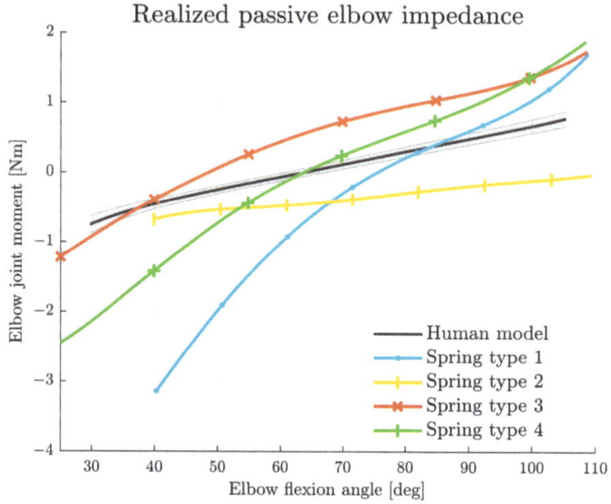

Figure 3. The realised passive eJimp using different spring types compared to the average human eJimp model of 12 non-disabled male individuals, with data retrieved from [26]. The solid grey line represents the group average, and the shaded area shows the standard error of the mean.

Table 3. Passive eJimp characteristics of different spring types compared to human model.

	1st-Order Fit Nm/rad x + Nm	**z (M ± SD) deg**
Human model	$1.14x - 1.29$	66 ± 11
Spring type 1	$4.31x - 2.46$	74
Spring type 2	$1.29x + 1.45$	-
Spring type 3	$3.04x - 0.03$	48
Spring type 4	$3.67x - 1.24$	63

Abbreviations: z, zero-crossing (equilibrium point); M, mean; SD, standard deviation.

3.2. Activities of Daily Living

The achieved range of motion of the shoulder joints is larger than the shoulder range of motion of the DAROR set-up, and the elbow joint is similar to the range of motion of a human arm in the DAROR set-up [12]. This achieved range of motion is sufficient to reach the desired activities that are considered relevant for daily living for this application. To illustrate the sufficient range of motion of the dummy arm in the DAROR set-up, some static poses of the relevant activities of daily living are displayed in Table 4. Moreover, the dummy arm design allows for the attachment of an additional object at the wrist location, which is illustrated in the third column (c), where a 200 g mass is attached to the attachment point to simulate a lifted object in hand.

Table 4. Illustration of the range of motion of the dummy arm for a set of activities of daily living.

a. Tabletop Activities	b. Feeding Activities	c. Lifted Object (200 g)

3.3. Cost-Analysis

The material costs for the various components of the dummy arm are listed in Table 5. Here, labour costs are not included in this overview, as they are dependent on the available facilities. Moreover, the costs of a 3D-printed shoulder joint are considered. However, a milled aluminium shoulder joint will be more expensive, especially due to the involved labour and equipment costs. The approximate fabrication time is around 5 h in the case of a 3D-printed shoulder joint, and for milled aluminium joints, an additional 5–10 h should be taken into account, depending on proficiency and the available equipment.

Table 5. Material costs of the various components. If the costs depend on the material properties, the total weight or length is reported. Dimensions of the different sub-components can be retrieved from the CAD files in Supplementary Materials S1. Costs are rounded to the nearest EUR 0.50.

Material	Costs
Bearings	EUR 13.50
Frame	EUR 6.00 (550 mm)
Mounting materials	EUR 15.00
Axes	EUR 1.00 (200 mm)
Weights	EUR 12.50 (1.9 kg)
Springs	EUR 11.00
Belt and pulleys	EUR 13.00
PLA filament	EUR 6.50 (280 g)
Total	EUR 78.50

4. Discussion

This work aims to present the design of a relatively simple, reproducible, and cost-effective dummy arm that can be used as a verification and development tool for (motorised) upper-extremity exoskeletons. The dummy arm mimics the human arm mechanics with adjustable mass, CoM, and eJimp characteristics. The forearm design simulates the shape of a human forearm and allows for the attachment of additional weights at the wrist to simulate (objects lifted in) the hand.

4.1. Limitations

Some simplifications are made when comparing the dummy arm with a human arm. First of all, it is assumed that the joints are 'ideal' joints (e.g., having a fixed centre of rotation), whereas, in reality, human joints have complex surface geometries and show shifts in their joint axis of rotation during motion [30]. This is especially apparent for the shoulder joint, as the elevation or suppression of the scapula occurs in most elevation movements. Joint misalignment might exert excessive interaction forces on the dummy arm, especially as the dummy arm joints give no slack. Therefore, there is little compensation possible against misalignment. This has practical implications, as the 3D-printed parts at the elbow joint may fatigue or fail when subjected to substantial joint misalignment. However, as opposed to human joints, the dummy arm joint centres are more easily visually aligned with the joints of the DAROR set-up.

Since the dummy arm is not instrumented with sensors like that in the work of [5,19], no feedback on the possible effects of joint misalignment between the dummy arm and exoskeleton could be obtained. Including sensors in the design would allow us to investigate influences of joint misalignment. However, adding sensors adds complexity and costs to the design of the dummy arm. Therefore, depending on the research interest, it must be considered whether it is worth including these components, or whether a simpler version might suffice. Moreover, since human joints are not 'ideal' like the simplified joints of the dummy arm, the measured interaction forces in a phantom limb might deviate from real-world scenarios with a human limb.

The generalisability of the dummy arm design to other upper-extremity exoskeletons has not yet been investigated. However, by providing our CAD design in the Supplementary Materials, we enable others to make the necessary adjustments for various exoskeleton applications.

4.2. Recommendations

Although beyond the scope of the current study, it is of interest to investigate whether the realised eJimp characteristics of the dummy arm spring types are representative of people with neuromuscular disorders who cope with elevated eJimp. Different spring characteristics could approximate the human eJimp more closely. This requires more investigation. Potentially, the torque-angle profile of people with DMD is less likely to behave linearly. As mentioned above, the dummy arm eJimp profile can be further tuned by adjusting the shape of a cam (wrapping cam mechanism). Examples of this can be found in the work of [28], who used a serial wrapping cam mechanism to design an upper-limb assistive device. A more general description of the design and use of a wrapping cam mechanism can be found in the work of [27].

Although not of interest within the current application, several directions exist in which to improve the dummy arm. For instance, the force transmission between the dummy arm and the exoskeleton could be considered. In reality, due to the soft-tissue deformation of the skin and subcutaneous tissue, the transmission of the forces to the arm is different than when a rigid structure is used. A possible solution would be to use a soft outer layer for the dummy arm, similar to what is done by the authors of [1]. Moreover, since humans have no ideal joints, it might be worth investigating, depending on the research interest, making the joints more human-like to allow for small shifts in the joint axes of rotation. However, adding soft tissue, or more human-like joints, adds complexity and costs to the design of the dummy arm.

To reduce the costs of the design, the shoulder brackets could also be 3D-printed. Nevertheless, care should be taken with the load demands on the material in combination with the material selection and manufacturing techniques. Otherwise, a redesign of the brackets to strengthen these structures is recommended. After our initial PLA 3D-printed shoulder bracket failed, we decided to mill the shoulder from aluminium.

The advantage of using a modular dummy arm is that it can be adjusted to the anthropometrics for different populations, such as children or people with pathologies affecting the upper extremity (e.g., DMD, spinal muscular atrophy, amyotrophic lateral sclerosis, stroke). By making use of a more patient-like dummy arm, the burden on patients for the development of exoskeletons can be decreased, as the tests can be very exhaustive and demanding for patients. Testing time with vulnerable patients is also limited, while a dummy arm is always available. Therefore, using a patient-like dummy arm allows for faster assistive or rehabilitative exoskeleton development. However, the translation of the exoskeleton performance with the dummy arm to the intended target population should be further investigated.

Lastly, it is challenging to objectively compare the performance of exoskeletons with respect to each other, especially as the performance metric (e.g., perceived level of provided support) can become very subjective and differ across the reachable workspace. Using standardised dummy arm and objective evaluation methods allows for a more objective comparison of exoskeleton performance, which is relevant for the patients as well for the developers, as it can point to possible areas of improvement.

5. Conclusions

The current dummy arm successfully mimicked the characteristics of the human arm and was proven to be a helpful development and verification tool for the developed investigational DAROR exoskeleton. Since the design is relatively simple, reproducible, and cost-effective, it promises to serve more arm exoskeleton applications, and is therefore shared. The design allows for easy adjustment of the mechanical properties (e.g., mass, CoM), outer dimensions, and eJimp torque-angle profiles to fit the specific characteristics of a (vulnerable) intended target population. Using such dummy limbs, like the dummy arm, enhances the fast iteration and objective evaluation of new control strategies, and can save time, labour, and the burden of voluntary participants, who may be exposed to safety risks in an early stage of exoskeleton development.

Supplementary Materials: The CAD design of the dummy arm can be downloaded at https://doi.org/10.4121/3c4fa57d-3fc6-4e27-933c-00a02a6e5a33. A supporting video is available at https://www.mdpi.com/article/10.3390/biomimetics9100579/s1.

Author Contributions: Conceptualisation, methodology, validation, data curation, and visualisation, S.J.F.; hardware design and realisation, B.J.v.d.B.; writing—original draft preparation, review and editing, S.J.F. and B.J.v.d.B.; writing—review, editing, and supervision, J.H. All authors have read and agreed to the published version of the manuscript.

Funding: This work is part of the research program Wearable Robotics, with the project number P16-05, funded by the Dutch Research Council (NWO), Duchenne Parent Project, Spieren voor Spieren, Festo, Yumen Bionics, Baat Medical, and the FSHD society.

Institutional Review Board Statement: Not applicable.

Data Availability Statement: The original contributions presented in this study are included in the article and supplementary material, further inquiries can be directed to the corresponding author/s.

Acknowledgments: We would like to thank the technical staff of the employee workshop of the Technical University at Delft for fabricating the shoulder joint brackets. Moreover, we would like to thank intern Kevin Laban, who assisted in the data collection of the dummy arm eJimp characteristics.

Conflicts of Interest: The authors declare no conflicts of interest. The funders had no role in the design of this study; in the collection, analyses, or interpretation of the data; in the writing of the manuscript; or in the decision to publish the results.

Abbreviations

The following abbreviations are used in this manuscript:

pJimp	passive joint impedance
eJimp	passive elbow joint impedance
CoM	centre of mass
DAROR	Duchenne ARm Orthosis
GH	glenohumeral
CAD	Computer Aided Design

References

1. Barrutia, W.S.; Bratt, J.; Ferris, D.P. A Human Lower Limb Mechanical Phantom for the Testing of Knee Exoskeletons. *IEEE Trans. Neural Syst. Rehabil. Eng.* **2023**, *31*, 2497–2506. [CrossRef] [PubMed]
2. Gull, M.A.; Bai, S.; Bak, T. A review on design of upper limb exoskeletons. *Robotics* **2020**, *9*, 16. [CrossRef]
3. Dollar, A.M.; Herr, H. Lower extremity exoskeletons and active orthoses: Challenges and state-of-the-art. *IEEE Trans. Robot.* **2008**, *24*, 144–158. [CrossRef]
4. De Looze, M.P.; Bosch, T.; Krause, F.; Stadler, K.S.; O'Sullivan, L.W. Exoskeletons for industrial application and their potential effects on physical work load. *Ergonomics* **2016**, *56*, 671–681. [CrossRef] [PubMed]
5. Dežman, M.; Massardi, S.; Pinto-Fernandez, D.; Grosu, V.; Rodriguez-Guerrero, C.; Babič, J.; Torricelli, D. A mechatronic leg replica to benchmark human–exoskeleton physical interactions. *Bioinspiration Biomim.* **2023**, *18*, 036009. [CrossRef] [PubMed]
6. Koopman, A.S.; Kingma, I.; Faber, G.S.; de Looze, M.P.; van Dieën, J.H. Effects of a passive exoskeleton on the mechanical loading of the low back in static holding tasks. *J. Biomech.* **2019**, *83*, 97–103. [CrossRef] [PubMed]
7. Van Dijk, W.; Van de Wijdeven, T.; Holscher, M.; Barents, R.; Könemann, R.; Krause, F.; Koerhuis, C.L. Exobuddy-A non-anthropomorphic quasi-passive exoskeleton for load carrying assistance. In Proceedings of the 2018 7th IEEE International Conference on Biomedical Robotics and Biomechatronics (Biorob), Enschede, The Netherlands, 26–29 August 2018; pp. 336–341. [CrossRef]
8. Plaza, A.; Hernandez, M.; Puyuelo, G.; Garces, E.; Garcia, E. Lower-Limb Medical and Rehabilitation Exoskeletons: A Review of the Current Designs. *IEEE Rev. Biomed. Eng.* **2023**, *16*, 278–291. [CrossRef] [PubMed]
9. Gandolla, M.; Antonietti, A.; Longatelli, V.; Pedrocchi, A. The Effectiveness of Wearable Upper Limb Assistive Devices in Degenerative Neuromuscular Diseases: A Systematic Review and Meta-Analysis. *Front. Bioeng. Biotechnol.* **2020**, *7*, 1–16. [CrossRef] [PubMed]
10. Chandler, R.F.; Clauser, C.E.C.; McConville, J.T.J.J.T.; Reynolds, H.M.; Young, J.W.; Yough, J. *Investigation of Inertial Properties of the Human Body*; Technical Report March; U.S. Department of Transportation: Washington, DC, USA, 1975.
11. Clauser, C.E.; McConville, J.T.; Young, J.W. *Weight, Volume, and Center of Mass of Segments of the Human Body*; Technical Report; Aerospace Medical Research Laboratory: Wright-Paterson Air Force Base, OH, USA, 1969.
12. Filius, S.J.; Keemink, A.Q.L.; Meijneke, C.; Gregor, W.; Kuenen, T.; Janssen, M.; Harlaar, J.; Kooij, H.V.D. *A 4DOF Impedance-Based Exoskeleton with Weight and Elbow Stiffness Compensation: For Duchenne Brooke Scale 4*; Department of BioMechanical Engineering, Delft University of Technology: Delft, The Netherlands, (Unpublished).
13. Maggioni, S.; Melendez-Calderon, A.; Van Asseldonk, E.; Klamroth-Marganska, V.; Lünenburger, L.; Riener, R.; Van Der Kooij, H. Robot-aided assessment of lower extremity functions: A review. *J. Neuroeng. Rehabil.* **2016**, *13*, 1–25. [CrossRef] [PubMed]
14. Filius, S.J.; Harlaar, J.; Alberts, L.; Houwen-van Opstal, S.; van der Kooij, H.; Janssen, M.M. Design requirements of upper extremity supports for daily use in Duchenne muscular dystrophy with severe muscle weakness. *J. Rehabil. Assist. Technol. Eng.* **2024**, *11*, 1–18. [CrossRef] [PubMed]
15. Noda, T.; Teramae, T.; Ugurlu, B.; Morimoto, J. Development of an upper limb exoskeleton powered via pneumatic electric hybrid actuators with bowden cable. In Proceedings of the 2014 IEEE/RSJ International Conference on Intelligent Robots and Systems, Chicago, IL, USA, 14–18 September 2014; pp. 3573–3578. [CrossRef]
16. Madani, T.; Daachi, B.; Djouani, K. Non-singular terminal sliding mode controller: Application to an actuated exoskeleton. *Mechatronics* **2016**, *33*, 136–145. [CrossRef]
17. Dávila-Vilchis, J.M.; LAZ-Avilés; Ávila Vilchis, J.C.; Vilchis-González, A.H. Design Methodology for Soft Wearable Devices—The MOSAR Case. *Appl. Sci.* **2019**, *9*, 4727. [CrossRef]
18. Toth, A.; Pilissy, T.; Bauer, M.O.; Al-Absi, G.; David, S.; Fazekas, G. Testing the Limit Range of Motion Safety Function of Upper Limb Rehabilitation Robots with an Anthropometrically Adjustable and Sensorized Dummy Limb. In Proceedings of the 2022 International Conference on Rehabilitation Robotics (ICORR), Rotterdam, The Netherlands, 25–29 July 2022; pp. 25–29. [CrossRef]
19. Bessler-Etten, J.; Schaake, L.; Prange-Lasonder, G.B.; Buurke, J.H. Assessing effects of exoskeleton misalignment on knee joint load during swing using an instrumented leg simulator. *J. Neuroeng. Rehabil.* **2022**, *19*, 13. [CrossRef]
20. Contini, R. Body Segment Parameters, Part II; Technical Report. Available online: https://citeseerx.ist.psu.edu/document?repid=rep1&type=pdf&doi=eb1971ff961ba567256c52fa89ad2be5e25ec9f9#page=5 (accessed on 15 January 2023).

21. Dempster, W.T.; Gaughran, G.R.L. Properties of body segments based on size and weight. *Am. J. Anat.* **1967**, *120*, 33–54. [CrossRef]
22. Drillis, R.; Contini, R.; Bluestein, M. *Body Segment Parameters A survey of Measurement Techniques;* Technical Report; Artifical Limbs: Washington, DC, USA, 1964.
23. Wu, G.; Van Der Helm, F.C.; Veeger, H.E.; Makhsous, M.; Van Roy, P.; Anglin, C.; Nagels, J.; Karduna, A.R.; McQuade, K.; Wang, X.; et al. ISB recommendation on definitions of joint coordinate systems of various joints for the reporting of human joint motion—Part II: Shoulder, elbow, wrist and hand. *J. Biomech.* **2005**, *38*, 981–992. [CrossRef] [PubMed]
24. Doorenbosch, C.A.; Harlaar, J.; Veeger, D.H. The globe system: An unambiguous description of shoulder positions in daily life movements. *J. Rehabil. Res. Dev.* **2003**, *40*, 147–155. [CrossRef] [PubMed]
25. Stienen, A.H.; Keemink, A.Q. Visualization of shoulder range of motion for clinical diagnostics and device development. In Proceedings of the 2015 IEEE International Conference on Rehabilitation Robotics (ICORR), Singapore, 11–14 August 2015; pp. 816–821. [CrossRef]
26. Filius, S.; Janssen, M.; van der Kooij, H.; Harlaar, J. Comparison of Lower Arm Weight and Passive Elbow Joint Impedance Compensation Strategies in Non-Disabled Participants. In Proceedings of the 2023 International Conference on Rehabilitation Robotics (ICORR), Singapore, 24–28 September 2023; pp. 1–6. [CrossRef]
27. Tidwell, P.H. Wrapping Cam Mechanisms. Ph.D. Thesis, Virginia Polytechnic Institute and State University, Blacksburg, VA, USA, 1995.
28. Schroeder, J.S.; Perry, J.C.; Beyerlein, S.W. Application of Wrapping Cams in an Unpowered, Upper-Limb Assistive Device. Ph.D. Thesis, University of Idaho, Moscow, ID, USA, 2016.
29. Story, A. Right Hand Reference. 2020. Available online: https://grabcad.com/library/right-hand-reference-1 (accessed on 10 January 2023).
30. Schiele, A.; Van Der Helm, F.C. Kinematic design to improve ergonomics in human machine interaction. *IEEE Trans. Neural Syst. Rehabil. Eng.* **2006**, *14*, 456–469. [CrossRef] [PubMed]

biomimetics

Article

Design and Assessment of Bird-Inspired 3D-Printed Models to Evaluate Grasp Mechanics

Pavan Senthil, Om Vishanagra, John Sparkman, Peter Smith and Albert Manero *

Limbitless Solutions, University of Central Florida, 12703 Research Parkway, Suite 100, Orlando, FL 32826, USA; p.senthil@limbitless-solutions.org (P.S.); o.vishanagra@limbitless-solutions.org (O.V.)
* Correspondence: albert@limbitless-solutions.org

Abstract: Adapting grasp-specialized biomechanical structures into current research with 3D-printed prostheses may improve robotic dexterity in grasping a wider variety of objects. Claw variations across various bird species lend biomechanical advantages for grasping motions related to perching, climbing, and hunting. Designs inspired by bird claws provide improvements beyond a human-inspired structure for specific grasping applications to offer a solution for mitigating a cause of the high rejection rate for upper-limb prostheses. This research focuses on the design and manufacturing of two robotic test devices with different toe arrangements. The first, anisodactyl (three toes at the front, one at the back), is commonly found in birds of prey such as falcons and hawks. The second, zygodactyl (two toes at the front, two at the back), is commonly found in climbing birds such as woodpeckers and parrots. The evaluation methods for these models included a qualitative variable-object grasp assessment. The results highlighted design features that suggest an improved grasp: a small and central palm, curved distal digit components, and a symmetrical digit arrangement. A quantitative grip force test demonstrated that the single digit, the anisodactyl claw, and the zygodactyl claw designs support loads up to 64.3 N, 86.1 N, and 74.1 N, respectively. These loads exceed the minimum mechanical load capabilities for prosthetic devices. The developed designs offer insights into how biomimicry can be harnessed to optimize the grasping functionality of upper-limb prostheses.

Keywords: biomimicry; prostheses; 3D printing

check for updates

Citation: Senthil, P.; Vishanagra, O.; Sparkman, J.; Smith, P.; Manero, A. Design and Assessment of Bird-Inspired 3D-Printed Models to Evaluate Grasp Mechanics. *Biomimetics* **2024**, *9*, 195. https://doi.org/10.3390/biomimetics9040195

Academic Editor: Rafhael Milanezi de Andrade

Received: 12 January 2024
Revised: 13 March 2024
Accepted: 22 March 2024
Published: 26 March 2024

1. Introduction

1.1. Motivation

A key area of development in the state of upper-limb prostheses is to improve prosthetic functionality [1]. A driving cause of high rejection or abandonment rates, estimated to be between 20 and 40 percent in recent decades, is prosthetic users finding their devices unreliable in interacting with objects in their day-to-day lives [1–3]. Issues with optimizing functionality are also linked with other commonly cited discomforts. Reducing prosthetic weight and retained socket heat typically requires designers to compromise on hardware used to deliver more robust performance [4,5]. Functionality, as defined by Light et al. [6], is the extent to which a device can adapt to perform certain tasks. Specifically for upper-limb prostheses, functionality can be classified at various levels: reliable grasp at the level of the hand, rotation at the level of the wrist and forearm, and movement in various heights at the level of the elbow and shoulder [6]. This research focuses on discovering methods to improve the most basic functionality, the grasp.

1.2. Background

Methods to address the issue of poor grasp prosthetic performance, without needing heavier or energy-intensive hardware components, are a priority. This research broadened the lens utilizing animal-based biomimicry. In general, biomimicry is defined as "innovation inspired by nature" [7–9]. This concept has been applied to several engineering applications. One field where biomimicry is commonly applied is the development of materials. Spider silk is known for its relatively high tensile strength and flexibility, which has prompted research on synthetically mimicking the unique material properties for mechanical testing [10–12]. The invention of Velcro® was inspired by Xanthium strumarium (Cockleburr), which has spiky fruits called burrs that easily hook onto animal fur to spread [13]. The designer mimicked the structure of these burrs to design a material with hooks and loops that would securely latch onto itself while being easily removable [14,15]. Biomimicry also has applications in robotics [16,17]. To develop a robot suited for running on irregular terrain, researchers Clark et al. [16] drew inspiration from the morphology and material properties of a cockroach leg to design a robotic leg that would emulate a similar behavior. Biomimicry inspired by amphibians, such as sea snakes and salamanders, was applied to create robots that perform undulating motions to maneuver through the water [17]. The use of birds specifically as biomimetic inspiration for robotics has also been explored for a variety of applications. Robotics for flight is a popular topic of research. For mechanical wing systems, the behavior of birds in adapting to their wing morphologies has been mimicked to maximize aerodynamics under different flight conditions [18]. The leg and claw structures of birds are also a common inspiration for designing landing and takeoff mechanisms. Feliu-Talegon et al. [19] utilized bird-inspired structures to stabilize and facilitate the takeoff action of flapping-wing robots.

The concept of biomimetic design has been applied to prostheses, with most upper-limb prosthetic devices modeled after the human hand and arm. The research by Varol et al. [20] describes different biomimetic approaches to designing a human-inspired prosthetic. One approach follows the biomechanics of the human hand closely to model the overall structure of the hand and designate the placement of the actuators [20]. A second approach looks to modify certain characteristics to make aspects of the device more practical for robotic systems. These changes include the placement of the actuators in the hand rather than the forearm and setting the opening of the arm as the active motion [20]. Other examples of biomimicry aiding the development of upper-limb robotics include inspiration for sensors. Stretchable sensors for soft robotics have been inspired by the structure of mechanoreceptors in plant surfaces and the skin of humans and animals [21]. The structure of bird claws has also been used to design soft robotic grippers. The research by Wang and Yu simulated air pressure-powered grasping devices modeled after features in bird and human claws [22].

Applications of biomimicry also go beyond simple replication and rather utilize the underlying mechanisms of that process [8,20,23–25]. For prosthetic design, improving the ability of the device to reliably grasp different types of objects will be beneficial to the user. This investigation aims to apply biomimicry, not exclusively human hands, specialized for grasp-related purposes.

1.3. Applications

The biomimetic model in this research investigated specifically how bird claws are suited for grasp-related applications and how they could improve the functionality of upper-limb prostheses. Across the many species, different birds have evolved unique adaptations to their claw structures that are suited for their environment and lifestyles. In identifying natural processes suited for prostheses, this investigation highlighted the claw structures of birds specialized for climbing or grasping. These structures are termed anisodactyl and zygodactyl claws and are displayed in Figure 1. Anisodactyl represents a configuration with three digits in the front and one at the back, and the zygodactyl represents a configuration with two in the front and two at the back [26].

The anisodactyl claw is commonly found in birds that perch and hunt their prey, for which the three digits in front offer greater shock absorption upon impact with structures or prey when approaching them at high velocities [27,28]. The outer digits on the front three act as stabilizers and provide additional reach for these actions [26]. Examples of birds exhibiting this pattern include birds of prey such as Falconiformes (falcons) and Accipitriformes (eagles and hawks) [26,28]. The zygodactyl claw is commonly found in birds that climb; the back two digits offer support and stability for a bird perched on vertical surfaces and allow improved maneuverability in this axis [26,29]. Piciformes (woodpeckers) and Psittaciformes (parrots) are among the many examples of birds exhibiting this claw configuration and behavior [29]. The objective was to design a representation of these digit patterns and to evaluate them for their grasping and mechanical abilities. To identify how to construct and evaluate these models, this research reviewed existing applications in the literature mimicking features from birds through computer-aided design (CAD) and 3D printing.

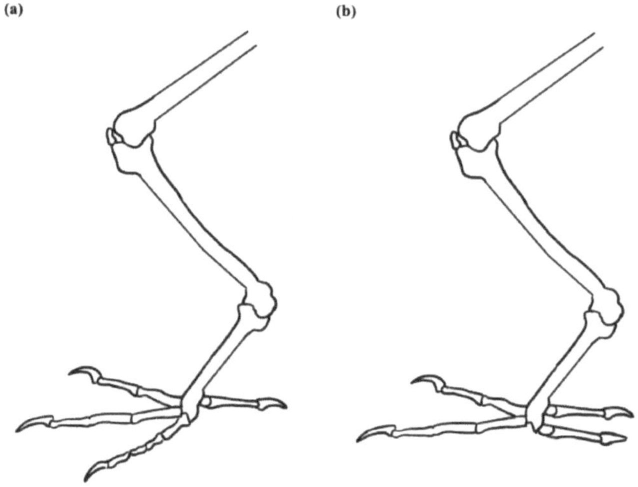

Figure 1. Representation of the two bird claw configurations found in nature: (**a**) anisodactyl; (**b**) zygodactyl.

One such example is a study conducted at Stanford University, where researchers Roderick et al. [27,30] created a 3D-printed drone modification based on the leg and claw structures of birds of prey. The drones use these additions to land on irregular surfaces such as tree branches and carry various payloads. This allows them to rest and conserve battery during long-distance flights and to carry supplies for disaster responses. The mechanism of the drone attachment also goes beyond a similar morphological mimic of the bird's legs and bases the features of actuation on a behavioral analysis of a bird performing a perching motion. Additional examples in the literature applying structures and mechanisms from bird claws in drone perching applications further corroborate the success of this aspect of biomimicry and aim to build on the grasping ability [26,31,32]. A study by Zhu et al. [26] tested some of the grasping and mechanical features of their developed claw model to assess its performance reliability. The design aims to transition the specific tree branch grasping application of the drone claw models to grasp a wider variety of objects. The model by Zhu et al. [26] could then be implemented for other applications and industries. This study tested two mechanisms of actuation of the claw: a standard model with the flexor digitorum longus and the extensor digitorum longus tendons and the other with a flexor digitorum longus tendon and torsion springs. Both types were evaluated based on their ability to adapt their grip to objects of varying sizes and incrementing weights.

A similar study by Nabi et al. [31] explores the utility of designing a robotic hand inspired by a pigeon claw, rather than a human hand, for industries that benefit from robotic automation. Specific features identified to streamline mechanical design include fewer actuators, reduction in material via fewer digits and joints, and a smaller palm size to improve the ability to maneuver smaller objects [31]. By expanding the initial scope of 3D-printed bird-inspired actuators, the literature supports how grasp-type devices modeled after bird claws may serve as an ideal inspiration for improving grasping devices.

To evaluate these grasping advantages, this research aimed to develop similar bird-inspired models for comparison with a traditional five-fingered prosthetic device. The Limbitless Solutions® prosthetic hand was used as the model of comparison [33,34] and built upon the Roderick et al. [27,30] research framework. The biomimetic models of this design drew structural inspiration from the existing literature utilizing the biomechanical features from bird claws along with certain parts from the Limbitless Solutions® five-fingered hand to translate the model from drone landing to prosthetic testing application. The design and assembly features of the biomimetic models and prosthetic hand were kept comparable to allow the testing to be more standardized. The biomimetic models were better standardized for evaluation alongside the five-fingered prosthetic. The Limbitless Solutions® prosthetic hand was developed in the same lab as the bird-inspired models, allowing all devices to be constructed with similar manufacturing techniques. The developed models were not intended to serve as functional prostheses; instead, these claw models served as a framework to assess how biomimicry could offer insights into potential improvements to the current state of upper-limb prostheses. The tests conducted on the devices involved a mixture of the evaluations used for biomimetic models in the literature along with existing biomechanical load testing standards for prostheses. The aim was to better understand how the device would function under those conditions. Through both qualitative and quantitative analyses, the grasp performance of the anisodactyl and zygodactyl claws was compared to that of the transitional five-fingered prosthetic. The results of these comparisons inform specific features that could be implemented in future prosthetic design to improve grasp.

2. Materials and Methods

2.1. Design

This work focused on developing a configurable multi-fingered mechanical model that mimics the anatomy and physiology of bird claws. Development primarily focused on two models with four digits, the anisodactyl and zygodactyl bird claw, characterized by key differences in digit arrangement. The anisodactyl and zygodactyl bird claw models are seen in Figure 2. Each digit on the frontal portion of the anisodactyl claw has an approximate angle of 45 degrees between the three digits, respectively. The outermost digits on the frontal portion of the palm and the rear digit have an approximate angle of 135 degrees.

Development of the zygodactyl base design adopted a similar approach. With a two-in-front and two-in-back configuration, each digit had a spacing of approximately 90 degrees. This created a symmetrical bird claw as seen in Figure 2b. This research designed a fully symmetrical bird claw design, a slight modification from a traditional zygodactyl bird claw seen in nature. Zygodactyl claws in nature (Figure 1b) do not follow a symmetrical digit arrangement compared to our zygodactyl model, which has equidistant digits. Biomimicry does not require the exact replication of biological elements found in nature and rather focuses on the functional aspects of the design [8,23,24]. This model was adapted to mimic the more symmetrical quality of the zygodactyl claw to evenly support a grasped object on all sides.

In nature, birds with these digit patterns often have differently sized digits on the same claw similar to how the fingers on a human hand differ proportionally [28]. These variations occur between different species exhibiting the same digit arrangement and even between members of the same species. The designed models kept the sizes consistent for digits by using the same phalanges and talons for each. The digit size variations may also depend on the dexterity of an actual bird claw for full effect, which a robotic application may not be able to truly replicate. Consistent digit sizes also facilitate the design and prototyping processes.

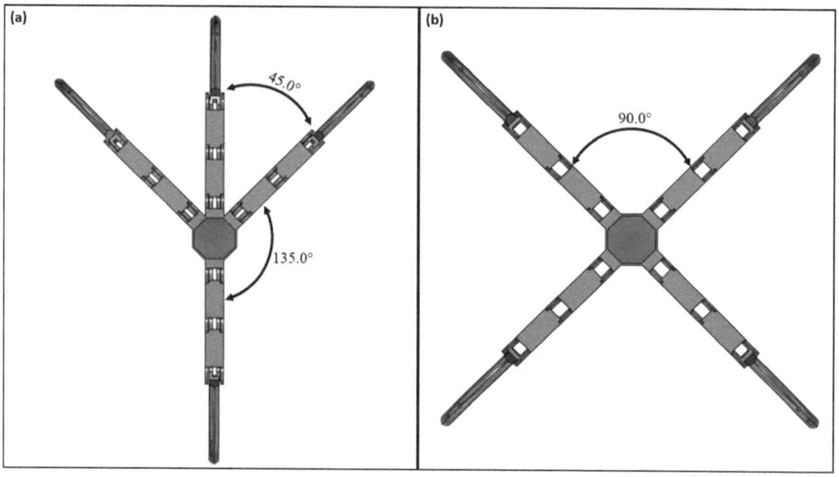

Figure 2. 2D depiction of the assembled bird claws and digit configuration angle: (**a**) anisodactyl claw; (**b**) zygodactyl claw.

A preliminary digital model of the bird claw was designed in CAD software. This software allows users to conduct motion studies that depict the ideal movement of 3D structures moving in conjunction with one another. Ideally, the actuation of the claw will result in the proximal phalanges moving first, followed by the distal phalanges and talons relative to the claw base. The motion study feature was used to visualize the contraction of a bird claw starting from the open claw position to a fully contracted grasp as seen in Figure 3. Analysis of the motion study revealed crucial structural and alignment impediments with the bird claw design, which were modified as needed to optimize claw functionality. Following the completion of the digital CAD model, each component (base, phalanges, and talon) of the bird claw was 3D-printed with acrylonitrile butadiene styrene (ABS) plastic using fused deposition modeling (FDM) 3D printing. FDM 3D printing deposits melt plastic layer-by-layer in the geometry of the defined shape until a full 3D model is created [35,36]. In addition to printing the 3D model with the desired material (ABS plastic), Supplementary Materials is applied to reinforce susceptible geometries against warping during printing. These vulnerable geometries may encompass overhangs and cavities within the 3D model. This study used soluble support material, Stratasys SR-30, to facilitate the 3D printing process for each claw component. These models were 3D-printed with the use of a Fortus 250mc 3D printer and post-processed in an SCA-1200HT support cleaning apparatus. A support cleaning apparatus was used to dissolve any residual support material that remained on the 3D model after the manual removal of supports.

The bird claw was designed in three components: anisodactyl/zygodactyl base, phalange, and talon. Each component is shown in greater detail in Figure 4. A fully assembled claw consists of 1 base (either the anisodactyl or zygodactyl), 8 phalanges, and 4 talons, excluding the various joint and tendon parts that connect the components together.

Multiple design elements were derived from the Limbitless Solutions® five-fingered prosthetic hand [33,34] and the Stanford drone model [27,30]. The foundational framework of the bird claw drew inspiration from the Stanford drone model, which primarily influenced the design of the phalange and talon [27]. This design provided a model of a tested bird claw that was developed to assess bird claw functionality in grasp and perching-related functions. These applications matched the goal of this investigation to assess the grasp and grip functions of bird claws, which can be referenced to improve current upper-limb prosthetic devices. Additional elements of the bird claw were modified in comparison to the drone model such as the joints. The Stanford team 3D-printed each joint of the phalange and talon as one unified digit. In contrast, this model created each phalange and talon separately, assembling the digit using finger posts and torsion springs as seen from the top view in Figure 5a. The choice to use finger posts was derived from the Limbitless Solutions® five-fingered prosthetic hand to improve cohesion between a previously tested prosthetic hand and the bird claw model [33,34]. For mechanical testing purposes, developing a bird claw model that utilizes the same components found in a traditional prosthetic is crucial for direct comparison of functionality. Torsion springs were implemented into the joints of the digit to return the digit to its initial resting position after a full contraction is achieved. The resting configuration of the bird claw is characterized by an open palm structure, with all four digits in a relaxed position.

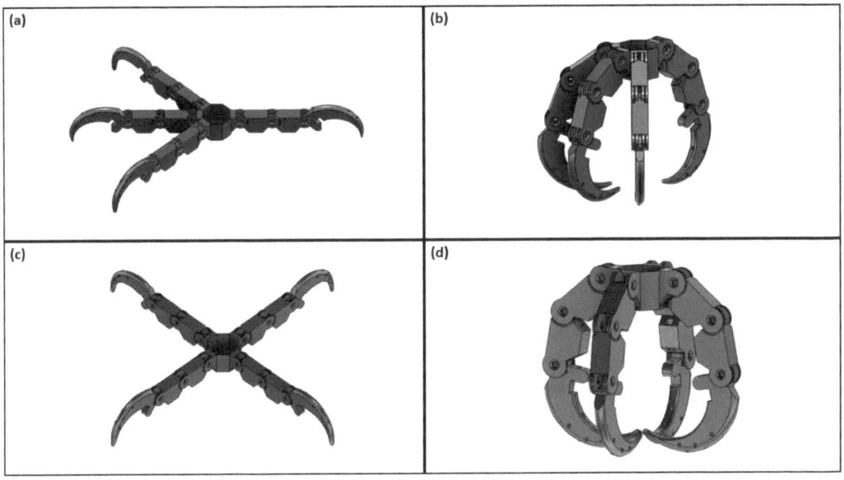

Figure 3. 3D representation of an ideal grasp for both claws using the motion study analysis: (**a**) anisodactyl open claw; (**b**) anisodactyl closed claw; (**c**) zygodactyl open claw; (**d**) zygodactyl closed claw.

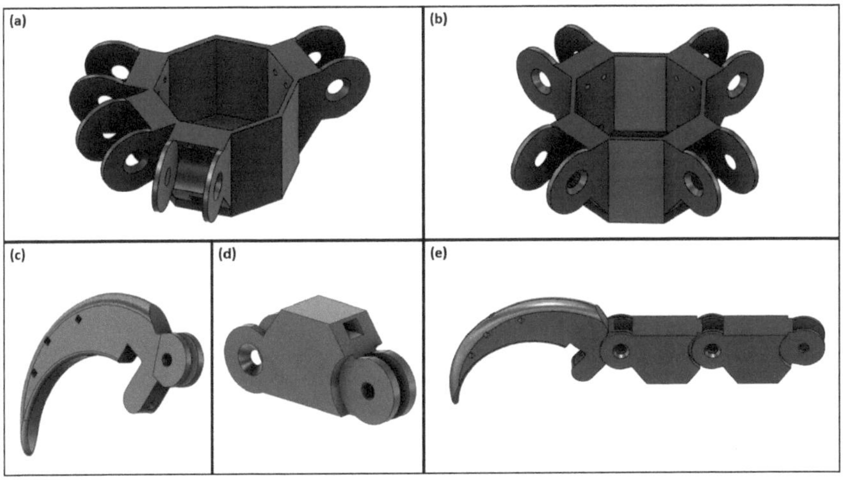

Figure 4. 3D claw components used for assembly of a complete bird claw: (**a**) anisodactyl claw base; (**b**) zygodactyl claw base; (**c**) talon; (**d**) phalange; (**e**) fully assembled digit.

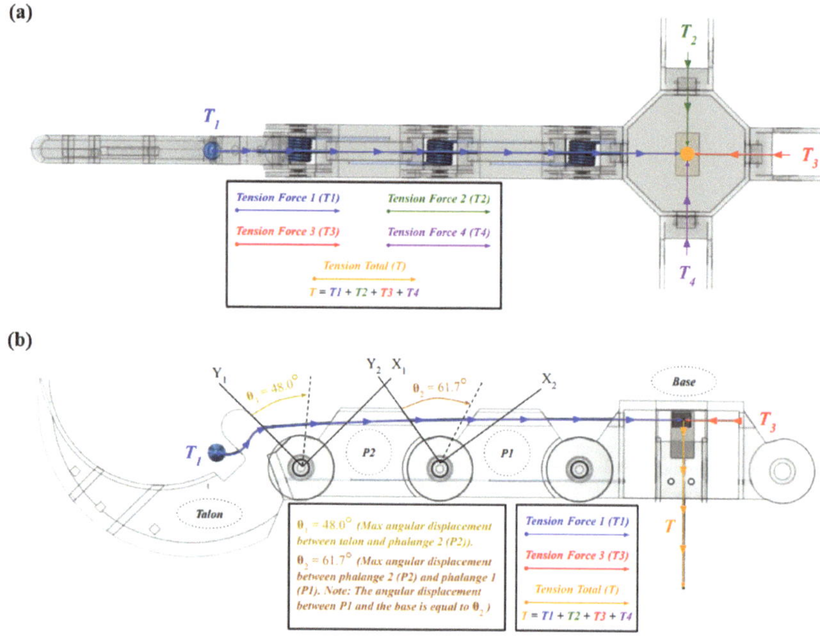

Figure 5. Single digit free body diagram: (**a**) top view; (**b**) side view.

During the digit assembly process, both target phalanges are aligned to ensure their joints overlap. A torsion spring is secured between each joint to supply torsion force, and a finger post is inserted into the central overlapping hole to connect the two parts. The use of a finger post and torsion spring allows the digit to mimic the movement of a pin joint, which allows each phalange to rotate about a central axis while being restricted to one degree of freedom. A 0.61 mm stainless steel cable was used to mimic the tendons of the bird claw. Creating a four-bar linkage by connecting the actuator to the talon with the intermediate phalanges in between allows the digit to curl without interference [37]. Figure 5a displays the junction point that comprises the four-bar linkage where all four digit cables meet together. Once threaded through the digit, each cable was crimped at the distal portion of the talon as well as the main cable in the bird claw base. The cables threaded through each of the 4 digits were crimped together with the main cable, which was then coiled around a spool attached to a MG996R servo motor. The main cable serves as the central connection point for the four-digit cables at the base, ensuring consistent contraction of each digit. The main cable was fed through the base and secured to the spool at the top of the claw housing, controlled by the servo motor, which rotates to actuate the digits. To create a grasping motion, the servo motor winds in a clockwise motion, reeling the cable in to curl the digit inward. To release the grasp, the servo rotates counterclockwise to relax the wound cables, and the torsion springs return the claw to its starting position. Figure 5b highlights the maximum angular displacement of each digit component during actuation. Maximum angular displacement represents how far each digit component can rotate before parts of their body begin to interfere with one another. The talon and the second phalange are capable of rotating 48.0 degrees relative to one another before a component of the talon begins to contact the face of the second phalange. These exact values were determined with the same motion analysis study used to analyze the ideal actuation of the fully assembled claw. These values reflect the ideal actuation of a singular digit. The angular displacements are significantly reduced during a grasp due to interference caused by other digits.

In addition to the components required to actuate the bird claw, various devices are required to control the servo motor and house the various bird claw parts. The housing encloses the servo motor, cable spool, a two-button breadboard circuit, and a detachable swivel base as seen in Figure 6. The two-button breadboard circuit consists of an Arduino Nano programmed using the Arduino Integrated Development Environment to control the rotation of the servo motor through two buttons. Pressing the first button would rotate the servo motor 180 degrees clockwise to wind the cable in to curl the claw digits, and pressing the other button would reset the motion by rotating the servo 180 degrees in the other direction. The housing design also includes a custom swivel base which allows the different claw configurations to be easily inserted and removed for mechanical testing and repairs.

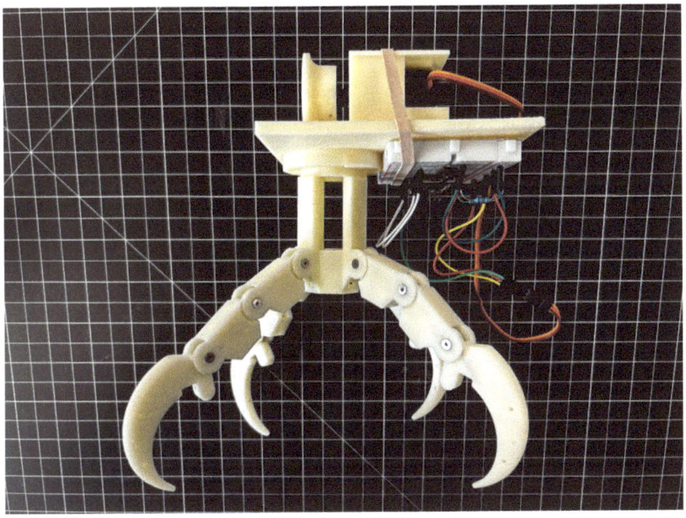

Figure 6. Fully assembled zygodactyl claw positioned within the prosthesis model housing, including the MG996R servo motor and two-button breadboard circuit.

2.2. Testing Methods

2.2.1. Qualitative Assessment

With the constructed anisodactyl and zygodactyl models, physical evaluations were executed to identify the biomechanical advantages of designing the grasping devices after two different four-fingered bird claws versus the traditional five-fingered human hand. The first phase involved a qualitative experiment assessing the grasp interactions of the three different devices: the anisodactyl claw, zygodactyl claw, and Limbitless Solutions® prosthetic. This experiment utilized lightweight objects of various shapes and sizes to assess how each device established a grip for grasping varying topologies. This experimental procedure was modeled after the methods of Zhu et al. [26], who tested the efficacy of two different anisodactyl claws adapted to objects establishing a secure grip. The objects chosen for this experiment are displayed in Figure 7: plane, cubic, spherical, wide rectangular, and irregular shapes. The first four shapes were created using Styrofoam™ to minimize potential grasping differences due to object weight. The final irregular object, represented by a rubber duck, was introduced to assess the device's approach to a more complex grasping of an uneven geometry. The procedure for this evaluation required the device to pick up the object from a resting position, hold it for at least five seconds to ensure a successful grasp, and then release the object. Evaluating device interactions with each of these objects offered insights into whether the biomimicked designs held any advantages in grasping specific shapes and sizes.

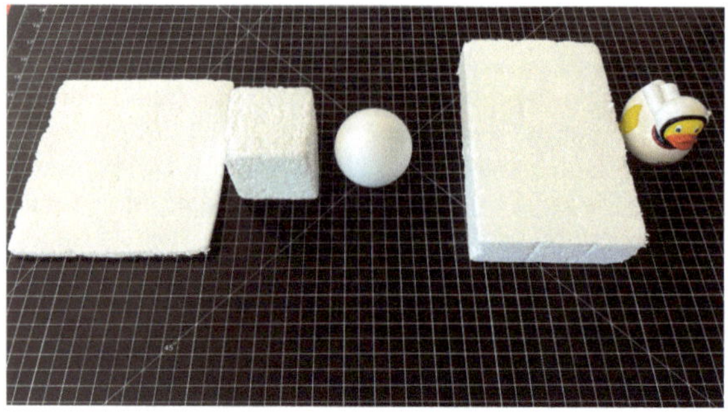

Figure 7. Objects used in qualitative assessment, from left to right: thin (220 mm × 135.5 mm × 48.5 mm), cubic (75.0 mm × 75.0 mm × 75.0 mm), spherical (75.0 mm diameter), wide rectangular (225.0 mm × 135.5 mm × 48.5 mm), and irregular (L_{max} 74.5 mm × W_{max} 180.0 mm × H_{max} 66.5 mm).

2.2.2. Quantitative Assessment

In addition to assessing object grasping, performance loading was evaluated. Specifically, the biomimicry designs were evaluated under prosthetic testing conditions for single-digit flexion and full grasp tests [38,39]. The results of tests under these conditions indicated the biomechanical advantages or disadvantages of structuring a prosthetic device with the bio-inspired designs. Modifications under consideration were the reduction in number of digits, the display of digits relative to the palm, and the shape and sizes of the phalanges and talon components [38,40]. The components under evaluation included the 3D-printed parts of the base and digits and the assembly parts including the finger posts, torsion springs, cable, and crimps. The circuit and motor driving the actuation were omitted from the load testing since the focus of this assessment was to test the ability of the plastic and connecting components to withstand the specified loads. The electromechanical actuation system's primary purpose was to drive the grasp and release of the claw for the quantitative assessment, and the MG966R servo motor used is not necessarily reflective of the actuator found in a functional prosthetic device.

The single-digit flexion test, informed by the testing procedure carried out by Mio et al. [38], evaluated the load capacity for a single digit [38,39]. The digit extension test conducted by Mio et al. follows guidelines outlined by the plastic testing standards ISO 178 and ASTM 790 [41,42]. The existing standards for prosthetic devices typically have the finger fully extended to apply a load at the most distal part of the digit, serving as a correlate for the pinch force capacity of a single finger [38,39]. The minimum load to meet this test's requirement is 30N [38,39]. The test conducted on the biomimicry models was modified to start with the digit in a curled position. The deviation was minor, due to the geometric differences between the talon and prosthetic finger design. This modification was made since the developed bird-inspired models do not use pinching as the primary grasp method, whereas the original testing procedure was designed to measure the prosthesis's pinch ability [38,40]. The force applied to the digit in a curled starting position was more similar to the type of force the digit would encounter when grasping an object. The single digit was mounted to the testing platform and secured as shown in Figure 8. The cabling running through the channels in the digit was pulled taut and crimped at the level of the base and talon to retain the digit in its curled position. An upward force was applied to the talon to represent the force encountered by a grasped object due to gravity. The loaded cable was looped over the t-slot bar at the top of the testing platform and pulled vertically. A handheld digital force gauge was used to simultaneously pull the cable and measure the force in newtons until the digit failed. Failure in this case was defined as the inability of the digit to retain its original position. The failure condition, defined as extreme digit displacement, caused the cable, finger posts, or 3D-printed plastic to break or deform.

The full grasp test, following the prosthetic testing standards outlined in Mio et al. [38], assesses the ability of the full hand to support a load. This portion of the work by Mio et al.

follows the test procedures outlined in ISO 22523 [43]. The minimum requirement for this standard is 60 N [38,39]. The device was supplied with an object, and a force was applied to the object to represent the weight of a grasped object. The applied force was measured until the device's failure, similar to the failure criteria of the single-digit flexion test. For the grasping object, a sphere was chosen due to its grasping reliability by both the zygodactyl and anisodactyl claw patterns. The sphere was adapted by removing its lower half and flattening the bottom so it would rest better on the base and mounting clamp. The object was placed in the grasp of the four-fingered device and enveloped by a Velcro® band to serve as an attachment point for the cable used for force application, depicted in Figure 9a. For both devices, a consistent increase in load was applied to the object until the failure of the device to retain the object within its grasp. The testing platform for the anisodactyl claw is displayed in Figure 9b and the zygodactyl claw in Figure 9c.

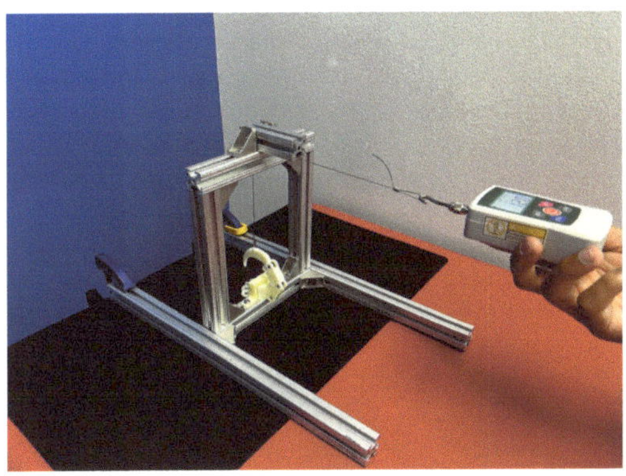

Figure 8. Single-digit flexion testing apparatus displaying the secured digit and the force application measured through the handheld force gauge.

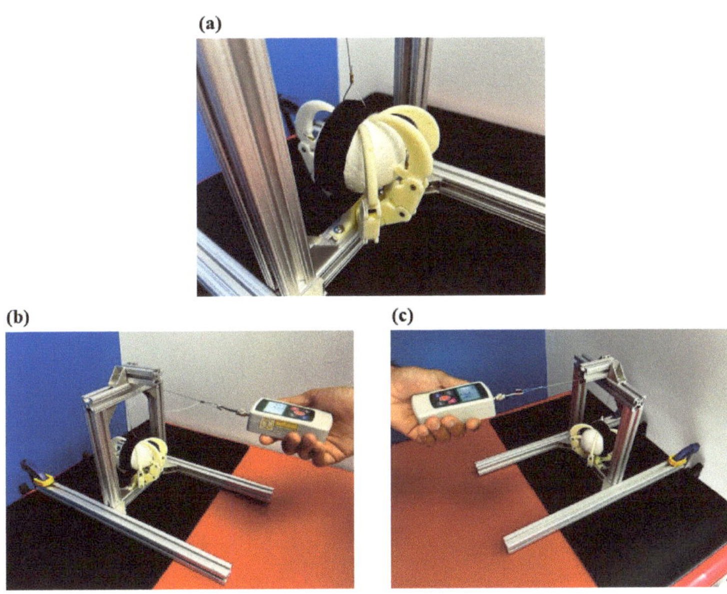

Figure 9. Full-claw flexion testing apparatus: (**a**) close-up view of the test object placed in the grasp of the anisodactyl claw; (**b**) load test conducted on the anisodactyl claw; (**c**) load test conducted on the zygodactyl claw.

3. Results and Discussion

3.1. Qualitative Grasp Test Results

In observing how each device interacted with the various objects as noted in Figure 10, key features from each of the devices that allowed for an improved grasp were identified. All three devices executed the intended function of contracting the more proximal joints followed by the distal joints relative to the base when grasping [44]. This was evident by the pattern in which the devices initiated contact with the objects. The phalanges closest to the palm made contact with the object, allowing the other sections to wrap around the object to establish more points of contact along the object. This enabled the digits to distribute the force rather than rely on a pinch-type grasp where a greater amount of force is concentrated at fewer points of contact.

A structure on both the biomimicry claws that emphasized this grasping method was the smaller, more central palm size. This feature improved opportunities for the more proximal portions of the digits to contact the object. Existing bird-inspired grasping devices with small palm structures similarly experience increased contact points between the digits and grasped objects [26]. The five-fingered hand, which has a much larger palm not located as central to the device, relied on grasping with more distal portions of the fingers in certain cases when not all of the proximal parts of the fingers could reach the object. As seen in the sphere column of Figure 10, the more central palm in both the anisodactyl and zygodactyl bird claw created a more secure grip around the object as opposed to the more human hand. These differences in grip were due to fewer points of contact seen in the Limbitless Solutions® hand. The primary grasp force is developed between the index and middle fingers and thumb, leaving a large portion of the palm exposed during a grasping motion. These factors led the grasp to be more pliable and the object easier to dislodge in the five-fingered hand. Although the palm of an actual human hand is similarly large, its composition allows it to deform and wrap securely around a grasped object. A similar effect may not be feasible to achieve for a low-cost prosthetic comprising rigid materials. Therefore, reducing the palm size to allow the digits more access to larger objects appears to be one solution for improving the grasp while maintaining a simpler design.

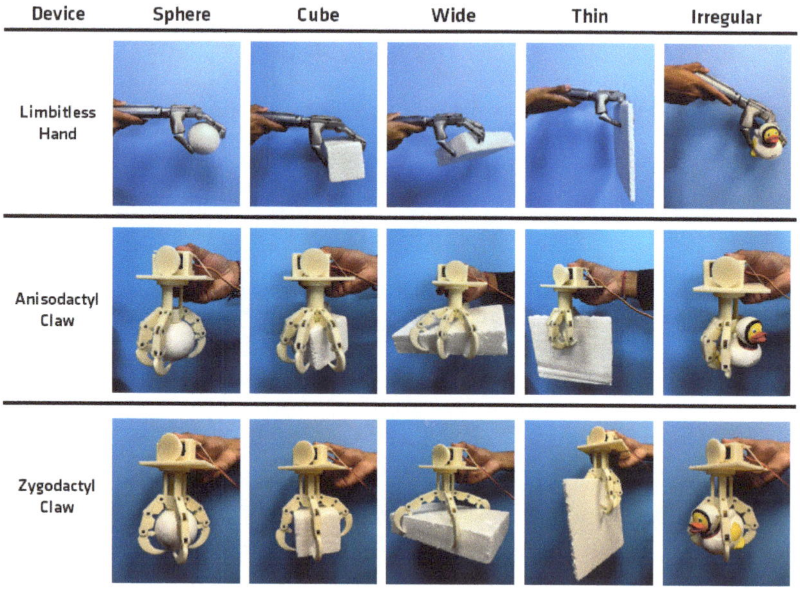

Figure 10. Qualitative grasp test results displaying the grasp interaction between each device and object type.

Another feature of the biomimicry claws that supplemented a more secure grasp was the structure of the most distal digital component: the talon. As the talon was curved, it offered additional reach in the curled position of the digit to make contact with the sides of wider objects. The five-fingered hand encountered difficulty in grasping all the way across the wide object and instead relied on making contact with adjacent sides, leading to a less stable grasp. Prosthesis with multi-positional thumbs may be able to change grip configurations to mitigate these limitations. The curved talon also served as an enclosure for smaller objects such as the sphere and irregular shapes by converging at the base of the object. The talon encountered challenges when grasping the thin object where the curved structure often forced the sharper digit tip to embed into the material to be able to establish a grasp. Predatory birds often use their talons to puncture prey. Grasping in this fashion may be problematic when handling fragile objects. In cases when the talon tips did not embed into the object, the contact from the relatively thin sections of the talon was precarious. The narrow contact afforded by talon-shaped digits is also observed in the study by Zhu et al., where the bird-inspired models faced similar challenges with thin geometries [26]. In comparison, the flatter, more linear fingertip of the five-fingered hand performed well in establishing a better grip on the thin object with a better pinch grip pattern. The pinch grip of the five-fingered hand offers a more reliable grasp for thin objects, whereas the encompassing grip of the bird claw models is generally preferred for larger objects [44]. A pinch grasp still demonstrates success in grasping larger objects; however, the encapsulating grasp of the bird claw models may be better suited for grasping larger objects in comparison to the five-fingered hand.

When comparing the two claw types, certain advantages exist between the anisodactyl and zygodactyl structures. The anisodactyl claw with its separated front and back digits is well suited to grasp objects that typically only require support on two sides. The rectangular shape benefits from this grasp due to its different length and width proportions. This also applies to cylindrical shapes, which corroborates why much of the literature focusing on bird claws for drone applications mimics an anisodactyl structure perching on cylindrical tree branches [27,30]. The zygodactyl claw, which was adapted to be radially symmetrical, establishes a more secure grasp with objects with more uniform sides such as the sphere and the block. The symmetrical display creates points of contact more evenly across the faces of the object. The larger spaces between the front and back digits in the anisodactyl claw create more of an opportunity for these types of objects to be dislodged from the grasp. A similar grasp feature was observed in the five-fingered hand due to the digits being positioned in the front and side of the palm, making it easier to grasp cylindrical objects with more direct contact from the digits as opposed to the zygodactyl model.

3.2. Quantitative Grasp Test Results

For the single-finger digit flexion test, the force was applied by pulling on the cable attached to the talon segment of the claw in its curled position. Recording the applied force with the handheld digital force gauge, the single digit consistently supported loads past the standard of 30 N for a mechanical prosthetic finger [38,39]. Across the five load assessments conducted for the single digit, the average load was 61.0 N, with a tolerance of 2.2 N for one standard deviation from the average. The maximum load the digit could support before failure was measured at 64.3 N, with failure occurring due to the cable in the curled digit snapping. Repeated testing indicated that the break generally occurred at the level of the crimp at the talon preventing the cable from passing through this channel. Supporting the cable at the talon crimp may help withstand further loads for a single digit. This test indicated that a 3D-printed digit structured after the morphology of a bird claw matched and even exceeded mechanical load requirements for prosthetic digits. The average load and tolerance for the five trials also demonstrated that the results were fairly consistent between trials. When observing the digit as it was being pulled, the curvature of the talon allowed the phalanges to straighten while retaining the position established where the pull cable was attached. Verification of the biomechanical load-bearing capacity of this specific structure is important in discovering whether it has any apparent weaknesses that would disqualify it for adaptation into prosthetic design or whether it holds any novel advantages, despite this current prototype not acting as a functional prosthetic.

For the full grasp test, separate tests were conducted for both the anisodactyl and zygodactyl configurations, and results were compared to existing prosthetic standards to determine whether the structure of either had advantages in supporting more load. The expectation for a full grasp test is to support 60 N of pulling force [38,39].

The first test with the anisodactyl model demonstrated this configuration supported a maximum of 86.1 N as an assembly. This result also exceeds the maximum load supported by any single digit (64.3 N), which is expected as there are additional digits to distribute the applied force [45]. However, the average for the five trials was lower at 76.8 N, with a tolerance of 9.0 N. Compared to the values for the single digit test, the results were not as consistent for the full anisodactyl grasp. The failure condition for this test occurred when the outer sets of digits on the front three digits partially relaxed from the deformation of the cable retaining the curled position of the claw. The digit relaxation allowed the object to escape the grasp through the gap between the front and back sets of digits. An explanation for the larger deviation in results may be in the difference in applying loads with an object rather than directly to the fingertip. The specified failure condition of the object slipping likely introduced some additional variability that may not occur when the load is applied directly via a cable and not the objects for the single digit.

The zygodactyl grasp test took several attempts to reach a maximum load that surpassed the 60 N threshold for a full grasp test. In the first two pull tests, it was observed that the cable adjacent to the crimp at the level of the talon tended to snap or deform, much like the failure condition for the single-digit extension test and the anisodactyl full grasp test. One possible explanation for the difference in performance between the two configurations is the lack of digits aligned with their respective cables in the zygodactyl testing setup. This may have caused the force to be applied at an angle to the digits and the cables holding them in place, resulting in the force overcoming the crimp at the talon with more ease. Tightening the cables in the claw to better position the crimps to avoid them adopting an unnatural angle helped overcome this issue, with this trial meeting a maximum pull force of 74.1 N to exceed the 60 N threshold. Repeat tests with this consideration for the zygodactyl claw demonstrated more consistent results in meeting the 60 N threshold. The average for the five trials was 58.0 N, with a tolerance of 10.2 N. Though the maximum values exceeded the 60 N requirement, the average is lower due to the premature failures observed in the first two tests. The results of each test are displayed in Table 1.

Table 1. Summary of the quantitative load testing results.

Test Type	Source [38,41–43]	Minimum Expected Load	Average Measured Load	Maximum Measured Load
Single-Digit Extension	Mio et al. (2019) ISO 178, ASTM 790	30 N	61.0 N ± 2.2 N	64.3 N
Anisodactyl Claw Grip	Mio et al. (2019) ISO 22523	60 N	76.8 N ± 9.0 N	86.1 N
Zygodactyl Claw Grip	Mio et al. (2019) ISO 22523	60 N	58.0 N ± 10.2 N	74.1 N

In comparing the results of the three different tests, it is observed that the maximum applied loads for the single digit significantly exceeded the minimum requirement. However, the maximum loads for the full claws were only higher by approximately 10–20 N. The device's assembly components, rather than the 3D-printed plastics, may have been the driving factor in the load the design could support. Adding more digits did not proportionally increase the supported load in comparison to the single digit since the failure condition included the cable or crimp deforming or snapping. Therefore, even one digit failing for the anisodactyl and zygodactyl patterns would likely lead to the object escaping the grasp. For the zygodactyl claw, the symmetrical arrangement of the digits may have been a disadvantage in this aspect since the loss of a digit on one end leads to an imbalance in the device's support of the grasped object. This imbalance is also applicable for the single digit in the rear of the anisodactyl claw. In contrast, the anisodactyl claw may still retain an object if the digits in the front supplement the failure of a neighboring digit. This may explain why the average and maximum measured loads for the anisodactyl evaluation were higher than the zygodactyl.

3.3. Limitations

A key limitation in the qualitative grasp testing portion of the comparison between the biomimicry and five-fingered models was that the five-fingered hand had a silicone segment at the fingertip, whereas the talons of the biomimicry models did not. Talons are often used by birds of prey when hunting for kinetic puncturing of objects. The purpose of the use of silicone on the prosthesis is to create a larger surface area and higher friction force between the object and the fingertip. Silicone was not included in the design of the talon tips of the biomimicry claws to more closely mimic the biological structure of a talon, although this difference between the three devices may contribute to the grasp performances.

While the reference quantitative testing method from Mio et al. [38] used a Zwick Roell Z0.5 multi-test machine to apply a consistently increasing load to their models, this study relied on a human to apply an increase in load. This may cause the results to be more sensitive to inconsistencies in the rate applied. The average loads for the full grasp test reflect this limitation, especially for the zygodactyl full grasp evaluation. However, the testing procedure primarily evaluated whether the models were able to handle the minimum load requirements rather than understanding the exact load that leads to failure. Further development of this research may focus on the use of industrial load testing equipment to corroborate the preliminary grip force results with a more sensitive testing procedure.

4. Conclusions

To develop novel biomimetic solutions to address the issue of high prosthetic rejection, a biomimetic grasping assembly was developed with inspiration from bird claws to test for features that would be advantageous in improving grasping ability. The biomimetic sources, the anisodactyl and zygodactyl configurations found in bird claws, were chosen for their alignment in nature with grasp-related activities. Design inspiration was derived from existing applications in the literature, adapting a 3D-printed bird claw for drone landing attachments and upper-limb human-hand-based prosthetic design. A testing platform was created for both claw configurations to emulate the structural and behavioral characteristics of a bird claw in a 3D-printed device. The constructed devices were evaluated both qualitatively for how they interact with various objects and quantitatively for how much load they can support relative to the standards for upper-limb prosthetic devices. These evaluations indicated certain features that improve the grasping ability without compromising the mechanical load-bearing capabilities of the device.

1. The cable attachment points and torsion springs placement along the digits allowed for rotation of the most proximal sections of the digits inward followed by the more distal components. This pattern appeared more natural and was advantageous in securely grasping larger objects by wrapping the digits around them rather than pinching them by initiating contact with the fingertips.
2. The palm design of the bird-inspired device, with a smaller and more central palm relative to the digits, compared to the human-hand prosthetic improved the ability of both the biomimetic claws to grasp and hold the various objects. This design feature enabled the proximal portions of the digits to have a better opportunity to establish contact with the object to improve the overall grasp.

3. The curved talon at the distal end of the digit held benefits in offering support for wider objects. For objects small enough for the digits to wrap around them, the curved talons also acted as an enclosure for the object by converging at the base to reduce the chances of accidentally dropping it. A disadvantage of this feature was the difficulty in securely holding thin objects because of the thin contact point and reduced application of force due to the curved tip angle.
4. Although a pinch-type actuation for the five-fingered hand was observed to grasp larger geometries with less security, it demonstrated greater success in holding thin objects as observed in the object interactions with the five-fingered device. The flatter fingertip shape of the human-inspired five-finger hand further contributes to handling thin objects relative to the talon structure of the bird models by increasing surface area at the points of contact.
5. In comparing the anisodactyl and zygodactyl configurations, each digit setup had grasp advantages for certain object types. The anisodactyl's 3×1 structure allows a wider grasp for more cylindrical objects. The central front and back digits create the primary grasp, the outer two digits in the front supply greater reach, and the greater angle between the front and back digits allows for the object to not be obstructed by another digit. The more symmetrical 2×2 feature of the zygodactyl allows an advantage when grasping smaller, more symmetrical objects as it more evenly supports the object on multiple sides and reduces the space between the digits for the object to potentially slip out.
6. Mechanical load testing results indicate that the robustness of the assembly is validated by testing standards traditionally applied for upper-limb prostheses and plastic properties. The maximum recorded load values were 64.3 N, 86.1 N, and 74.1 N for the single digit, anisodactyl claw, and the zygodactyl claw models, respectively. These values exceed the load metrics of 30 N for a single digit and 60 N for a full grasp set for prosthetic testing. The low averages for the full grasp tests across five trials warrant further testing with a mechanized force application procedure to further validate the initial results.

Future Research in Upper-Limb Prostheses

These features have the potential to be implemented in the design of current prosthetic devices to make them more functional and reliable. Models may benefit from a design that strikes a balance between the features of the current human hand prosthetic and the developed biomimetic features. On current prosthetic hands, examples of this design approach may include resizing the palm to have it take up less space and positioning it more central to the displayed digits. It is also important to consider the existing advantages of the five-fingered hand, notably the pinch grip pattern that demonstrates improved grasping ability for thin objects, and ensure that any design modifications preserve those as well. Human-inspired hands are anticipated to be less task-specified and required to be adaptable to a variety of situations.

Adding a slight curve to the curvature of the fingertips may allow the fingers to establish a pinch-type grasp for thinner objects while still offering some additional reach. This testing enabled an adaptable research platform to be designed, enabling a variety of digit configurations to be feasible for testing. Modifications may be made to the current base to allow the digit to rotate around the central palm structure and enable different angular spacing between each digit.

In the effort to develop an optimized model to mitigate prosthetic rejection, future work would also benefit from exploring techniques to create more lightweight devices. New models with the mentioned modification may be constructed with different materials and evaluated to determine the properties that allow for a durable yet lightweight device. These mechanical evaluations may be assessed with finite element analysis (FEA), a simulation study run on physical models to assess how a model behaves under certain stresses. Next-generation models closer to a functional upper-limb prosthetic would also benefit from involving a wider variety of shapes in the qualitative evaluation; while the current testing evaluated the grasping ability of relatively elementary shapes for the initial models, updated designs would involve objects encountered in daily activities with more complex geometries to better inform how the device may perform for a prosthetic user.

This research primarily focused on the assessment of the anisodactyl and zygodactyl bird claw configurations; however, many different bird claw configurations exist in nature, including the heterodactyl, syndactyl, and pamprodactyl, each with their unique digit configurations and adaptations [26]. Future research investigating the utility of biomimicry from further types of bird claws in prosthetic design would be of use to uncover the most optimal digit arrangements and spacing angles for prosthetic use.

Further design modifications may also address a limitation between the existing prosthetic hand model and the 3D-printed test devices. The inclusion of silicone along the contact points may provide increased performance for all varieties. As observed in the qualitative grasp tests, the increased friction between the silicone and the object's surface allowed the prosthetic hand to successfully grasp and hold objects even with fewer contact points. Including silicone along the major contact points of the developed bird claw models, notably at the interior side of the phalanges and the talon tip, may improve overall grip. Further increases in the prosthesis contact patterns may translate to improved functionality with limited weight increases. Further evaluating the performance, weight increase, and complexity for assembly may offer insights into optimization for the prosthetic version and the bird-claw biomimetic device.

Supplementary Materials: The following supporting information can be downloaded at: www.mdpi.com/xxx/s1, which includes the mechanical testing statistical analysis.

Author Contributions: Conceptualization, A.M. and J.S.; hardware, P.S. (Pavan Senthil), O.V. and J.S.; writing—original draft preparation, P.S. (Pavan Senthil) and O.V.; testing, P.S. (Pavan Senthil) and O.V.; supervision, A.M., P.S. (Peter Smith), and J.S.; writing—review and editing, all. All authors have read and agreed to the published version of the manuscript.

Funding: This research was partially funded by the UCF 2023 Summer Undergraduate Research Fellowship.

Institutional Review Board Statement: Not applicable.

Informed Consent Statement: Not applicable.

Data Availability Statement: Data is contained within the article or Supplementary Materials.

Acknowledgments: The authors would like to thank the University of Central Florida 2023 Summer Undergraduate Research Fellowship program and the interdisciplinary student research team and alumni at Limbitless Solutions®, notably Benjamin Mathew, Dominique Courbin, Julia Sheiber, and Jacob Benjamin. The researchers appreciate the philanthropic support for undergraduate research from the Paul B. Hunter & Constance D. Hunter Charitable Foundation, the Don & Lorraine Freeberg Foundation, and the Albert E. & Birdie W. Einstein Foundation.

Conflicts of Interest: The Limbitless Solutions® program has received financial and in-kind support from the 3D printer manufacturer Stratasys LTD® and 3D design software manufacturer Autodesk®. The authors want to disclose that the funders had no role in the design of the overarching research study; in the collection, analyses, or interpretation of data; in the writing of the manuscript; or in the decision to publish the results.

Abbreviations

The following abbreviations are used in this manuscript:

AM	Additive manufacturing
FDM	Fused deposition modeling
ABS	Acrylonitrile butadiene styrene
CAD	Computer-aided design
FEA	Finite element analysis

References

1. Resnik, L. Development and testing of new upper-limb prosthetic devices: Research designs for usability testing. *J. Rehabil. Res. Dev.* **2011**, *48*, 697. [CrossRef] [PubMed]
2. Biddiss, E.; Chau, T. Upper-limb prosthetics: Critical factors in device abandonment. *Am. J. Phys. Med. Rehabil.* **2007**, *86*, 977–987. [CrossRef] [PubMed]
3. Espinosa, M.; Nathan-Roberts, D. Understanding prosthetic abandonment. *Proc. Hum. Factors Ergon. Soc. Annu. Meet.* **2019**, *63*, 1644–1648. [CrossRef]
4. Segil, J.L.; Pulver, B.; Huddle, S.; Weir, R.F.f.; Sliker, L. The point digit II: Mechanical design and testing of a ratcheting prosthetic finger. *Mil. Med.* **2021**, *186*, 674–680. [CrossRef] [PubMed]
5. Wagner, L.V.; Bagley, A.M.; James, M.A. Reasons for prosthetic rejection by children with unilateral congenital transverse forearm total deficiency. *JPO J. Prosthetics Orthot.* **2007**, *19*, 51–54. [CrossRef]
6. Light, C.M.; Chappell, P.H.; Kyberd, P.J.; Ellis, B.S. A critical review of functionality assessment in natural and prosthetic hands. *Br. J. Occup. Ther.* **1999**, *62*, 7–12. [CrossRef]
7. Seankennedy. Biomimicry/Bimimetics: General Principles and Practical Examples. SCQ, 2004. Available online: https://www.scq.ubc.ca/biomimicrybimimetics-general-principles-and-practical-examples/ (accessed on 18 October 2023).
8. Volstad, N.L.; Boks, C. On the use of biomimicry as a useful tool for the industrial designer: Biomimicry as a tool for the industrial designer. *Sustain. Dev.* **2012**, *20*, 189–199. [CrossRef]
9. Marshall, A.; Lozeva, S. Questioning the theory and practice of biomimicry. *Int. J. Des. Nat. Ecodyn.* **2009**, *4*, 1–10. [CrossRef]
10. Gosline, J.M.; Guerette, P.A.; Ortlepp, C.S.; Savage, K.N. The mechanical design of spider silks: From fibroin sequence to mechanical function. *J. Exp. Biol.* **1999**, *202*, 3295–3303. [CrossRef]
11. Römer, L.; Scheibel, T. The elaborate structure of spider silk: Structure and function of a natural high performance fiber. *Prion* **2008**, *2*, 154–161. [CrossRef]
12. Vollrath, F. Strength and structure of spiders' silks. *Rev. Mol. Biotechnol.* **2000**, *74*, 67–83. [CrossRef] [PubMed]
13. Mazzolai, B.; Beccai, L.; Mattoli, V. Plants as model in biomimetics and biorobotics: New perspectives. *Front. Bioeng. Biotechnol.* **2014**, *2*, 2. [CrossRef] [PubMed]
14. Lurie-Luke, E. Product and technology innovation: What can biomimicry inspire? *Biotechnol. Adv.* **2014**, *32*, 1494–1505. [CrossRef] [PubMed]
15. Hapsari, F.N.; Purwaningsih, R.; Azzahra, F.; Sari, D.P. Velcro product design with biomimicry approaches. *IOP Conf. Ser. Earth Environ. Sci.* **2022**, *1111*, 012057. [CrossRef]
16. Clark, J.; Cham, J.; Bailey, S.; Froehlich, E.; Nahata, P.; Full, R.; Cutkosky, M. Biomimetic design and fabrication of a hexapedal running robot. In Proceedings of the 2001 ICRA IEEE International Conference on Robotics and Automation (Cat. No.01CH37164), Seoul, Republic of Korea, 21–26 May 2001; Volume 4, pp. 3643–3649. [CrossRef]
17. Guetta, O.; Shachaf, D.; Katz, R.; Zarrouk, D. A novel wave-like crawling robot has excellent swimming capabilities. *Bioinspir. Biomim.* **2023**, *18*, 026006. [CrossRef] [PubMed]
18. Dharmdas, A.; Patil, A.Y.; Baig, A.; Hosmani, O.Z.; Mathad, S.N.; Patil, M.B.; Kumar, R.; Kotturshettar, B.B.; Fattah, I.M.R. An experimental and simulation study of the active camber morphing concept on airfoils using bio-inspired structures. *Biomimetics* **2023**, *8*, 251. [CrossRef] [PubMed]
19. Feliu-Talegon, D.; Rafee Nekoo, S.; Suarez, A.; Acosta, J.A.; Ollero, A. Modeling and under-actuated control of stabilization before take-off phase for flapping-wing robots. In Proceedings of the ROBOT2022: Fifth Iberian Robotics Conference, Zaragoza, Spain, 23–25 November 2022; Lecture Notes in Networks and Systems; Tardioli, D., Matellán, V., Heredia, G., Silva, M.F., Marques, L., Eds.; Springer: Cham, Switzerland, 2023; pp. 376–388. [CrossRef]
20. Varol, H.A.; Dalley, S.A.; Wiste, T.E.; Goldfarb, M. Biomimicry and the design of multigrasp transradial prostheses. In *The Human Hand as an Inspiration for Robot Hand Development*; Springer Tracts in Advanced Robotics; Balasubramanian, R., Santos, V.J., Eds.; Springer International Publishing: Cham, Switzerland, 2014; pp. 431–451. [CrossRef]
21. Li, S.; Bai, H.; Shepherd, R.F.; Zhao, H. Bio-inspired design and additive manufacturing of soft materials, machines, robots, and haptic interfaces. *Angew. Chem. Int. Ed.* **2019**, *58*, 11182–11204. [CrossRef]
22. Wang, B.; Yu, T. Numerical investigation of novel 3D-SPA for gripping analysis in multi-environment. *Int. J. Mech. Sci.* **2023**, *240*, 107916. [CrossRef]
23. Perera, A.S.; Coppens, M.O. Re-designing materials for biomedical applications: From biomimicry to nature-inspired chemical engineering. *Philos. Trans. R. Soc. A Math. Phys. Eng. Sci.* **2019**, *377*, 20180268. [CrossRef]

24. Zhang, D.; Xu, J.; Liu, X.; Zhang, Q.; Cong, Q.; Chen, T.; Liu, C. Advanced bionic attachment equipment inspired by the attachment performance of aquatic organisms: A review. *Biomimetics* **2023**, *8*, 85. [CrossRef]
25. Rao, R. Biomimicry in architecture. *Int. J. Adv. Res. Civil Struct. Environ. Infrastruct. Eng. Dev.* **2014**, *1*, 101–107.
26. Zhu, Y.; He, X.; Zhang, P.; Guo, G.; Zhang, X. Perching and grasping mechanism inspired by a bird's claw. *Machines* **2022**, *10*, 656. [CrossRef]
27. Roderick, W.R.; Chin, D.D.; Cutkosky, M.R.; Lentink, D. Birds land reliably on complex surfaces by adapting their foot-surface interactions upon contact. *eLife* **2019**, *8*, e46415. [CrossRef] [PubMed]
28. Csermely, D.; Rossi, O. Bird claws and bird of prey talons: Where is the difference? *Ital. J. Zool.* **2006**, *73*, 43–53. [CrossRef]
29. Leblanc, K.; Pintore, R.; Galvão, A.; Heitz, E.; Provini, P. Foot adaptation to climbing in ovenbirds and woodcreepers (Furnariida). *J. Anat.* **2023**, *242*, 607–626. [CrossRef] [PubMed]
30. Roderick, W.R.T.; Cutkosky, M.R.; Lentink, D. Bird-inspired dynamic grasping and perching in arboreal environments. *Sci. Robot.* **2021**, *6*, eabj7562. [CrossRef] [PubMed]
31. Nabi, F.G.; Sundaraj, K.; Vijean, V.; Shafiq, M.; Planiappan, R.; Talib, I.; Rehman, H.U. A novel design of robotic hand based on bird claw model. *J. Phys. Conf. Ser.* **2021**, *1997*, 012034. [CrossRef]
32. Nadan, P.M.; Anthony, T.M.; Michael, D.M.; Pflueger, J.B.; Sethi, M.S.; Shimazu, K.N.; Tieu, M.; Lee, C.L. A bird-inspired perching landing gear system. *J. Mech. Robot.* **2019**, *11*, 061002. [CrossRef]
33. Manero, A.; Smith, P.; Sparkman, J.; Dombrowski, M.; Courbin, D.; Kester, A.; Womack, I.; Chi, A. Implementation of 3D printing technology in the field of prosthetics: Past, present, and future. *Int. J. Environ. Res. Public Health* **2019**, *16*, 1641. [CrossRef]
34. Manero, A.; Sparkman, J.; Dombrowski, M.; Smith, P.; Senthil, P.; Smith, S.; Rivera, V.; Chi, A. Evolving 3D-Printing Strategies for Structural and Cosmetic Components in Upper Limb Prosthesis. *Prosthesis* **2023**, *5*, 167–181. [CrossRef]
35. Gibson, I.; Rosen, D.; Stucker, B. Introduction and basic principles. In *Additive Manufacturing Technologies: 3D Printing, Rapid Prototyping, and Direct Digital Manufacturing*; Gibson, I., Rosen, D., Stucker, B., Eds.; Springer: New York, NY, USA, 2015; pp. 1–18. [CrossRef]
36. Dudek, P. FDM 3D printing technology in manufacturing composite elements. *Arch. Metall. Mater.* **2013**, *58*, 1415–1418. [CrossRef]
37. Backus, S.B.; Sustaita, D.; Odhner, L.U.; Dollar, A.M. Mechanical analysis of avian feet: Multiarticular muscles in grasping and perching. *R. Soc. Open Sci.* **2015**, *2*, 140350. [CrossRef] [PubMed]
38. Mio, R.; Sanchez, M.; Valverde, Q.; Lara, J.; Rumiche, F. Mechanical testing methods for body-powered upper-limb prostheses: A case study. *Adv. Sci. Technol. Eng. Syst. J.* **2019**, *4*, 61–68. [CrossRef]
39. Smit, G.; Plettenburg, D.H.; van der Helm, F.C.T. The lightweight delft cylinder hand: First multi-articulating hand that meets the basic user requirements. *IEEE Trans. Neural Syst. Rehabil. Eng.* **2015**, *23*, 431–440. [CrossRef] [PubMed]
40. Joyce, T.; Unsworth, A. A test procedure for artificial finger joints. *Proc. Inst. Mech. Eng. Part H J. Eng. Med.* **2002**, *216*, 105–110. [CrossRef] [PubMed]
41. *ISO 178:2019*; Plastics—Determination of Flexural Properties. International Organization for Standardization: Geneva, Switzerland, 2019.
42. *ASTM D790-17*; Standard Test Methods for Flexural Properties of Unreinforced and Reinforced Plastics and Electrical Insulating Materials. American Society for Testing and Materials: Geneva, Switzerland, 2019.
43. *ISO 22523:2006*; External Limb Prostheses and External Orthoses—Requirements and Test Methods. International Organization for Standardization: Geneva, Switzerland, 2006.
44. Kashef, S.R.; Amini, S.; Akbarzadeh, A. Robotic hand: A review on linkage-driven finger mechanisms of prosthetic hands and evaluation of the performance criteria. *Mech. Mach. Theory* **2020**, *145*, 103677. [CrossRef]
45. Kargov, A.; Pylatiuk, C.; Martin, J.; Schulz, S.; Döderlein, L. A comparison of the grip force distribution in natural hands and in prosthetic hands. *Disabil. Rehabil.* **2004**, *26*, 705–711. [CrossRef]

MDPI AG
Grosspeteranlage 5
4052 Basel
Switzerland
Tel.: +41 61 683 77 34

Biomimetics Editorial Office
E-mail: biomimetics@mdpi.com
www.mdpi.com/journal/biomimetics

www.ingramcontent.com/pod-product-compliance
Lightning Source LLC
Chambersburg PA
CBHW041352210926
51850CB00013B/3062
* 9 7 8 3 7 2 5 8 3 2 8 2 8 *